T0331615

Who's Bigger?

Where Historical Figures *Really* Rank

Is Napoleon bigger than Hitler? Washington bigger than Lincoln? Picasso bigger than Einstein?

Quantitative analysts are finding homes in cultural domains from finance to politics. What about history? In this fascinating book, Steven Skiena and Charles B. Ward bring quantitative analysis to bear on ranking and comparing historical reputations. They evaluate each person by aggregating the traces of millions of opinions, similar to how Google ranks web pages.

Did you know:

- Women remain significantly underrepresented in the historical record compared to men and have long required substantially greater achievement levels to be equally noted for posterity?
- The long-term prominence of Elvis Presley rivals that of the most famous classical composers? Roll over Beethoven, and tell Tchaikovsky the news!
- If you've got a spare billion dollars, and want to be remembered forever, your best investment is to get a university named after you?

Along the way, the authors present the rankings of more than one thousand of history's most significant people in all areas of human endeavor, including science, politics, and entertainment. Anyone interested in history or biography can see where their favorite figures place in the grand scheme of things. While revisiting old historical friends and making new ones, you will come to understand the forces that shape historical recognition in a whole new light.

About the Authors

Steven Skiena serves as Distinguished Teaching Professor of Computer Science at Stony Brook University. He is the author of four well-regarded books: *The Algorithm Design Manual* (2008); *Calculated Bets: Computers, Gambling, and Mathematical Modeling to Win* (2001); *Programming Challenges* (with Miguel Revilla, 2003); and *Computational Discrete Mathematics* (with Sriram Pemmaraju, 2003). Skiena heads the Data Science Laboratory at Stony Brook, using large-scale text analysis to chart the frequency, sentiment, and relationships among millions of people, places, and things. This technology forms the foundation of General Sentiment (http://www.generalsentiment.com), where he serves as co-founder and chief scientist. His news analysis has been applied to research projects from financial forecasting to presidential election analysis. The rankings underlying *Who's Bigger?* derive from this analysis.

Charles B. Ward works as an engineer on the ranking team at Google. He is the author of more than a dozen scholarly papers, including research in text analysis, computational social science, computational biology, and graph theory. Ward worked as a lead developer with the Data Science Laboratory during his four years of postdoctoral studies at Stony Brook University. In his spare time, he is an accomplished solo pianist. He is also an authority on historical strategy games, and spends perhaps too much time playing and designing them.

WHO'S BIGGER?

Where Historical Figures
Really *Rank*

STEVEN SKIENA
Stony Brook University, New York

CHARLES B. WARD

Shaftesbury Road, Cambridge CB2 8EA, United Kingdom

One Liberty Plaza, 20th Floor, New York, NY 10006, USA

477 Williamstown Road, Port Melbourne, VIC 3207, Australia

314–321, 3rd Floor, Plot 3, Splendor Forum, Jasola District Centre, New Delhi – 110025, India

103 Penang Road, #05–06/07, Visioncrest Commercial, Singapore 238467

Cambridge University Press is part of Cambridge University Press & Assessment, a department of the University of Cambridge.

We share the University's mission to contribute to society through the pursuit of education, learning and research at the highest international levels of excellence.

www.cambridge.org
Information on this title: www.cambridge.org/9781107041370

First published 2014

A catalogue record for this publication is available from the British Library

Library of Congress Cataloging-in-Publication data
Skiena, Steven S., author.
Who's bigger? : where historical figures really rank / Steven Skiena, Charles B. Ward.
pages cm
Includes bibliographical references and index
ISBN 978-1-107-04137-0 (hardback)
1. Quantitative research. I. Ward, Charles, 1984– author. II. Title.
Q180.55.Q36S55 2013
920.02–dc23 2013032173 CIP

ISBN 978-1-107-04137-0 Hardback

Additional resources for this publication at http://www.whoisbigger.com

Contents

Rationale

Quantitative analysts ("quants") are finding new homes in the social and cultural domains. Finance? Hedge funds employing quantitative trading strategies now rule Wall Street, to the extent that more than 80 percent of all buy-sell decisions are made by algorithms, not people. Sports? The best-selling book *Moneyball* detailed how successful major league teams now hire general managers who value statistical analysis above gut instinct – and get to be played by Brad Pitt in the movie version of the story. Politics? Prediction markets and the meta-analysis of polling data work so reliably that it hardly seems necessary to hold the actual election.

In this book, we bring quantitative analysis to bear on ranking and comparing historical reputations. Who's bigger: Washington or Lincoln? Hitler or Napoleon? Picasso or Michelangelo? Charles Dickens or Jane Austen? Did you realize that:

- Although *Paul Revere (1735–1818)* [627] and *Betsy Ross (1752–1836)* [2430] are well known to all American schoolchildren, they fell into complete obscurity for several generations after their contributions to the American Revolution. Their rediscoveries, completely independent of their actual achievements, tell us much about the capricious forces of history.
- Women remain significantly underrepresented in the historical record compared to men. We can prove that women have long required substantially greater achievement levels (analogous to about 4 IQ points in the mean) than men to get equally noted for posterity.

- Got a spare billion dollars, and want to be remembered forever? Your best investment is to get a university named after you.
- The long-term historical significance of *Elvis Presley (1935–1977)* [69] rivals that of the most famous classical composers. Roll over *Beethoven* [27], and tell *Tchaikovsky* [63] the news!

This is all good fun, but our quantitative approach also enables us to address important cultural questions in a rigorous manner:

- Are the historical figures discussed in school texts *really* worth teaching?
- How good are committees at recognizing historical greatness?
- Why do reputations change posthumously? Have copyright laws affected the literary canon, determining which authors get revived and which are forgotten?

We believe this book will be of interest to a broad readership and that anyone who likes history or biography will be curious about our rankings, eager to see where their favorite figures place in the grand scheme of things. Our book provides an opportunity to revisit old historical friends and make new ones. The reader will come to understand the forces that shape historical recognition in quite a different way after reading it. We detail the difference between fame and significance, the cultural biases in how we think about the past, and how rapidly reputations decay with time.

We also expect our share of right-brained readers, attracted by our quantitative methods and intrigued by our analysis. No mathematical or computational background is necessary to understand our rankings or conclusions. But we include a technical discussion for readers interested in the details of our methods.

We anticipate that professional historians may be skeptical of computer scientists intruding on their turf. But what we are really trying to do here is study what shapes the process of historical recollection. We predict that many scholars will be taken with the power of our methods, and will seek to employ these tools in their own studies. We have collaborated on research projects with political scientists and sociologists: watching them change from aloof to curious and then finally excited. We are making our massive datasets available to the academic community (and the curious public) through our website, `http://www.whoisbigger.com`. Computational

Social Science is a new and growing area of academic scholarship, and we hope to bring the field greater visibility and acceptance.

Textual Notes

We employ certain textual conventions in this book that may confuse readers, until they understand our rationale. We explain them here briefly.

More than 1,000 historical figures are mentioned by name within the body of this book. At each such reference, each person's name will be accompanied by their dates of birth and death, as well as their significance rank as assessed by our methods. Thus, Brad Pitt is *Brad Pitt (1963–)* [1845] and Aristotle is *Aristotle (384–322 B.C.)* [8]. We sort people from most significant to least significant, and assign ranks by counting down from the top. Thus the most significant person gets rank 1, and smaller numbers are better than bigger ones. Embedding our ranking with every reference will help you appreciate each individual's historical magnitude, and implicitly encourage you to evaluate the performance of our methods.

Describing first-hand experiences and opinions within a coauthored text is somewhat problematic. The royal "We" isn't always accurate here, since we are distinct individuals: one raised a Southern gentleman while the other was toilet-trained in the Bronx. And so we have adopted the following convention. We both will lay claim to "I," distinguishing which I when necessary. Thus "I (Steve)" will be used to represent bits of Bronx cheer. But we will mostly be "we," since we developed the methods, analysis, and contents of this text together, as a team.

Acknowledgments

I (Steve) dedicate this book to my second daughter, Abby. Fate has determined that she will never get to experience things like the fifth grade ahead of her older sister. Abby: don't worry, because you will get your own opportunities to make history. You are second only chronologically, but equal where it counts, in our hearts. And to my wife Renee: you know you will always be the most significant person in my book.

I (Charles) dedicate this book to my parents and Srivani. One could not wish for better parents or a better partner.

We acknowledge the help of several people, who should be famous even if they have not gotten there yet. Bala Mundiam built much of the infrastructure underlying our data collection and display, making this project possible. Ajeesh Elikkottil and Dhruv Matani updated our rankings engine. Vincent Tseui and Wenbin Lin built the `www.whoisbigger.com` site into what it is today. Qi Zhou contributed background research, data curation, and other bits of analysis. Goutham Bhat polished our Ngram data to perfection. The entire Data Science Laboratory (most recently, Rami al-Rfou, Bryan Perozzi, and Yanqing Chen) contributed to general discussions and generated some of the data underlying our analysis. We thank the National Science Foundation for their interest in our work, which was partially supported by NSF Grant IIS-1017181.

We thank the dozen book agents who passed on this project, driving us safely back into the arms of Lauren Cowles at Cambridge University Press. It has always been a pleasure to work with her. We particularly appreciate her help in shaping this manuscript. We also thank other Cambridge people who helped to get our book into your hands: Rachel Ewen, Josh Penney, Melissanne Scheid, Katy Strong, and Karen Verde. Finally, we thank other friends who read preliminary versions of our book, including Michael Holloway, Theo Pavlidis, Scott Perlman, Marshall Poe, Arnout van de Rijt, Eran Shor, and Jerry and Suzanne Trajan.

Sir *Walter Raleigh (1552–1618)* [313] wrote his history of the world while in prison, awaiting execution. We wrote ours at Stony Brook University. On balance, we feel more blessed by circumstances. In particular, we have enjoyed working with social scientists here to help make sense of our data, including Arnout van de Rijt, Eran Shor, Leonie Huddy, Matt Lebo, and Ellen Key.

PART I

Quantitative History

1

History's Most Significant People

People love lists: the Ten Commandments, the Seven Deadly Sins, and the Four Beatles. But they are fascinated by *rankings*, which are lists organized according to some measure of value or merit. Who were the most important women in history? The best writers or most influential artists? Our least illustrious presidents? Who's bigger: John, Paul, George, or Ringo?

This is a book about measuring the "significance" of historical figures. We do not answer these questions as historians might, through a principled assessment of their individual achievements. Instead, we evaluate each person by aggregating the traces of millions of opinions in a rigorous and principled manner. We rank historical figures just as Google ranks web pages, by integrating a diverse set of measurements about their reputation into a single consensus value.

Significance is related to fame but measures something different. Forgotten U.S. president *Chester A. Arthur (1829–1886)* [499] is more historically significant than young pop singer *Justin Bieber (1994–)* [8633], even though he may have a less devoted following and lower contemporary name recognition. Significance is the result of social and cultural forces acting on the mass of an individual's achievement. We think you will be impressed by the extent to which our results capture what you think of as "historical significance." And our computational, data-centric analysis provides new ways to understand and interpret the past.

1.1 People as Memes

We will be interested in the concept of people as *memes*, simple ideas that reproduce when spread from mind to mind. Memes were introduced by *Richard Dawkins (1941–)* [1630] in his book *The Selfish Gene* [Dawkins, 1990]. He observed that ideas undergo the same processes of natural selection and modification as that of biological species, and hence can be studied using the same tools of evolutionary theory.

For example, the "teenaged pop star" meme that is *Justin Bieber (1994–)* [8633] reproduces every time someone reads his Wikipedia page, or he makes news for some performance or gossip-worthy transgression. It weakens whenever a newly grown-up fan removes his poster from the bedroom wall. The Bieber meme will continue to thrive until some future star comes to occupy his particular environmental niche.

Many historical figures reduce to small stories of who they are and why they are known. The meme of *Betsy Ross (1752–1836)* [2430] as the "woman who first sewed the American flag" is an excellent example. It does not really matter whether she actually did sew the first flag (the evidence isn't very strong here) but catching this meme is valuable as a cultural reference in American colonial history and the evolution of gender roles.

Thinking about historical figures as memes turns the processes of fame into a legitimate area of study. We can think of people as occupying niches in history, analogous to how species thrive in particular ecological systems. Sometimes cultural niches disappear, along with memories of all those who occupied them. Historical figures are always in danger of being displaced, whenever stronger but analogous memes rise up to replace them.

Our historical significance measures can be thought of as a quantitative tool to measure the strength of historical memes. We will use this tool to highlight the forces at work in building popular history.

1.2 Our 100

So we have a ranking for you. Figures 1.1 and 1.2 present our ranking of the 100 most significant historical figures according to our computational methods.

Rank	Name	Dates	Description
1	Jesus	(7 B.C.–A.D. 30)	Central figure of Christianity
2	Napoleon	(1769–1821)	Emperor of France (Battle of Waterloo)
3	Muhammad	(570–632)	Prophet and founder of Islam
4	William Shakespeare	(1564–1616)	English playwright (*Hamlet*)
5	Abraham Lincoln	(1809–1865)	16th U.S. president (U.S. Civil War)
6	George Washington	(1732–1799)	1st U.S. president (American Revolution)
7	Adolf Hitler	(1889–1945)	Fuehrer of Nazi Germany (World War II)
8	Aristotle	(384–322 B.C.)	Greek philosopher and polymath
9	Alexander the Great	(356–323 B.C.)	Greek king and conqueror of the known world
10	Thomas Jefferson	(1743–1826)	3rd U.S. president (Decl. of Independence)
11	Henry VIII	(1491–1547)	King of England (six wives)
12	Charles Darwin	(1809–1882)	Scientist (Theory of Evolution)
13	Elizabeth I	(1533–1603)	Queen of England (The Virgin Queen)
14	Karl Marx	(1818–1883)	Philosopher ("Communist Manifesto")
15	Julius Caesar	(100–44 B.C.)	Roman general and statesman ("Et tu, Brute?")
16	Queen Victoria	(1819–1901)	Queen of Britain (Victorian Era)
17	Martin Luther	(1483–1546)	Protestant Reformation (95 Theses)
18	Joseph Stalin	(1878–1953)	Premier of USSR (World War II)
19	Albert Einstein	(1879–1955)	Theoretical physicist (Relativity)
20	Christopher Columbus	(1451–1506)	Explorer, discoverer of the New World
21	Isaac Newton	(1643–1727)	Scientist (Theory of Gravity)
22	Charlemagne	(742–814)	First Holy Roman Emperor ("Father of Europe")
23	Theodore Roosevelt	(1858–1919)	26th U.S. President (Progressive Movement)
24	Wolfgang Amadeus Mozart	(1756–1791)	Austrian composer (*Don Giovanni*)
25	Plato	(427–347 B.C.)	Greek philosopher (*Republic*)
26	Louis XIV	(1638–1715)	King of France ("The Sun King")
27	Ludwig van Beethoven	(1770–1827)	German composer ("Ode to Joy")
28	Ulysses S. Grant	(1822–1885)	18th U.S. president and Civil War general
29	Leonardo da Vinci	(1452–1519)	Italian artist and polymath ("Mona Lisa")
30	Augustus	(63 B.C.–A.D. 14)	First Emperor of Rome (Pax Romana)
31	Carl Linnaeus	(1707–1778)	Swedish biologist (Father of Taxonomy)
32	Ronald Reagan	(1911–2004)	40th U.S. president (Conservative Revolution)
33	Charles Dickens	(1812–1870)	English novelist (*David Copperfield*)
34	Paul the Apostle	(A.D. 5–A.D. 67)	Christian apostle and missionary
35	Benjamin Franklin	(1706–1790)	Founding father/scientist (captured lightning)
36	George W. Bush	(1946–)	43rd U.S. president (Iraq War)
37	Winston Churchill	(1874–1965)	Prime minister of Britain (World War II)
38	Genghis Khan	(1162–1227)	Founder of the Mongol Empire
39	Charles I	(1600–1649)	King of England (English Civil War)
40	Thomas Edison	(1847–1931)	Inventor (light bulb, phonograph)
41	James I	(1566–1625)	King of England (King James Bible)
42	Friedrich Nietzsche	(1844–1900)	German philosopher ("God is dead")
43	Franklin D. Roosevelt	(1882–1945)	32nd U.S. President (New Deal, World War II)
44	Sigmund Freud	(1856–1939)	Neurologist and creator of psychoanalysis
45	Alexander Hamilton	(1755–1804)	U.S. Founding Father (National Bank)
46	Mohandas Karamchand Gandhi	(1869–1948)	Indian nationalist leader (Nonviolence)
47	Woodrow Wilson	(1856–1924)	28th U.S. president (World War I)
48	Johann Sebastian Bach	(1685–1750)	Classical composer (Well-Tempered Clavier)
49	Galileo Galilei	(1564–1642)	Italian physicist and astronomer
50	Oliver Cromwell	(1599–1658)	Lord Protector of England (English Civil War)

FIGURE 1.1. The 100 Most Historically Significant Figures (1–50).

Rank	Name	Dates	Description
51	James Madison	(1751–1836)	4th U.S. president (War of 1812)
52	Gautama Buddha	(563–483 B.C.)	Central figure of Buddhism
53	Mark Twain	(1835–1910)	American author (*Huckleberry Finn*)
54	Edgar Allan Poe	(1809–1849)	American author ("The Raven")
55	Joseph Smith	(1805–1844)	American religious leader (Mormonism)
56	Adam Smith	(1723–1790)	Economist (*The Wealth of Nations*)
57	David	(1040–970 B.C.)	Biblical King of Israel (Jerusalem)
58	George III	(1738–1820)	King of England (American Revolution)
59	Immanuel Kant	(1724–1804)	German philosopher (*Critique of Pure Reason*)
60	James Cook	(1728–1779)	Explorer and discoverer of Hawaii, Australia
61	John Adams	(1735–1826)	Founding Father and 2nd U.S. President
62	Richard Wagner	(1813–1883)	German composer (*Der Ring des Nibelungen*)
63	Pyotr Ilyich Tchaikovsky	(1840–1893)	Russian composer (*1812 Overture*)
64	Voltaire	(1694–1778)	French Enlightenment philosopher (*Candide*)
65	Saint Peter	(?–?)	Early Christian leader
66	Andrew Jackson	(1767–1845)	7th U.S. president ("Old Hickory")
67	Constantine the Great	(272–337)	Emperor of Rome (First Christian emperor)
68	Socrates	(469–399 B.C.)	Greek philosopher and teacher (Hemlock)
69	Elvis Presley	(1935–1977)	The "king of rock and roll"
70	William the Conqueror	(1027–1087)	King of England (Norman Conquest)
71	John F. Kennedy	(1917–1963)	35th U.S. president (Cuban Missile Crisis)
72	Augustine of Hippo	(354–430)	Early Christian theologian ("The City of God")
73	Vincent van Gogh	(1853–1890)	Post-impressionist painter ("Starry Night")
74	Nicolaus Copernicus	(1473–1543)	Astronomer (Heliocentric cosmology)
75	Vladimir Lenin	(1870–1924)	Soviet revolutionary and Premier of USSR
76	Robert E. Lee	(1807–1870)	Confederate General (U.S. Civil War)
77	Oscar Wilde	(1854–1900)	Irish author and poet (*Dorian Gray*)
78	Charles II	(1630–1685)	King of England (Post-Cromwell)
79	Cicero	(106–43 B.C.)	Roman statesman and orator (*On the Republic*)
80	Jean-Jacques Rousseau	(1712–1778)	Philosopher (*On the Social Contract*)
81	Francis Bacon	(1561–1626)	English scientist (Scientific method)
82	Richard Nixon	(1913–1994)	37th U.S. president (Watergate)
83	Louis XVI	(1754–1793)	King of France (executed in French Revolution)
84	Charles V	(1500–1558)	Holy Roman Emperor (Counter-Reformation)
85	King Arthur	(?–?)	Mythical 6th-century King of Britain
86	Michelangelo	(1475–1564)	Italian sculptor and Renaissance man (David)
87	Philip II	(1527–1598)	King of Spain (Spanish Armada)
88	Johann Wolfgang von Goethe	(1749–1832)	German writer and polymath (*Faust*)
89	Ali	(598–661)	Early Caliph and a central figure of Sufism
90	Thomas Aquinas	(1225–1274)	Italian theologian ("Summa theologiae")
91	Pope John Paul II	(1920–2005)	20th-century Polish Pope (Solidarity)
92	René Descartes	(1596–1650)	French philosopher ("I think, therefore I am")
93	Nikola Tesla	(1856–1943)	Inventor (alternating current)
94	Harry S. Truman	(1884–1972)	33rd U.S. president (Korean War)
95	Joan of Arc	(1412–1431)	French military leader and saint
96	Dante Alighieri	(1265–1321)	Italian poet (*Divine Comedy*)
97	Otto von Bismarck	(1815–1898)	1st chancellor and unifier of modern Germany
98	Grover Cleveland	(1837–1908)	22nd and 24th U.S. president
99	John Calvin	(1509–1564)	French Protestant theologian (Calvinism)
100	John Locke	(1632–1704)	English Enlightenment philosopher (Tabula rasa)

FIGURE 1.2. The 100 Most Historically Significant Figures (51–100).

Please study our rankings for a while. We are confident that you will have at least a nodding familiarity with most of these people. Grade yourself on how many of our choices you have heard of: knowing 70 is a C, 80 earns a B, and 90 will get you on an A. We are pretty sure we have a lot of A students/readers out there, but if you're not yet one of them, consider this book your opportunity to meet some new people.[1]

We don't expect you will agree with everyone chosen for the top 100, or exactly where they are placed. But we trust you will agree that most selections are reasonable: a mix of famous people including the major pillars of Western civilization. A quarter of them are philosophers or major religious figures, plus eight scientists/inventors, thirteen giants in literature and music, and three of the greatest artists of all time.

The success of our ranking methods is best established by the banality of our results. You should be reassured by your familiarity with our top 100, instead of being startled by our claims: say, if we promoted *Francis Scott Key (1777–1843)* [1050] as a critical historical figure. Our methods summarize the knowledge of all the authors and readers of the English-language Wikipedia, to order historical figures consistent with the general views of this community. By definition, you should see the names here that you expect to see.

1.3 Other People's Rankings

Historical judgment is subjective. Scholars continue to argue about the causes of wars and other great events. Political and cultural biases come into play, and the past is always being reinterpreted. There is no replacement for the critical process of highly trained scholars to the workings of the humanities and social sciences. And yet, we can learn important things about the past by studying its traces using computational methods.

We are by no means the first people to publish rankings of the most significant people in history. Over the course of this project, we have uncovered more than three dozen published rankings of the (typically) top 100 people in one historical domain or another. But we believe that we are the first to do so using a rigorous statistical methodology, which avoids some of the vagaries of individual human opinion. To better understand the strengths and limitations of our algorithmic methods, we will compare our rankings to two prominent published rankings of historical figures.

[1] Short descriptions of each member of our top 100 appear in Appendix C, in case you want to become more familiar with someone.

1.3.1 MICHAEL HART'S *THE 100*

The 100 [Hart, 1992] is probably the best known ranking of historic figures by influence. It has sold more than half a million copies since the first edition in 1978. I (Steve) owned one of those copies back in high school, which no doubt stimulated my interest in both history and ranking.

Hart himself is a curious character, with graduate degrees in physics, astronomy, law, and computer science. His writings embrace a variety of controversial topics, pegging *Edward de Vere (1550–1604)* [1603] as the author of *William Shakespeare's (1564–1616)* [4] plays and supporting racial/ethnic separation both in the United States and abroad. Still, his biographies in *The 100* make informed and stimulating reading. We will study his rankings from the revised 1992 edition of the book.

Hart's top 100 and our own share many historical figures in common. What is more enlightening is to study where our rankings sharply differ. We start by identifying the ten people in his 100 who are ranked lowest by our methods.

Bottom of the Hart 100

Us	Hart	Person	Dates	Description
47910	82	Gregory Goodwin Pincus	(1903–1967)	American biologist (oral contraceptive pill)
7233	37	William T. G. Morton	(1819–1868)	Dentist and pioneer of anesthesia
6950	7	Cai Lun	(A.D. 50–121)	Chinese inventor (paper)
5746	96	Menes	(?–?)	First pharaoh of ancient Egypt
5663	61	Nikolaus Otto	(1832–1891)	German inventor (internal combustion engine)
4724	85	Emperor Wen of Sui	(541–604)	Founder of China's Sui Dynasty
3005	47	Louis Daguerre	(1787–1851)	French inventor of photography
2751	83	Mani	(216–276)	Prophet and the founder of Manichaeism
2732	71	Wilhelm Röntgen	(1845–1923)	German physicist (X-rays)
1835	92	Mencius	(372–289 B.C.)	Chinese thinker (Confucianism)

The least significant member of Hart's list is *Gregory Goodwin Pincus (1903–1967)* [47910], who is promoted as the father of the oral contraceptive pill. The Pill has indeed changed the world, but we think he has honored the wrong man (or, in particular, woman) here. *Carl Djerassi (1923–)* [47277] was the scientist who developed the compound (Norethirsterone) that became the first practical oral contraceptive. *Margaret Sanger (1879–1966)* [2672] was the activist who established Planned Parenthood, and was responsible for the funding that Pincus used to validate Djerassi's compound. We rank Sanger as a far more significant figure than Pincus.

Hart's rankings glorify technological achievement, but his heart lies with the underdog. We would contest his choice of the seminal figure in several other areas as well:

- Hart credits *Nikolaus Otto (1832–1891)* [5663], inventor of the four-stroke internal combustion engine, as the pioneer of the automobile. But we more highly rank *Gottlieb Daimler (1834–1900)* [1461] and *Karl Benz (1844–1929)* [840], who actually built the first cars.
- Hart credits *Louis Daguerre (1787–1851)* [3005] as the pioneer of photography, but he was just one of several inventors with diverse chemical processes for recording images, like his rival *William Fox Talbot (1800–1877)* [2650]. Our choice for the real father of photography was *George Eastman (1854–1932)* [1584], whose invention of roll film and the Eastman Kodak camera led the way to the modern photographic era.
- Hart recognizes *Menes* [5746], the first pharaoh of the first dynasty. Legend credits him with uniting Upper and Lower Egypt, but there is little evidence of his existence in the historical record. Instead, we identify *Ramesses II (1302–1213 B.C.)* [293] as the most significant pharaoh, who ruled Egypt for 66 years during its time of greatest power.

Hart omitted several of our top 100 from his rankings who prove to be much stronger vessels. We are happy to find room for *Abraham Lincoln (1809–1865)* [5], *Henry VIII (1491–1547)* [11], and *Wolfgang Amadeus Mozart (1756–1791)* [24] ahead of anesthesia pioneer *William T. G. Morton (1819–1868)* [7233] or *Werner Heisenberg (1901–1976)* [1659], a great scientist but one who doesn't crack our rankings of the top five modern physicists.

Missing from the Hart 100

Us	Hart	Person	Dates	Description
4		William Shakespeare	(1564–1616)	English playwright (*Hamlet*)
5		Abraham Lincoln	(1809–1865)	16th U.S. president (U.S. Civil War)
11		Henry VIII	(1491–1547)	King of England (six wives)
16		Queen Victoria	(1819–1901)	Queen of Britain (Victorian Era)
23		Theodore Roosevelt	(1858–1919)	26th U.S. President (Progressive Movement)
24		Wolfgang Amadeus Mozart	(1756–1791)	Austrian composer (*Don Giovanni*)
26		Louis XIV	(1638–1715)	King of France ("The Sun King")
28		Ulysses S. Grant	(1822–1885)	18th U.S. president and Civil War general
29		Leonardo da Vinci	(1452–1519)	Italian artist and polymath ("Mona Lisa")
31		Carl Linnaeus	(1707–1778)	Swedish biologist (Father of Taxonomy)

1.3.2 *LIFE* MAGAZINE'S 100 MOST INFLUENTIAL FIGURES OF THE MILLENNIUM

The year 2000 provoked many backward glances at mankind's achievements over the past one thousand years, particularly a popular ranking from *Life* Magazine (2000). They neglect figures from ancient times and the early Middle Ages, but we only have twenty-ish figures from these periods ourselves, leaving enough shared people for a reasonable comparison with our rankings.

The relative order of *Life*'s rankings correlate better with ours (0.54) than Hart's rankings did (0.31), so we respect their choices more. Still, there are revealing differences.

Life managed to find room for the "Four-Minute Miler" *Roger Bannister (1929–)* [11095], Chinese landscape painter *Fan Kuan (1020–1030)* [35313], and medieval music theorist *Guido of Arezzo (991–1033)* [6215]. All were at the expense of *George Washington (1732–1799)* [6], *Joseph Stalin (1878–1953)* [18], *Winston Churchill (1874–1965)* [37], and others in our top 100.

Bottom of the *Life* 100

Us	Life	Person	Dates	Description
35313	59	Fan Kuan	(1020–1030)	Chinese landscape painter
14490	67	Cao Xueqin	(1715–1763)	Chinese classical writer
11095	92	Roger Bannister	(1929–)	English athlete (four-minute mile)
7177	65	Hiram Stevens Maxim	(1840–1916)	American inventor (Maxim gun)
6215	62	Guido of Arezzo	(991–1033)	Medieval music theorist
3774	99	Kwame Nkrumah	(1909–1972)	Founding leader of Ghana
3302	72	Simone de Beauvoir	(1908–1986)	French existentialist philosopher
3005	79	Louis Daguerre	(1787–1851)	French inventor of photography
2799	96	Jacques Cousteau	(1910–1997)	French oceanographic explorer
2061	45	Zhu Xi	(1130–1200)	Chinese Confucian scholar

Missing from the *Life* 100

Us	Life	Person	Dates	Description
6		George Washington	(1732–1799)	1st U.S. president (American Revolution)
11		Henry VIII	(1491–1547)	King of England (six wives)
18		Joseph Stalin	(1878–1953)	Premier of USSR (World War II)
23		Theodore Roosevelt	(1858–1919)	26th U.S. president (Progressive Movement)
24		Wolfgang Amadeus Mozart	(1756–1791)	Austrian composer (*Don Giovanni*)
28		Ulysses S. Grant	(1822–1885)	18th U.S. president and Civil War general
32		Ronald Reagan	(1911–2004)	40th U.S. president (Conservative Revolution)
33		Charles Dickens	(1812–1870)	English novelist (*David Copperfield*)
35		Benjamin Franklin	(1706–1790)	Founding father/scientist (captured lightning)
36		George W. Bush	(1946–)	43rd U.S. president (Iraq War)

We think these comparisons with the Hart and *Life* rankings validate the basically sensible nature of our historical significance analysis. We have performed far more substantial evaluations (see Section 2.5) which demonstrate that (a) our rankings show an excellent correlation with published rankings by human experts, and (b) we correlate better with these experts than they do among themselves. This supports the argument that our rankings are better than those produced by human experts. We will also show that our significance rankings can be used to predict the prices of such diverse commodities as celebrity autographs, baseball cards, and modern paintings.

1.4 History vs. Historiography

We are not historians, and this is not a history book. *History* is the study of past events, with the emphasis on what happened and why. We have nothing new to say in this book that will help historians reconstruct past events, or understand the driving forces that made them unfold.

Certain authorities reasonably declare *Gavrilo Princip (1894–1918)* [2814] to be the most significant historical figure of the twentieth century. *Who?* Princip was the Bosnian/Serbian nationalist who assassinated *Archduke Franz Ferdinand of Austria (1863–1914)* [541]. This event precipitated World War I, which redrew the political boundaries of Europe and sowed the seeds for World War II. Without Princip, none of this might ever have happened. Historians are concerned with such arguments, as they struggle to understand the root causes of events that shaped our world. Our analysis cannot really help them in their quest.

By contrast, *historiography* is the study of the field of history itself. We are interested in what makes the lives of various historical personages worth recording and passing on to future generations. Why is it that the reader probably could place *John J. Pershing (1860–1948)* [764] or *Jacqueline Kennedy Onassis (1929–1994)* [1387] much more easily than Mr. Princip? There are larger patterns at work here that make this phenomenon worth studying.

Indeed, throughout this book we will see several forces acting on our collective memory to determine which figures get preserved for posterity:

- *Founder Effects* – Among the most enduring figures are those recognized as first in some way: say, the first European to discover America (*Christopher Columbus (1451–1506)* [20]) or the first president (*George Washington (1732–1799)* [6]). Both turn out to be arguable distinctions: read about *Leif Ericson (970–1020)* [563] or *John Hanson (1721–1783)* [4024]. But once canonized as first, they cannot ever be dislodged.

  Founder figures, like explorers and inventors, last longer than political or cultural leaders whose reputations are not anchored to a single enduring achievement. These effects are strong enough to overcome the Anglo-centric bias of Wikipedia: Russian cosmonaut *Yuri Gagarin (1934–1968)* [2324] beat the first American into space (*Alan Shepard (1923–1998)* [6750]) by only three weeks, yet by 4,446 spots in historical rank.
- *Role Replacement* – Conversely, other cultural niches are refilled on a regular basis, condemning earlier figures to reduced significance with time. Olympic champions are replaced every four years, as are losing presidential candidates. The meme of "greatest living saint" has transferred from *Albert Schweitzer (1875–1965)* [1512] to *Mother Teresa (1910–1997)* [820] to the Dalai Lama, no doubt to be recycled again soon after his passing.
- *Narrative Simplification* – As history recedes into the past, the story gets simplified, with supporting players granted a smaller fraction of the tale. Western outlaw *Jesse James (1847–1882)* [413] used to share almost equal billing with his brother *Frank James (1843–1915)* [4608], who is now barely remembered as a member of the gang. Early contributors to the Civil Rights movement have become much less visible in the shadow of *Martin Luther King (1929–1968)* [221], a trend we anticipate will only increase with time.
- *Storification* – Details with little objective impact on stature help fix certain historical figures in the collective memory. Untimely death, particularly martyrdom, proves a great preservative. Others hang on a single saying, like *Nathan Hales's (1755–1776)* [1090] "regret that I have only one life to give for my country" or *Patrick Henry's (1736–1799)* [472] "Give me Liberty, or give me Death!"

> Memes must be easy to spread, necessitating capsule summaries that make them easy to remember. Most of the tables of ranked figures we provide in this book include brief (45-character) descriptions of each person, but it is amazing how often that suffices for their unambiguous placement in the canon.

We intend this book as both a survey of historical figures for a general audience and a genuine academic contribution to historiography. We will expose you, the reader, to more than a thousand of the historical memes most worth knowing. Our algorithmic rankings help focus attention where it should properly be focused. Any person interested in livestock production would miss the point if they study how to raise geese and Angora rabbits at the expense of cows, pigs, and chickens. Our rankings help answer the question "Where's the beef?" in history in a rigorous and effective way. The memes most worth knowing are usually most *interesting* to read about, so you should look forward to making their acquaintance (or re-acquaintance) here.

That said, we believe there is generally a high correlation between the strength of a given historical meme (what we will call *significance* in this book) and its "true" importance as would be measured by historians. Several times we will put our measure to the test against objective standards in particular domains: history textbooks, Hall of Fame electoral results, statistical evidence of sports performance, published rankings of notable people, and more. In all these experiments, our historical significance rankings do an excellent job of capturing the perceived importance of members in the target group.

This multipronged approach to assessment is important, because it validates our significance scores as a tool for scholarly research on large-scale historical/historiological forces. We employ it to study questions like the effectiveness of human decision processes, or measuring the degree to which women have been excluded from the historical record. *Culturomics* is a new paradigm employing massive datasets to answer questions in the social sciences and humanities. We look forward to watching future scholars make even better use of our data and analytical methods.

1.5 Overview

By now, we hope you are convinced that our historical significance rankings are generally sensible, and worth further investigation. To help you, our book is organized into three parts.

The first, *Quantitative History*, introduces our analytical methods and explains how our algorithmic rankings were constructed. We introduce the datasets and ranking methods necessary to make reasoned statistical judgments about historical significance, tools that we will use to order the individuals marching in the vast parade of history. We then deploy our analysis to study several "Big Picture" questions in history and popular culture. Who belongs in children's textbooks? How accurately do experts recognize the future historical significance of then-contemporary figures? Are women and other groups sufficiently represented in the historical record?

The second part of our book, *Historical Rankings*, presents rankings of people within specific spheres of influence, for example, world leaders, athletes, artists, or outlaws. It is quite revealing to identify the top dog in every domain, with results that will enlighten most readers and infuriate others. Our rankings provide an efficient way to review and expand your knowledge of the world. Indeed, the 1,000 names we drop in this book make it a crash course in historical Who's Who.

At the end of the book are two appendices that provide resources to help you continue your quantitative investigations into history. Our website `http://www.whoisbigger.com` is the main electronic supplement to this book, which includes our latest rankings for all historical figures. There is even an App available for game players to test their knowledge.

Ready? Let our journey into the past begin....

2

Ranking Historical Figures

Our rankings of historical significance will be used to rationally assess the reputation of figures in the historical canon. But how can algorithms do this in a fair and sensible way? We start by reviewing how people approach ranking problems, to better understand the strengths and limitations of our computational methods.

2.1 Traditional Ranking Methodologies

The rankings most prominent in popular culture are produced in several ways. Perhaps most popular is the *expert poll*, used to create the Associated Press Top 25 College Football Rankings. Every week a set of experts (here coaches and sportswriters) independently rank the top teams. These ratings are combined using a point system, and the cream skimmed off after sorting teams by points yields the Top 25. Expert polls prove effective in clearly defined domains followed by many knowledgeable individuals. They tend to be fairly conservative, however, and are often based on second-hand judgments. Do we really believe that active college football coaches have enough time, or even a good enough cable television plan, to watch all their peers play?

Still, polls of professional historians might be used to rank historical figures. Indeed, historian polls ranking the greatness of U.S. presidents are conducted regularly as a sort of academic parlor game; an early example was the 1948 poll by *Arthur M. Schlesinger (1888–1965)* [27072] of Harvard University. These polls reflect interesting changes in historical reputation over time. Opinions of *Dwight D. Eisenhower (1890–1969)* [110] and *Ronald Reagan (1911–2004)* [32] have generally been rising in recent years, at the expense of *Grover Cleveland (1837–1908)* [98] and *Herbert Hoover (1874–1964)* [183]. But these polls are subject to political and contemporary biases. Indeed, in our charged political climate, could *anyone* give an unbiased judgment on the relative merits of Republican *George W. Bush (1946–)* [36] versus Democrat *Barack Obama (1961–)* [111]?

Further, expert polls do not scale well. There are simply too many historic personages and too few experts, particularly those with the broad knowledge to compare inventors to poets, politicians to entertainers, and athletes to philosophers.

An alternate approach might poll the public instead of the experts. The *wisdom of crowds* hypothesis [Surowiecki, 2004] dictates that large, diverse groups can make better decisions than individual experts. How much you believe this reflects your faith in democracy: does the voting public generally do a good job in identifying the right candidate for political office? The Internet makes it much easier to conduct public polls on a massive scale, now employed by sports leagues to select their All-Star teams. Technology also enables more passive means of gathering popular sentiment, such as analyzing the content of search engine queries. That more people google *Selena Gomez (1992–)* [5365] than *Lefty Gomez (1908–1989)* [20494] tells you something meaningful about their relative fame. Indeed, the Internet Movie Database (IMDb) *STARmeter* ranks celebrities by the frequency of recent search queries, and seems quite effective at distinguishing who's hot from who's not.

But an accurate public poll requires a large audience of knowledgeable and enthusiastic participants. This seems more likely with movie stars than nineteenth-century poets. Further, search engine queries track notoriety far better than historical significance; otherwise the greatness of a political leader would rise whenever they got caught in bed with a mistress.

Yet another approach to ranking identifies a single salient statistic, and uses it to order things accordingly. In the arts, this statistic is often (ironically) sales: the top-40 music countdown, the *New York Times* best-selling books, and the highest-grossing films of the year. University reputations are measured by the average SAT scores of the incoming class, or the percentage of student applicants accepted.

But single statistics paint a one-dimensional picture of reputation. Only an investor would argue that the highest-grossing movies are the best films. Indeed, the Oscars are awarded through an expert poll. Critical rankings of the best books by quality or influence tend to be quite independent of sales. Individual statistics can be easy to game, producing misleading or self-serving results. Publishers have been known to buy multiple copies of their own product just to prop up its rank on the best seller list, and universities have become increasingly adept at omitting suspect classes of students from their self-reported "average" SAT.

A final approach mixes several statistical measures in a witch's brew to yield a single authoritative ranking. Representative here are the *U.S. News and World Report* 2011 rankings of American universities, which combine factors like faculty/student ratio, graduation rates, and expert polls into a single score. Such meta-analyses can be compelling, but they must be performed with a level of rigor to have genuine meaning. *U.S. News* needs its university rankings to change substantially every year, so its customers must buy new editions. It manages this by tinkering with its rating formula. However, when done with the proper data, integrity, and care, such meta-analysis can produce good results.

We take a meta-analysis approach in this book to rank historical significance. We combine a diverse set of data sources to determine the most famous people in the English-speaking world. Our cast of characters includes all the people appearing in Wikipedia as of a specific date: 843,790 strong, or roughly the population of San Francisco. Our computational analysis ranks them by significance in a satisfying and convincing way. As we showed back in in Figure 1.1, *Jesus (7 B.C.–A.D. 30)* [1] is the most significant figure in world history. The least significant, *Sagusa Ryusei (1954–)* [843790],[1] is a Yoshinkan aikido master whose claim to fame is being the student of a better-known master.

2.2 Measurements of Historical Reputation

Our research on statistical ranking methods has developed ways to make informed judgments about relative significance in Anglo-American culture.

[1] In case you have not caught on to our convention, each person mentioned in this book is annotated with their birth/death dates and significance ranking.

Here we explain what is going on under the hood. There are three major aspects of our methodology: (1) raw measurements of historical reputation, i.e., data, (2) statistical factor analysis to combine them into a single score, and (3) corrections for the decay in reputation that occurs with the passing of time. We deal with each of these in turn in this chapter.

2.2.1 WIKIPEDIA

Let's talk about data. Historically significant figures leave statistical evidence of their presence behind, if one knows where to look for it. We use several data sources to fuel our ranking algorithms. Most important is Wikipedia [Wales, 2009], the web-based, collaborative, multilingual encyclopedia project supported by the nonprofit Wikimedia Foundation. But we use it quite differently from the typical reader.

Wikipedia is enormous, featuring more than three million articles in its English edition alone. It got that way by being *open*. Anyone on the Internet can contribute what they wish to Wikipedia. You are free to start a new article or edit any existing one. Perhaps you are an expert on Polynesian butterflies, and catch a minor but gnawing error in its coverage of a particular species. You have the power to fix it, and make the world a better place. The same would be true for any person knowledgeable about actor *Humphrey Bogart (1899–1957)* [1291], cellist *Pablo Casals (1876–1973)* [3112], or football player *Bubba Smith (1945–2011)* [102599].

This open collaborative model might seem an invitation to abuse, but the Wikipedia community has developed effective mechanisms to harness users' enthusiasm while enforcing proper behavior. Studies have shown that the general accuracy of Wikipedia is comparable to traditional centrally edited encyclopedias. In particular, the journal *Nature* has reported that Wikipedia's scientific articles had a similar rate of "serious errors" as the authoritative *Encyclopedia Britannica*, and came close to its general level of accuracy [Giles, 2005].

We second their opinion. Yes, there are corporate flacks lurking who clean up the sordid reputations of their employers [Hafner, 2007]. Yes, there are true believers willing to rewrite history at the behest of political or spiritual leaders. And yes, jokers or conspiracy theorists can use Wikipedia to claim anyone did anything. For example, respected editor

John Seigenthaler (1927–) [12815] was understandably perturbed to read about his involvement in killing both *John F. Kennedy (1917–1963)* [71] and *Robert F. Kennedy (1925–1968)* [1355] [Seelye, 2005].

But such errors typically disappear instantly after discovery, and are but a dirty drop in the vast ocean of truth that has been properly reported. The deeper we have delved into analyzing Wikipedia, the more we have been impressed with the general sensibility and correctness of this amazing human artifact. And because we work in volume, our algorithms will not be particularly affected by any specific factual errors.

The openness of Wikipedia proves critical to our analysis here, because the entire bundle is freely available for download – all the text, all the metadata – everything. This means we can write computer programs that use Wikipedia as *data*, inputs to fuel our models of historical reputation. Implicit in the structure and contents of Wikipedia are several quantitative measures of the "fame" or "achievement" of each entity. We will use these as input to our reputation model.

Our statistical manipulations get somewhat technical, so certain details have been relegated to Appendix A. But understanding the basic model is helpful for making sense of our results, and an interesting exercise to boot. So let's get started.

2.2.2 PAGERANK

Web search engines such as Google and Bing are spectacularly successful examples of the power of ranking technology. Responding instantly to almost any question we ask, they rank billions of documents from most to least relevant so accurately that surfers rarely look beyond the top few results.

The original secret sauce underlying Google's search engine is a clever and powerful algorithm called *PageRank* [Brin and Page, 1998]. PageRank ignores the textual content of web pages to focus only on the structure of the hyperlinks between them. The fundamental idea is simple: significant pages have more links to them than lesser pages. Further, the more significant a page is, the stronger a recommendation it lends to any page *it* links to. Having a large stable of contacts recommending you for a job is great, but it is even better when one of them served as the president of the United States.

A Random Walk Along the Network

PageRank is best understood in the context of a random walk along a network. Suppose we start from an arbitrary web page and then randomly select an outgoing link from the set of possibilities. This puts you on a neighboring web page. Now jump to a random neighbor of this web page, and repeat for as many random hops as you please. Each website's PageRank is essentially the probability that you will arrive there after a long series of random clicks.

To get a feel for PageRank, let's take a random walk along Wikipedia pages and see how many recognizable figures we stumble upon. To start, we went to Wikipedia and clicked on the "Random article" button. The article turned out to be *Marfa Sobakina (1552–1571)* [279246], and her story was not a happy one. The third wife of *Ivan the Terrible (1530–1584)* [297], she was poisoned and died a few days after her marriage.

Although herself obscure, she leads us directly to *Ivan*, a signature figure of his age. One random step away is his son, *Tsarevich Ivan Ivanovich of Russia (1554–1581)* [50371]. This is another unhappy story: *Ivan* killed him after beating his pregnant wife into a miscarriage. The son leads us to *Ivan*'s first wife, *Anastasia of Russia (1530–1560)* [19255], also possibly poisoned. Our random walk starts out as a trip through Russian genealogy, but after seven more random hops we land on *Henry VIII (1491–1547)* [11]. Now we are back in the big leagues again. *Henry VIII* was the child of *Henry VII (1457–1509)* [179], who defeated *Richard III (1452–1485)* [344] at the Battle of Bosworth Field to claim the English throne. And, of course, *Richard III* was the subject of an eponymous play by *William Shakespeare (1564–1616)* [4].

Wikipedia is a hyperlinked collection of web pages just like the rest of the web, so we can use PageRank to compute the probability that random Wikipedia surfing will lead to a specific page. Unlike the rest of the web, however, each article in Wikipedia can be considered as a proxy for its subject. That the Wikipedia page for *George Washington (1732–1799)* [6] has a higher PageRank than the page for former Met first baseman *Marv Throneberry (1933–1994)* [171634] is an informed judgment that the Father of Our Country was a more significant figure than Marvelous Marv.

PageRank in Practice

PageRank relies on the idea that if all roads lead to Rome, Rome must be a pretty important place. Additional links to your web page increase your PageRank, but quality matters. You should thank your brother-in-law for linking to your Wikipedia page, but it will be considerably less valuable than a link from a high PageRank fellow like *William Shakespeare (1564–1616)* [4] would be. On the other hand, linking from your page to the Bard shows you to be an erudite fellow, but it does you no good in the PageRank department. The paths *to* your page are what count. This is what makes PageRank hard to game: other people must link to your web page, and whatever shouting you do about yourself is irrelevant.

Adding and deleting links from a given network gives rise to different networks, some of which better reveal underlying significance using PageRank. Google implicitly adds links from every web page to a single super-page (called *G-d*), to ensure that random walks never get trapped in some small corner of the network. It also deletes any link from a page to itself, as not indicative of the significance of anything.

More important, Google deletes all links deemed "link spam," pages specifically created to be read by Google's web-crawling programs instead of people. Suppose I create a new spam website (say `www.gx928zq93.com`), containing multiple links to every page on another site I really care about, like `www.whoisbigger.com`. Left undetected, these links raise the PageRank of `www.whoisbigger.com`, and thus increase its prominence in search engine results. Generating/detecting link spam is a cat-and-mouse game, for the fidelity of any search engine requires meaningful edges in its PageRank graph. Fortunately for our purposes, Wikipedia is relatively free of link spam, because its team of editors quickly swoop down on obviously self-serving links.

How well does PageRank work at smoking out significant people? We examine people with the highest and lowest PageRank (PR1) drawn from a list of 1,000 recognizable figures.

The high-PageRank figures are all readily recognized as very accomplished individuals, which is why they generally peg our "Celebrity/Gravitas" meter.[2] The least familiar person in this table is probably *Carl*

[2] We will discuss exactly how we measure this tradeoff between popular renown (celebrity) and achievement (gravitas) in Section 2.3.

Highest PR1 Ranking

Person	PR1	Sig	Celeb/Grav
Napoleon	1	2	C G
George W. Bush	2	36	C G
Carl Linnaeus	3	31	C G
Jesus	4	1	C G
Barack Obama	5	111	C G
Aristotle	6	8	C G
William Shakespeare	7	4	C G
Elizabeth II	8	132	C G
Adolf Hitler	9	7	C G
Bill Clinton	10	115	C G

Lowest PR1 Ranking

Person	PR1	Sig	Celeb/Grav
Vijay	16693	4456	C G
Daniel Radcliffe	12373	7080	C G
Jesse McCartney	11850	4236	C G
Randy Orton	11093	4123	C G
Ashley Tisdale	10299	4445	C G
Edge	10066	2603	C G
Kane	9065	2229	C G
Rey Mysterio	8581	2740	C G
Nicole Scherzinger	7653	6058	C G
Big Show	7551	4221	C G

Linnaeus (1707–1778) [31], biology's "father of taxonomy" whose genus-species system is used to classify *Homo sapiens* and all other life on earth. He was a great scientist, but why is he *so* very highly regarded by PageRank? The Wikipedia pages of all the plant and animal species he classified link back to him, so thousands of life-forms contribute prominent paths to his page.

The low-PageRank figures on the right are all well-known people, primarily young actors, singers, and professional wrestlers. Almost all of their fame is due to celebrity. Being young, they have not yet had the opportunity to interact with too many other significant people, so their networks are not as well established as more senior folk. They are all extremely *popular*, just not well connected.

The Linnaeus example points out a possible weakness of PageRank: do we really want plants, institutions, and inanimate objects voting on who the most prominent people are? Figure 2.1 (left) shows part of the PageRank graph for *Barack Obama (1961–)* [111]. Although the direct links associated with Obama are all very reasonable, we can reach him in only two clicks from the Wikipedia page for Dinosaurs. Should extinct beasts contribute to the President's significance?

The same techniques that Google uses to deal with link spam can come to our aid. Suppose we compute a second PageRank on only the Wikipedia pages/vertices that correspond to people. We will call this new PageRank PR2. This computation would ignore any contribution from places, organizations, and lower organisms. Figure 2.1 (right) shows a sample of Obama's network when we restrict it to only people.

This second PageRank yields a different measure of significance. There is considerable overlap with the previous rankings, but also some subtle

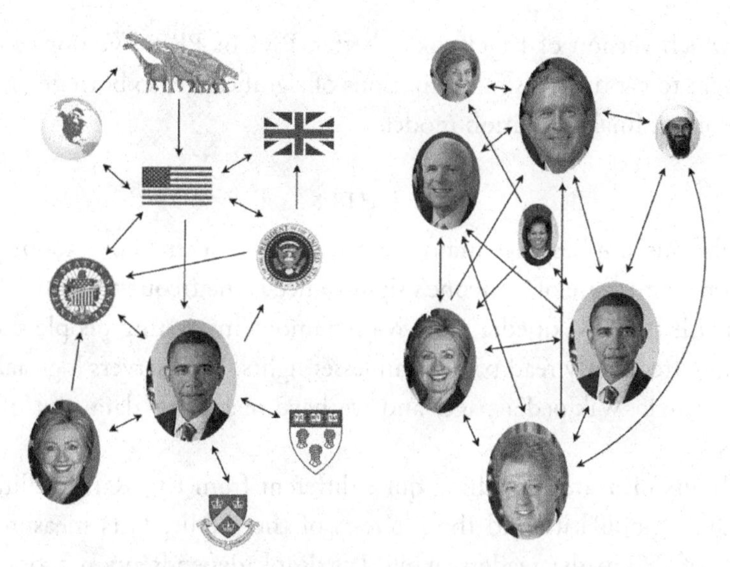

FIGURE 2.1. PageRank graphs for *Barack Obama,* taken over all Wikipedia pages (left) and then restricted to people (right).

changes. On the left, Jesus, Linnaeus, and *Aristotle (384–322 B.C.)* [8] are gone, replaced by three recent U.S. presidents – who clearly have direct connections from many important people. On the right, prominent explorers (*Amerigo Vespucci (1454–1512)* [407] and *John Cabot (1450–1498)* [314]) and open-source technologists (*Richard Stallman (1953–)* [4921] and *Jimmy Wales (1966–)* [3198]) lose ground when they must rely on direct connections from other individuals instead of organizations they founded and places they discovered.

Highest PR2 Ranking

Person	PR2	Sig	Celeb/Grav
George W. Bush	1	36	C G
Bill Clinton	2	115	C G
William Shakespeare	3	4	C G
Ronald Reagan	4	32	C G
Adolf Hitler	5	7	C G
Barack Obama	6	111	C G
Napoleon	7	2	C G
Richard Nixon	8	82	C G
Franklin D. Roosevelt	9	43	C G
Elizabeth II	10	132	C G

Lowest PR2 Ranking

Person	PR2	Sig	Celeb/Grav
Amerigo Vespucci	19812	407	C G
Richard Stallman	17040	4921	C G
The Notorious B.I.G.	16944	3199	C G
Morrissey	15345	3897	C G
John Cabot	14566	314	C G
Ashlee Simpson	14453	4744	C G
Ludacris	13576	3183	C G
Jimmy Wales	13572	3198	C G
Beck	13488	4304	C G
Henry Hudson	13390	353	C G

So which version of PageRank is better, PR1 or PR2? We don't know. Both seem to capture reasonable notions of significance, so both deserve to be part of our final reputation model.

2.2.3 HITS

Being famous implies that many people want to learn more about you. An alternate measure of someone's significance is the frequency with which readers visit their Wikipedia page. More famous/interesting people should have more frequently read pages than lesser lights. Web servers log each hit to every single Wikipedia page, and we have made this data part of our analysis.

Web hits measure something quite different from PageRank, telling us more about popularity and the interests of the public. Hits measure the number of Wikipedia readers, while PageRank depends upon actions by the authors of Wikipedia pages. These are different groups of people, so we should expect them to arrive at somewhat different opinions on matters of historical significance.

Highest Hits Ranking				Lowest Hits Ranking			
Person	Hits	Sig	Celeb/Grav	Person	Hits	Sig	Celeb/Grav
Eminem	3	823	C G	Gough Whitlam	11525	1114	C G
Lady Gaga	5	2502	C G	Paul Martin	9435	3224	C G
Adolf Hitler	6	7	C G	Charles S. Peirce	8564	225	C G
Lil Wayne	7	1803	C G	Brian Mulroney	8055	2596	C G
Katy Perry	9	4737	C G	Joseph Priestley	7809	330	C G
Rihanna	10	1185	C G	Lester B. Pearson	7603	1204	C G
Barack Obama	12	111	C G	Suharto	7599	1836	C G
Michael Jackson	13	180	C G	Anthony Burgess	7420	1430	C G
Kanye West	15	1494	C G	Thomas Henry Huxley	7294	365	C G
Miley Cyrus	16	2009	C G	William L. M. King	7163	668	C G

The high-hits figures[3] are largely contemporary entertainers (such as *Lady Gaga (1986–)* [2502] and *Rihanna (1988–)* [1185]), because young people turn to Wikipedia as a primary source of information. It is disconcerting to see *Adolf Hitler (1889–1945)* [7] appear as the only non-contemporary figure on the list, a frightening measure of the grip he retains on the public consciousness nearly seventy years after his death.

[3] The high-hits ranks do not count from 1 to 10 because certain artifacts like pages associated with unrecognizable people have been excluded from view.

Hit frequency varies over time, responding actively to news events. There is nothing like a juicy scandal to drive more readers to your Wikipedia page. The data we analyze here thus overreacts to people who were in the news during our sampling period. This explains the presence of *Michael Jackson (1958–2009)* [180] (who died shortly before our sampling) and why current president *Barack Obama (1961–)* [111] substantially outranks all his predecessors in hits.

The low-hits people on the right include earnest scholarly figures, like oxygen discoverer *Joseph Priestley (1733–1804)* [330] and philosopher *Charles Sanders Peirce (1839–1914)* [225]. We also get major political figures from important but quiet countries like Canada and Australia. The top figure here, Prime Minister *Gough Whitlam (1916–)* [1114], could be described as the *Richard Nixon (1913–1994)* [82] of Australia.

2.2.4 ARTICLE LENGTH

The length of an individual's Wikipedia article, as measured in words, provides a natural reflection of his or her fame: more significant people deserve longer articles than lesser folk. They generally have amassed more substantial achievements, requiring additional text to describe. Their activities also attract greater interest and scrutiny, meaning more people can contribute to building detailed, lengthy articles.

Article length in Wikipedia is not the same hard constraint that it is in printed texts, where publishers face a real financial cost for each additional printed page. Yet there are clear social pressures toward brevity and conciseness. The Wikipedia community aggressively edits, and indeed deletes, the articles of people whose significance does not measure up to its accepted standards. More than 100,000 people have suffered the ignominy of having their (usually autobiographical) articles removed from Wikipedia.

The people with the longest articles include controversial figures with intense followings like religious figures (*Joseph Smith (1805–1844)* [55] and *L. Ron Hubbard (1911–1986)* [1208]) and revolutionaries (*Che Guevara (1928–1967)* [457] and *Paul Robeson (1898–1976)* [1278]). These constituencies can take advantage of Wikipedia's open access policies to write about people they consider important. This is a real strength of the Wikipedia models, although it does introduce a potential bias into the data.

People with shorter articles than their significance suggests (as in the right-hand side of the table) include *Euclid* [149] and *Euripides (480–406 B.C.)* [348], ancient figures about whom little is known. This leaves little to write about. We also get a gaggle of contemporary celebrities whose accomplishments are too vapid for comment: consider *Vijay (1974–)* [4456], *Ja Rule (1976–)* [6330], and *Fergie (1975–)* [3327].

Highest Length Ranking				Lowest Length Ranking			
Person	Words	Sig	Celeb/Grav	Person	Words	Sig	Celeb/Grav
Adolf Hitler	5	7	C G	Euclid	38130	149	C G
Stanley Kubrick	12	1935	C G	Tony Hawk	36692	3787	C G
Elvis Presley	14	69	C G	Vijay	32752	4456	C G
Joseph Stalin	18	18	C G	Sean Hannity	32352	4566	C G
Joseph Smith	20	55	C G	Ja Rule	32149	6330	C G
L. Ron Hubbard	21	1208	C G	Fergie	30525	3327	C G
Che Guevara	25	457	C G	Euripides	27995	348	C G
Paul Robeson	27	1278	C G	Will Smith	26604	2774	C G
Janet Jackson	28	559	C G	John Travolta	25103	1692	C G
Michael Jackson	29	180	C G	Tim Berners-Lee	23895	3931	C G

The actual difference in length between long and short articles is not quite as large as the differences in rankings might suggest. The longest prominent article (Hitler) runs 29,341 words, while the shortest (Euclid) contains 1,542. As with PageRank, there are several possible ways to measure the length of an article. Do we count all words appearing within the confines of an article, or exclude references, tables, and other structures? Each approach yields somewhat different measures of significance. We use the meat of the text, without references and tables, as our measure of article length.

2.2.5 PAGE EDITS

The collaborative model underlying Wikipedia empowers thousands of authors to contribute their knowledge to the world. Each change to every article is retained in Wikipedia's database, providing a historical record of how the page evolved to its present state. We expect that famous/important people possess more refined articles than lesser personages, because a larger fraction of the editorial body has both motive and opportunity: information to contribute and the desire to do so.

The high-edit people include celebrities with intense popular followings, such as *Michael Jackson (1958–2009)* [180] and *Britney Spears (1981–)*

[689]. These people lead volatile lives, requiring fast-moving encyclopedia articles to keep up with them. Similarly, the frequent flow of events involving active presidents and sports stars triggers many updates to their articles.

Highest Edits Ranking | Lowest Edits Ranking

Person	Edits	Sig	Celeb/Grav	Person	Edits	Sig	Celeb/Grav
George W. Bush	1	36	C G	Tacitus	7248	300	C G
Michael Jackson	2	180	C G	Francis I	6836	351	C G
Jesus	3	1	C G	Toyotomi Hideyoshi	6767	515	C G
Britney Spears	4	689	C G	Plutarch	6697	258	C G
Adolf Hitler	5	7	C G	William Pitt the Younger	5977	409	C G
Barack Obama	6	111	C G	Josephus	5959	358	C G
Muhammad	7	3	C G	Peter Paul Rubens	5916	410	C G
Elvis Presley	8	69	C G	Friedrich Engels	5869	384	C G
Roger Federer	11	743	C G	Jerome	5550	320	C G
Mariah Carey	12	636	C G	George II	5418	340	C G

Religious figures trigger far more edits than would appear their due, particularly since *Jesus (7 B.C.–A.D. 30)* [1] and *Muhammad (570–632)* [3] both died more than one thousand years ago. These edits reflect jockeying between members of opposing theologies, eager to ensure that the public record reflects their personal beliefs. Thus, page edit frequency serves as a measure of attachment as well as fame. The open-edit model is based on good faith, but breaks down in the face of two equally determined groups deleting the changes the other side makes. As a result, Wikipedia has developed mechanisms to limit edits to trusted moderators after such flame wars catch fire. Hitler's high-edit status here also must be credited to such a battle.

The prominent low-edit people (right side of the table) are all long dead, gone more than 1,000 years on average. Perhaps the least familiar figure for Western readers is *Toyotomi Hideyoshi (1537–1598)* [515], the sixteenth-century ruler who unified Japan. Such historically important, academic figures are simply not generating fresh activity that requires updating their articles. It is difficult to imagine the man-on-the-street having much dirt to contribute about any of them.

There are several closely related measures of page edit frequency, including distinct editors, number of changes, and number of edit reversions. All serve as proxies for significance, but we use the total number of all revisions as our edit frequency measure.

2.3 Celebrity vs. Gravitas

There seems something inherently different about the reputation of contemporary celebrities (say pop singer *Britney Spears (1981–)* [689] or golfer *Tiger Woods (1975–)* [2101]) from that of lower-profile people backed by more hefty achievements: Nobel Prize–winning scientists like *James D. Watson (1928–)* [3619], Supreme Court justices like *Louis Brandeis (1856–1941)* [1134], or national leaders like Israel's *David Ben-Gurion (1886–1973)* [1269]. Can we use algorithmic measures to distinguish between popular fame and accomplishment-based recognition, so as to better understand the nature of someone's prominence?

We call these different notions of fame *celebrity* and *gravitas*. Our tools for discriminating between them follow naturally from *factor analysis*, the statistical method we employ to combine our input variables into a single quantity measuring significance. Factor analysis describes the commonality of variables in terms of a small set of unobserved variables, or "factors."

Suppose we have a large set of people's physical measurements: height, weight, shoe size, inseam length, waist size, neck size, and jacket length. Many of these variables will be correlated with each other: people with large waists will generally be of substantial weight, while shoe size and inseam length correlate better with height than weight. Indeed, two underlying factors do a pretty good job of explaining all these measurements: your girth (how wide are you?) and your span (how tall/long are you?). This is why height and weight measurements suffice at your doctor's office, if not quite for your tailor.

Factor analysis is best known for its application to intelligence (IQ) testing, the task for which it was developed by psychologist *Charles Spearman (1863–1945)* [15211] in 1904. Spearman showed that a single unobserved variable, his *g* factor for "general intelligence," underlies scores on many different types of intelligence tests.

We expected that factor analysis would extract a single underlying factor of individual significance. However, two independent factors better explained the data. *Gravitas*, our first factor, largely comes from (or "loads on," in statistical parlance) the two forms of PageRank. Gravitas seems to accurately reflect achievement-based recognition.

Celebrity, our second factor, loads more strongly on page hits, revisions, and article length. Our celebrity factor better captures the popular (some might say vulgar) notions of reputation. The wattage of singers, actors, and other entertainers is better measured by celebrity than gravitas. Finally, our combined measure of *fame* is given by the sum of these celebrity and gravitas factors.

To get a feel for the distinction between gravitas and celebrity, compare our highest figures by each factor. The high gravitas figures are clearly old-fashioned heavyweights, people of stature and accomplishment: kings, philosophers, and statesmen. The low gravitas figures are such complete celebrities that the top four walk this earth using only one name. They are professional wrestlers, actors, and singers. It is quite telling that the only two figures here displaying even a sliver of gravitas are *Britney Spears (1981–)* [689] and *Michael Jackson (1958–2009)* [180]. Both are among the Platonic ideals of modern celebrity.

Highest Gravitas Ranking

Person	Grav	Sig	Celeb/Grav
Napoleon	8	2	C G
Carl Linnaeus	13	31	C G
Plato	23	25	C G
Aristotle	27	8	C G
F. D. Roosevelt	30	43	C G
Plutarch	32	258	C G
Charles II	33	78	C G
Elizabeth II	35	132	C G
Queen Victoria	38	16	C G
William Shakespeare	42	4	C G
Pliny the Elder	43	212	C G
Tacitus	52	300	C G
Herodotus	58	123	C G
Charles V	61	84	C G
George V	64	235	C G

Highest Celebrity Ranking

Person	Celeb	Sig	Celeb/Grav
The Undertaker	2	2172	C G
Vijay	8	4456	C G
Edge	10	2603	C G
Kane	13	2229	C G
John Cena	16	2277	C G
Beyoncé Knowles	19	1519	C G
Triple H	26	1596	C G
Rey Mysterio	36	2740	C G
Britney Spears	37	689	C G
Ann Coulter	45	3376	C G
Jesse McCartney	48	4236	C G
Roger Federer	57	743	C G
Ashley Tisdale	60	4445	C G
Michael Jackson	75	180	C G
Dwayne Johnson	78	1446	C G

The celebrity-gravitas continuum provides a powerful tool for analyzing the reputation of historical personages, which we will employ throughout this book. For example, we can study how the distribution of celebrity and gravitas has changed over history. Gravitas has held basically constant over the past seventy years, while celebrity scores have exploded since 1990. As we would expect, figures who are still creating news are attracting far more hits, page revisions, etc. than their less active (and often dead) counterparts.

2.4 The Half-Life of Fame

> My name is Ozymandias, king of kings: Look on my works, ye mighty, and despair!
> *Percy Bysshe Shelley (1792–1822)* [329]

Time humbles all great men, as memories fade and their fame crumbles to dust. The glow of today's brightest stars will gradually diminish, and then fade to black. Where will *Justin Bieber (1994–)* [8633] rank after Justin Time: 50 years, 100 years, or 200 years from now? To what extent will he be remembered? How about *Barack Obama (1961–)* [111]? Estimating the speed with which glory fades is essential to compare the significance of current individuals with older figures long gone.

The need to compensate for temporal effects on reputation becomes apparent when we add celebrity and gravitas to create the measure we term *Fame*. The top 100 people ranked by Fame are presented in Figure 2.2. It is clear this list wildly overstates the significance of recent figures. We count 28 of the top 100 as still living at the time of this writing, plus another 18 no longer with us but active post-1970. Any merit criterion that puts *Britney Spears (1981–)* [689] in the same league as *Aristotle (384–322 B.C.)* [8] seems unworthy of further discussion.

Correcting for the influence of time in a principled way requires building a model to age contemporary figures, so as to predict how much of their reputation will remain years from now. There are two distinct processes at work here: first, the lapse from living memory inherent in the passage of generations, and second, a selection bias caused by the recent advent of Wikipedia. We will correct for each of these factors in turn, to complete our model for quantifying historical significance.

2.4.1 LIVING MEMORY

Each generation experiences events that become permanently seared into its collective consciousness. Where were you on September 11, 2001, at the time of the attack on the World Trade Center in New York? What were you thinking when you heard that *Barack Obama (1961–)* [111] had been elected president? I (Steve) am old enough to remember the explosion of the Space Shuttle *Challenger*, the shooting of *John Lennon (1940–1980)* [162],

Fame	Person	Fame	Person
1	George W. Bush	51	Fidel Castro
2	Jesus	52	Mariah Carey
3	Barack Obama	53	Wolfgang Amadeus Mozart
4	Adolf Hitler	54	John McCain
5	Napoleon	55	Lyndon B. Johnson
6	Ronald Reagan	56	Margaret Thatcher
7	Bill Clinton	57	Plato
8	Muhammad	58	Harry S. Truman
9	Michael Jackson	59	Vladimir Lenin
10	William Shakespeare	60	Dwight D. Eisenhower
11	Elvis Presley	61	Sigmund Freud
12	Joseph Stalin	62	Louis XIV
13	Abraham Lincoln	63	Ludwig van Beethoven
14	Albert Einstein	64	Ulysses S. Grant
15	George Washington	65	Benito Mussolini
16	Elizabeth II	66	Leonardo da Vinci
17	John F. Kennedy	67	Augustus
18	Pope John Paul II	68	Carl Linnaeus
19	Madonna	69	Al Gore
20	Aristotle	70	Mao Zedong
21	Richard Nixon	71	Woodrow Wilson
22	Franklin D. Roosevelt	72	Thomas Edison
23	Winston Churchill	73	Charles Dickens
24	Bob Dylan	74	Bill Gates
25	Alexander the Great	75	Roger Federer
26	Pope Benedict XVI	76	Paul the Apostle
27	Britney Spears	77	Gerald Ford
28	Thomas Jefferson	78	Benjamin Franklin
29	Theodore Roosevelt	79	Jimi Hendrix
30	John Lennon	80	Sarah Palin
31	Tony Blair	81	Marilyn Monroe
32	Saddam Hussein	82	Arnold Schwarzenegger
33	Henry VIII	83	Friedrich Nietzsche
34	Hillary Rodham Clinton	84	Genghis Khan
35	Paul McCartney	85	Charles I
36	Charles Darwin	86	James I
37	Elizabeth I	87	Osama bin Laden
38	John Kerry	88	Vladimir Putin
39	Karl Marx	89	Christina Aguilera
40	Jimmy Carter	90	Steven Spielberg
41	Julius Caesar	91	J. R. R. Tolkien
42	Queen Victoria	92	David Bowie
43	Mohandas K. Gandhi	93	Alexander Hamilton
44	Martin Luther	94	Diana
45	Martin Luther King	95	Michael Jordan
46	Christopher Columbus	96	Ernest Hemingway
47	Isaac Newton	97	Mary
48	George H. W. Bush	98	Noam Chomsky
49	Charlemagne	99	Che Guevara
50	Eminem	100	Nelson Mandela

FIGURE 2.2. The top 100 people by our measure of historical fame, without corrections for time. Far more contemporary figures appear here than for our final time-corrected significance measure, as shown in Figure 1.1.

and the resignation of *Richard Nixon (1913–1994)* [82]. I even vaguely recall seeing the funerals of *Robert F. Kennedy (1925–1968)* [1355] and *Martin Luther King (1929–1968)* [221] on TV as a child. I have a living connection to these people and events.

I also have a personal but more tenuous connection to the events of my parents' lifetimes. As a child, my father once saw *Babe Ruth (1895–1948)* [434] drive his big car through the neighborhood. My mother had a crush on comedian *Danny Kaye (1913–1987)* [4774], and saw him perform at the Paramount. Other historical figures left their mark on the movies, books, and music that I was exposed to growing up. I am aware of heroic physician *Thomas Anthony Dooley III (1927–1961)* [92953] and the wit of two-time failed presidential candidate *Adlai Stevenson (1900–1965)* [2641] only through discarded paperbacks picked up at used book sales during my youth.

I hear even fainter echos of the past through the grandfather I was named for, who died just two weeks before I was born. He was the first Skiena in America; a leather-worker. He built protective training headgear for all the fighters, and said that boxing champ *Joe Louis (1914–1981)* [2820] was a real gentleman. But before that, I have no record – just my great-grandfather's name written on a shipping manifest from the boat that took his son to a new life in America.

Contemporary figures will slip from active memory as personal contact with their achievements fades. We need to model how fast this will happen. It is natural to think of a "half-life" model of fame, analogous to that of nuclear decay, where only half as much stuff remains after (say) 50 years, one-quarter as much after 100 years, and so forth. But this model does not hold up to careful scrutiny. Consider figures from antiquity who are still of wide renown, such as *Jesus (7 B.C.–A.D. 30)* [1]. His fame would have been halved forty times over the past 2,000 years, meaning it must have burned a trillion times brighter at the start of the Christian era. It is inconceivable that Christ's fame could ever have been that much greater than it is now. Today the term "Jesus" accounts for roughly one out of every 10,000 words appearing in scanned books. How much more prominent could he possibly ever have been? There is not enough room for much upside here.

We conclude that the decay rate cannot be exponential, and further does not go on forever. At some point, living memory ceases, after which reputation has essentially been fixed in the historical record. In order to quantify these effects, we turn to the Google book Ngram dataset, which we discuss in full detail in Chapter 4. It counts the number of times each person, place, and thing appears in books published in a given year, and provides an excellent means of tracking the volume of discussion about individuals over a long period of time.

Examining this data[4] reveals two important facts about reputations over time. First, discussion of the average individual typically peaks between 60 and 70 years after their birth. The more famous the individual, the later in life that discussion peaks. Second, discussion decays from this peak relatively slowly. Thus, any significant individual will continue to be discussed long after his/her demise.

We can use these observations to construct a principled forecast of future reputational decay for contemporary figures. We equate an individual's significance rank to a given reference frequency in the Google Ngram dataset. Given this cohort's rate of decay, we can project their significance forward to the age of 170, providing equal grounds to compare past and present figures.

2.4.2 THE WIKIPEDIA GENERATION

Nevertheless, even after incorporating reputational decay into our model, contemporary figures still appeared to be overvalued. Do we live in an unrecognized era of Giants, or is some other force at work?

The problem becomes clear when we plot the number of people from each era significant enough to earn their own Wikipedia articles. In particular, Figure 2.3 shows the number of Wikipedia articles about people who died in each year since the 1770s. Death is the last major event of every person's life: rare are historical figures who have not done their best work by then.

That an increasing number of historically significant people are dying each year is not surprising. After all, the world has experienced continuous population increase since about 1350 – once the Bubonic plague and Hundred Years Wars finally ran their courses. The world first hit one billion

[4] Presented in Figure A.4 of the Appendix.

people around 1804, and two billion in 1927. Population growth continues so rapidly that it reached seven billion before this book hit the press [Wikipedia, 2011].

We juxtapose the per-year Wikipedia people counts in Figure 2.3 against a projection of how many people we should expect to see based on a linear combination of the U.S. and World populations. The English Wikipedia is heavily biased toward the inclusion of American figures, so world population alone was not sufficient to produce an accurate estimate. There is *amazingly* good agreement here. Over the past 250 years, the number of people ending Wikipedia-worthy lives are almost exactly what would be predicted by gross population statistics. This is strong evidence that Wikipedia is better than it has any right to be: there is no historical era that it systematically underrepresents.

But Figure 2.3 clearly shows three bumps corresponding to periods that are *overrepresented* according to this model. The first two bumps correspond to World Wars I and II respectively. Far more historical figures died in the horrors of war – both soldiers and civilians – than would have during times of comparative peace. But the third bump, the biggest of all, arises with the advent of Wikipedia.

Clearly, if you had the good fortune to die over the past fifteen years, you were significantly more likely to be immortalized in Wikipedia than those who lived in earlier times. Wikipedia is a product of the Internet

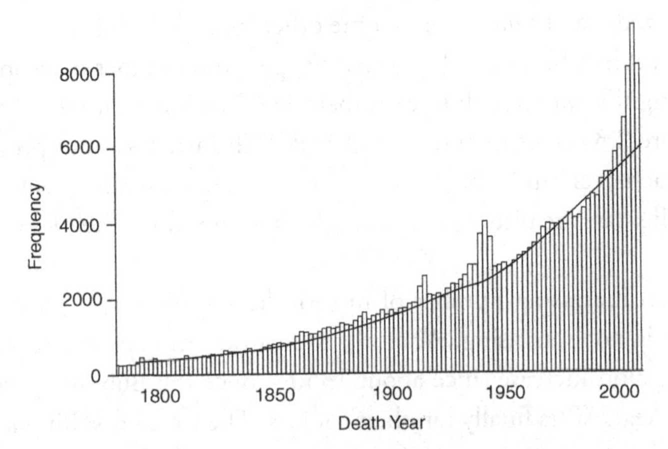

FIGURE 2.3. The number of Wikipedia people dying each year is well predicted by a linear combination of U.S. and world population.

era, and has been chronicling contemporary events with particular passion. Today's news represents the main source of updates to Wikipedia, with any significant event generating fresh Wikipedia pages only minutes after being reported. This recency bias exaggerates several of our measures for contemporary figures, such as page hits and revisions. The significance of active people must therefore be further reduced to account for this effect.

Starting around 1990, the average celebrity level of active figures increases dramatically (see Figure A.5 in the Appendix). This causes us to over-inflate the significance of recent figures, particularly those most associated with popular culture. We correct for this by explicitly subtracting a fixed amount of celebrity from each contemporary person as a function of their peak year of activity.

This gives us our final, corrected measure of significance, which we presented in Figure 1.1. We note, with satisfaction, that our algorithmic ranking corresponds well with what we think of as genuine historical significance. Modern presidents are now dropped down to plausible positions, and *Abraham Lincoln (1809–1865)* [5], *George Washington (1732–1799)* [6], and *Thomas Jefferson (1743–1826)* [10] now appropriately rise to the top of our rankings. Our most recent single-term presidents, *George H. W. Bush (1924–)* [363], *Jimmy Carter (1924–)* [462], and *Gerald Ford (1913–2006)* [230], fall into the bottom half of U.S. presidents, joining a more foggily remembered crowd. Other standards, such as the significance of Oscar winners for Best Actor (see Section 14.4), suggest that we have removed contemporary bias from our model fairly effectively.

Unless otherwise noted, we will use our corrected measure as our true measure of historical significance over the rest of this book. Let's see what we can do with it.

2.5 Evaluation Results

We will use our significance scores to draw conclusions about a variety of interesting cultural questions. Who belongs in history textbooks? Are women really underrepresented in Wikipedia and the historical record? How accurately do contemporary observers recognize the significance of their peers?

Our answers to these questions will mean something only if our significance rankings reliably measure the strength and durability of historical memes. For this reason, we have engaged in a rigorous evaluation process. We assembled a collection of 35 published rankings of people from a wide variety of domains, in history, sports, and entertainment. These standards include expert rankings, Internet opinion polls, and statistical formulae computed over sports statistics. The published standards we compare ourselves to include:

- *Actors and Actresses* – Expert rankings such as the American Film Institute's (AFI) Screen Legends ("100 years, 100 stars") and the IMDb STARmeter Top 100 of 2010.
- *Athletes* – Expert ratings and statistical rankings of athletes in baseball, basketball, and other sports, including the Associated Press Top 100 athletes of the twentieth century.
- *Musicians* – A collection of polls and expert rankings on musicians across genres, including *Rolling Stone* magazine's greatest singers of all time.
- *U.S. presidents* – The United States Presidency Center's expert poll rankings of U.S. presidents [Wikipedia, 2012a] and other similar rankings.
- *General Historical Rankings* – Popular books including Hart's *The 100*, and *1000 Years, 1000 People* [Gottlieb et al., 1998].
- The *Time 100* social networking influence index, based on Twitter followers and Facebook connections.

Each standard ranks between 25 and 250 people within some domain. These rankings serve as a gold standard to assess how well our computationally generated rankings correspond to expert evaluation of fame/significance.

For each standard, we compared the published rankings of people according to each variable, our original factor analysis–based scores, and finally the time-corrected version of our significance measure. Figure 2.4 summarizes these results.

Both forms of PageRank do well, with average correlations of 0.511 and 0.509 for the full and person-only networks, respectively. Our uncorrected factor analysis model of significance does even better. But our decayed significance score performs the best overall by far, with an average correlation of 0.554.

ListGroup	NL	ILA	NPR	PPR	PH	NR	AL	Celeb	Grav	Fame	Sig
Actors	3	0.412	0.499	0.556	0.273	0.327	0.283	0.178	0.411	0.466	**0.523**
Actresses	3	0.419	0.491	0.514	0.272	0.389	0.348	0.226	0.349	0.501	**0.546**
Authors	3	0.353	0.419	0.426	0.415	0.358	0.285	0.189	0.358	0.436	**0.458**
Directors	5	0.491	0.586	0.562	0.431	0.502	0.498	0.364	0.466	0.576	**0.608**
Musicians	3	N/A	0.648	0.621	0.572	0.569	0.473	0.416	0.413	0.618	**0.672**
Individual Sports	10	0.280	0.459	0.457	0.408	0.406	0.348	0.316	0.381	0.453	**0.463**
General Athletics	3	0.569	0.489	**0.571**	0.467	0.463	0.374	0.369	0.323	0.537	0.554
U.S. Presidents	5	0.909	0.576	0.490	0.625	0.549	0.472	0.386	0.532	0.580	**0.655**
General Historical	3	N/A	0.434	0.388	0.499	0.489	0.339	0.420	0.324	0.482	**0.511**
Overall	35	0.490	0.511	0.509	0.440	0.450	0.380	0.318	0.395	0.517	**0.554**

FIGURE 2.4. Rank correlations between variables and the published rankings in different categories. Abbreviations: NL (Number of Lists in group), ILA (Inter-list Agreement), NPR (Normal PageRank), PPR (Person PageRank), PH (Page Hits), NR (Number of Revisions), AL (Article Length), Celebrity, Gravitas, Fame (uncorrected Significance), Historical Significance. Bold signifies best overall correlation with validation lists in each category.

To put our score of 0.554 into context, realize that the published standards do not all agree with each other, meaning it is impossible to get a perfect score on this test. But we can measure how consistent the published standards are. The average interstandard correlation of 0.490 was substantially lower than that achieved by our corrected significance measure, suggesting that we are more accurate than the human standards themselves.

Beyond this, we have compared our historical significance scores against many other measures of meme strength, always with positive results. Subsequent sections will discuss our assessment studies including public polls (Section 3.8), frequency of appearance in books (Section 4.3), sports statistics (Section 6.1), autograph and baseball card prices (Section 6.7), and auction prices for modern art (Section 13.3.3).

With all of this evidence in hand, we assert that our significance measure is substantially superior to traditional ranking methods and produces results strong enough to support our conclusions in this book.

3

Who Belongs in Bonnie's Textbook?

Helping your children with their homework is a part of every parent's job description. It can be frustrating to watch your kid struggle with something as basic as elementary subtraction, but incredibly rewarding when they finally master borrowing ten under your tutelage.

Homework gets harder as students progress through grade levels and, like all parents, I (Steve) realized that the time would come when I would no longer be able to help my daughter Bonnie with her homework. But I didn't expect that moment would first arise in fifth grade U.S. history!

Bonnie's assignment was to match historical figures with descriptions of who they were and why they were important. Several names were well known to her from prior experience, such as *Christopher Columbus (1451–1506)* [20], *George Washington (1732–1799)* [6], and *Abraham Lincoln (1809–1865)* [5]. Others were new, but clearly worth learning about as she began her study of American history – people such as:

- *Henry Ford (1863–1947)* [148]: automobile industry pioneer.
- *Robert Fulton (1765–1815)* [954]: steamboat inventor.
- *William Lloyd Garrison (1805–1879)* [735]: abolitionist who fought against slavery.
- *Alexander Hamilton (1755–1804)* [45]: secretary of the Treasury who put the United States on a firm financial footing.
- *Robert E. Lee (1807–1870)* [76]: commander of the Confederate Army during the Civil War.

But Bonnie's homework also contained the names of several people who, frankly, I had never heard of before. And this homework was just practice, an aid to help her memorize these names for the upcoming exam. Were these figures *really* historically significant enough to justify this level of attention from students in such an early grade?

Bonnie's fifth grade history textbook contained a "Biographical Dictionary," a handy reference guide with capsule biographies of 250 figures highlighted in the text. A typical assigned task would be matching a person's name to the description of what they did. You give it a try. See how well you do matching the names and descriptions of the following people from Bonnie's textbook.

ID	Name
1	Mary Antin
2	Nat Love
3	George Shima
4	Benjamin Pap Singleton
5	Luzena Wilson

ID	Description
A	Cowboy and author of a popular autobiography
B	Entrepreneur who ran hotels and restaurants during the California Gold Rush
C	Immigrant from Japan who became known as the "Potato King" for his successful farming of potatoes in California
D	Immigrant from Russia who published a popular autobiography called *The Promised Land*
E	Leader of African American homesteaders known as Exodusters

The correct answers are given in the footnote.[1] By contrast, her text found room for only half of the U.S. presidents. So who *really* belongs in Bonnie's fifth grade history book? How can we tell? And what does this say about our educational system and society at large?

Our significance rankings provide the right tool to study such important questions. Further, Bonnie's history book provides an excellent testbed to establish how meaningful our historical significance rankings are. You, the reader, presumably passed fifth grade in style, and so should be familiar with most of the historical figures that appear here.

3.1 The Significance of Significance

This is the point in our book where many academic readers start to get a queasy feeling. They generally find the notion of measuring historical meme strength as interesting, even important, and respect our methodology for measuring it. But here we seem to be claiming that it has some objective meaning, and are using it to make judgment calls about historical interpretation.

[1] Correct answers: 1D, 2A, 3C, 4E, 5B.

There are several reasonable objections they may have to our line of thinking. First, we haven't really defined what we mean by significance: it is just some quantity that falls out of a formula. Second, the English-language Wikipedia is inherently culturally biased: favoring American figures above those of the rest of the world. The Wikipedia authors did not leave their prejudices at the door, so any results concerning minorities and gender reflect attitudes as well as accomplishments. Finally, meme strength is in no way synonymous with achievement or importance. A strong historical meme like the traitor *Benedict Arnold (1741–1801)* [222] might be easier to catch than that of *Jawaharlal Nehru (1889–1964)* [530], the political founder of modern India, but that doesn't mean Arnold has had even remotely comparable impact on world events or human society.

So let's tackle these issues head on. What is the significance of Significance?

We make the analogy between what we call Significance and various notions of intellectual capacity measured by intelligence tests and college-entry examinations like the Scholastic Aptitude Test (SAT). Indeed, our statistical factor–based methodology is exactly what Spearman used to identify his underlying "g" factor reflecting general intelligence.

- Are the questions in such intellectual aptitude tests culturally biased? Absolutely.
- Might differences in testing conditions, access to pretest training courses, and the evaluation methodology impose class, ethnic, and gender biases on outcomes? Certainly.
- Do test scores measure academic accomplishment, such as grade point average or class rank, or important intangibles like leadership, gumption, and creativity? Of course not.
- Should test scores be the only criteria used to make admissions decisions? Sure, if you don't care about selecting the best candidates.

But despite all this, do these tests provide useful data for making admissions decisions or understanding differences in educational performance? Of course they do. Judgment is always needed to properly interpret statistical data, and an understanding of its biases and limitations is important in making these interpretations. Reasonable people can arrive at different interpretations regarding the validity of an observation or the explanation for some phenomenon. But reasonable people often make better interpretations when presented with quantitative data.

The value of our significance rankings is in quantifying phenomena, providing evidence that, despite all its flaws and biases, correlates well with a lot of external references and standards. We will use it throughout this book to shed light on questions and observations we find interesting.

3.2 The People in Bonnie's Textbook

The most familiar names in Bonnie's textbook provide a strong reference point to calibrate your faith in our rankings. If we rank the people you know best in a reasonable order, this should build trust that carries over to the rest of the historical universe. Figure 3.1 pairs the 40 most significant people from the biographical dictionary of Bonnie's textbook with the 40 least significant, for your inspection.

The people on the left are quite an impressive bunch. By our calculations, all rank among the 200 or so most significant people who ever lived. Almost all should be familiar to you. Personally, I might fear being quizzed on exactly why *James I (1566–1625)* [41] made the list,[2] or whether *Francis Drake (1540–1596)* [150] was an explorer or military officer,[3] but the rest I would be willing to be tested on.

Fully 26 of the hundred most significant figures in world history appear in Bonnie's history text. Who ranks among the top 100 yet did not make her textbook? Excluding those who lived before Columbus plus political, scientific, and cultural figures of European extraction leaves only half a dozen people with a legitimate connection to the history of the United States: Presidents *Barack Obama (1961–)* [111] and *Grover Cleveland (1837–1908)* [98], scientist *Albert Einstein (1879–1955)* [19], writers *Mark Twain (1835–1910)* [53] and *Edgar Allan Poe (1809–1849)* [54], musician *Elvis Presley (1935–1977)* [69], and World War II ally/honorary citizen Sir *Winston Churchill (1874–1965)* [37]. At the top of its biographical

[2] Because British colonization of North America began during his eventful reign as king.

[3] Both. He led England to victory over the Spanish Armada and the second voyage (after *Ferdinand Magellan (1480–1521)* [311]) to circumnavigate the earth.

Most Significant

Person	Sig	Fame
Abraham Lincoln	5	13
George Washington	6	15
Adolf Hitler	7	4
Thomas Jefferson	10	28
Elizabeth I	13	37
Joseph Stalin	18	12
Christopher Columbus	20	46
Theodore Roosevelt	23	29
Ulysses S. Grant	28	64
Ronald Reagan	32	6
Benjamin Franklin	35	78
George W. Bush	36	1
Thomas Edison	40	72
James I	41	86
Franklin D. Roosevelt	43	22
Alexander Hamilton	45	93
Woodrow Wilson	47	71
James Madison	51	107
Joseph Smith	55	114
George III	58	118
John Adams	61	124
Andrew Jackson	66	131
John F. Kennedy	71	17
Robert E. Lee	76	159
Richard Nixon	82	21
Harry S. Truman	94	58
Benito Mussolini	101	65
Andrew Carnegie	104	248
Andrew Johnson	105	250
Alexander Graham Bell	106	210
Dwight D. Eisenhower	110	60
Thomas Paine	113	260
Bill Clinton	115	7
Marco Polo	147	352
Henry Ford	148	185
Francis Drake	150	361
John D. Rockefeller	172	403
Lyndon B. Johnson	184	55
Jefferson Davis	188	428
Samuel de Champlain	205	468

Least Significant

Person	Sig	Fame
Luzena Wilson	206506	370972
George Shima	144560	229435
Mary Antin	101158	137054
Venture Smith	97353	213273
Hendrick Theyanoguin	48999	122111
Dolores Huerta	44244	20035
Nampeyo	36510	70203
Joseph Cinqué	35465	93100
Nat Love	35100	81060
Peter Salem	25384	69519
William Harvey Carney	24221	66675
Tomochichi	23880	65783
Maya Lin	23823	8825
John Parker	23333	64394
Marjory Stoneman Douglas	22679	28056
Samoset	20841	58287
York	17907	50656
Belle Boyd	16296	44161
Samuel Prescott	16052	45995
Blanche Bruce	14221	43475
Eliza Lucas	14007	40587
Narcissa Whitman	13567	39411
Prince Hall	12142	35739
William Prescott	12010	35340
Juan Seguín	10797	31983
David Walker	10625	31540
Levi Coffin	10446	31066
James Armistead	10241	30456
Charles Goodnight	10172	30207
Hiram Rhodes Revels	9500	28207
Francis Cabot Lowell	8791	26367
Daniel Shays	8476	25434
Hiawatha	8362	18108
James W. Marshall	7740	23323
Madam C. J. Walker	7730	13050
William Dawes	7692	23190
Marcus Whitman	6599	19932
Massasoit	6590	19912
Bernardo de G. y. Madrid	6220	18822
Carrie Chapman Catt	6037	13378

FIGURE 3.1. The most historically significant members of Bonnie's textbook and the least significant.

dictionary, the editors of Bonnie's textbook have clearly identified the right people.

But it is baffling to confront the 40 *least* significant figures in her text, representing the bottom 15 percent of its biographical dictionary. The

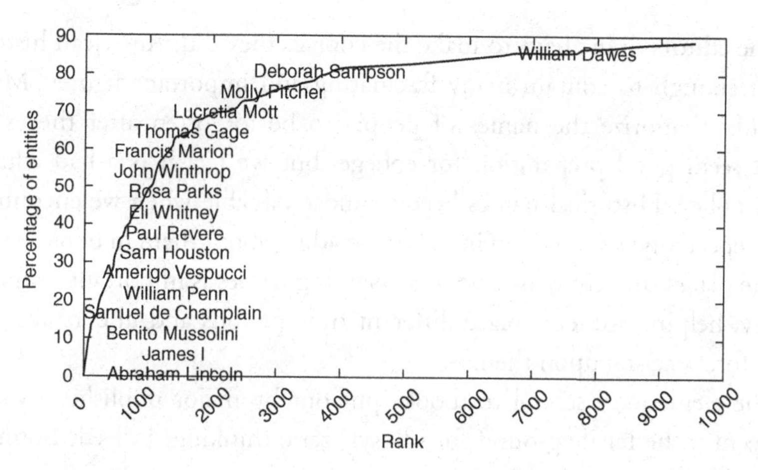

FIGURE 3.2. The percentage of historical figures of given or better rank in Bonnie's textbook.

ones I knew included historical footnotes like *William Dawes (1745–1799)* [7692], the other guy who went along on *Paul Revere's (1735–1818)* [627] ride.[4]

Others were ethnic firsts, like the first two black Southern senators during Reconstruction (*Blanche Bruce (1841–1898)* [14221] and *Hiram Rhodes Revels (1827–1901)* [9500]); both were chased from office after a single term. *Maya Lin (1959–)* [23823] was an undergraduate student who won the competition to design the Vietnam War Memorial on the Mall in Washington, DC. She made Bonnie's textbook ahead of any military or protest movement leader of that war. All the remaining names were a complete mystery to me.

I suspect that I am not alone here. Figure 3.2 displays the fraction of the figures in Bonnie's book achieving a given significance rank, along with representative figures at each level. Roughly 65 percent of them rank among the 2,000 most historically significant people. But then the significance quickly thins out. Even quintupling this range to the top 10,000 only fills another 20 percent of the textbook's roster, leaving the weakest 15 percent off this chart entirely.

[4] *William Dawes (1745–1799)* [7692] and *Samuel Prescott (1751–1777)* [16052] were Paul Revere's riding mates, who got his message through to Concord but not to *Henry Wadsworth Longfellow (1807–1882)* [349], whose poem is the source of Revere's legend.

The editors didn't have to make the choices they did. American history is rich enough to contain many fascinating and important figures. Making kids memorize the names of people to be forgotten after the exam might seem good preparation for college, but we feel it is a bad educational policy. Historical names become most valuable when we encounter them repeatedly: seeing them in movies, reading about them in books, visiting the places they lived and worked. Naming names is important to frame history: helping students place different time periods and understand the social forces acting upon them.

The elementary school textbooks put out by major publishers would not seem to be fertile ground for idiosyncratic thinking. Indeed, Bonnie's textbook[5] listed seven "Program Authors," seven more "Contributing Authors," eight "Content Consultants," and no fewer then 33 "Classroom Reviewers." So could this august panel have made better selections?

That is the question we will explore in the rest of this chapter. We partition the historical figures appearing in Bonnie's text into logical groups, and compare the book's selections with the most prominent excluded representatives. We believe that the text could have made better known choices for about 30 percent of the available slots.

In particular, we propose the following rules of thumb for future textbook authors. People of historical rank less than 2,000 or so are broadly known by the educated public. People of rank up to 5,000 might be historically significant enough to bring to the attention of a general audience in good conscience – if you have a particular point to make. But figures beyond that limit represent off-beat choices that should require considerable soul searching.

3.3 Political Leaders

The organization of the government of the United States into three separate but equal branches (executive, congressional, and judicial) provides one of the major lessons of elementary school–level social studies. Each branch offers a rich set of characters and stories. Have they been treated properly in Bonnie's textbook?

[5] *Social Studies: The United States* (Gold Edition), Pearson Scott Foresman, 2008.

3.3.1 PRESIDENTS

The names and faces of our presidents confront Americans more than any other historical figures. The full roster appears on commemorative dollar coins, restaurant placemats, and even animatronic robots at Disney World. *Barack Obama (1961–)* [111] is just the forty-fourth president, a tractable number for young people to master and understand. We both feel proud that we can enumerate all our presidents in order.

Indeed, the primary decision the editors of Bonnie's textbook faced was whether *all* the presidents should appear in the biographical dictionary. Even the lowest ranked president (*Chester A. Arthur (1829–1886)* [499]) would score in the most significant quarter of the figures appearing in the book.

The editors found room for 24 presidents, as shown in Figure 3.3. Given this constraint, they generally did a good job of selection. Obama's omission gets a pass because of the publication date of the text. The two weakest presidents appearing in the book (*Jimmy Carter (1924–)* [462] and *George*

Included

Person	Sig	Fame
Abraham Lincoln	5	13
George Washington	6	15
Thomas Jefferson	10	28
Theodore Roosevelt	23	29
Ulysses S. Grant	28	64
Ronald Reagan	32	6
George W. Bush	36	1
Franklin D. Roosevelt	43	22
Woodrow Wilson	47	71
James Madison	51	107
John Adams	61	124
Andrew Jackson	66	131
John F. Kennedy	71	17
Richard Nixon	82	21
Harry S. Truman	94	58
Andrew Johnson	105	250
Dwight D. Eisenhower	110	60
Bill Clinton	115	7
Lyndon B. Johnson	184	55
James Monroe	220	506
Gerald Ford	230	77
James K. Polk	240	557
George H. W. Bush	363	48
Jimmy Carter	462	40

Excluded

Person	Sig	Fame
Grover Cleveland	98	224
Barack Obama	111	3
John Quincy Adams	135	323
William Howard Taft	153	241
William McKinley	176	393
Herbert Hoover	183	151
James Buchanan	237	547
Warren G. Harding	242	283
Martin Van Buren	277	648
James A. Garfield	285	675
William Henry Harrison	288	683
Zachary Taylor	299	712
Rutherford B. Hayes	325	769
Benjamin Harrison	339	808
John Tyler	341	818
Calvin Coolidge	370	363
Franklin Pierce	427	1037
Millard Fillmore	446	1079
Chester A. Arthur	499	1267

FIGURE 3.3. U.S. presidents in Bonnie's history book, along with those excluded.

H. W. Bush (1924–) [363]) are justifiable selections as recent leaders still ranking among the top 50 individuals in terms of contemporary fame. The most obscure included presidents *James Monroe (1758–1831)* [220] and *James K. Polk (1795–1849)* [240]), are strongly associated with major events in our history (the Monroe Doctrine and the Mexican War), which are well covered in the text.

Of the omitted presidents, perhaps the strongest cases can be made for *Grover Cleveland (1837–1908)* [98] (who dominated American politics from 1884 to 1896) and *Herbert Hoover (1874–1964)* [183] (on whose watch the Great Depression began). But all in all, the editorial decisions concerning our presidents appear to be sound and well justified.

3.3.2 CONGRESSIONAL FIGURES

Congress, it is said, is filled by men who think they should be president. Excluding the 13 who eventually went on to the top job, Bonnie's textbook includes ten congressmen.

Congressmen in Bonnie's Textbook

Sig	Person	Dates	C/G	Description
188	Jefferson Davis	(1808–1889)	C G	President of the Confederacy (U.S. Civil War)
252	Henry Clay	(1777–1852)	C G	Senator ("The Great Compromiser")
399	John C. Calhoun	(1782–1850)	C G	U.S. vice president/senator (nullification)
453	Daniel Webster	(1782–1852)	C G	American statesman and Mass. U.S. senator
498	Sam Houston	(1793–1863)	C G	First president of Texas Republic
553	Joseph McCarthy	(1908–1957)	C G	Republican senator (McCarthyism)
1362	Richard Henry Lee	(1732–1794)	C G	Virginia statesman (Continental Congress)
2977	John Glenn	(1921–)	C G	Astronaut (first American to orbit Earth)
9500	Hiram Rhodes Revels	(1827–1901)	C G	First African American congressman
14221	Blanche Bruce	(1841–1898)	C G	First elected African American senator

Only four of these ten figures are remembered primarily as important figures in the representative branch of government: *Henry Clay* [252], *John C. Calhoun* [399], *Daniel Webster* [453], and *Joseph McCarthy* [553]. Interestingly, two (*Sam Houston* [498] and *Jefferson Davis* [188]) earned their primary fame as the presidents of short-lived independent republics (Texas and the Confederacy), and a third outside the Earth's atmosphere (astronaut *John Glenn* [2977]).

We would replace the three least significant congressmen in Bonnie's textbook by other high-ranking figures with more interesting stories to tell:

- *Aaron Burr (1756–1836)* [373] – a fascinating fellow whose mischief in the election of 1800, fatal duel with *Alexander Hamilton (1755–1804)* [45], and subsequent conspiracies more than outweighed his genuine contributions to establishing the republic.
- *William H. Seward (1801–1872)* [796] – the man most responsible for the purchase of Alaska as secretary of state under presidents Lincoln and Johnson.
- *William Jennings Bryan (1860–1925)* [698] – the dominant populist force in America during the Progressive Era, and pioneer of the modern presidential campaign.

None of these figures were selected primarily for their work in Congress, a deliberative body whose achievements are those of the whole rather than its individual members. Section 8.2 will discuss congressmen in much greater detail, and we invite you to flip over and find your own alternative selections.

3.3.3 JUDICIAL HISTORY

Three Supreme Court justices qualified for Bonnie's textbook: *John Marshall* [401], responsible for establishing the power of the court through the doctrine of judicial review in *Marbury v. Madison*, and the first black and female justices (*Thurgood Marshall* [1707] and *Sandra Day O'Connor* [5221]). All appear worthy of inclusion, with perhaps a quibble that O'Connor's historical achievements do not extend far beyond the precedent of her selection. Thurgood Marshall's contributions as litigator for civil rights perhaps exceed his as justice.

Supreme Court Justices in Bonnie's Textbook

Sig	Person	Dates	C/G	Description
401	John Marshall	(1755–1835)	C G	U.S. chief justice (judicial review)
1707	Thurgood Marshall	(1908–1993)	C G	First African American Supreme Court justice
5221	Sandra Day O'Connor	(1930–)	C G	First female Supreme Court justice

Earl Warren (1891–1974) [2034] is the most significant excluded justice. The Warren Court was instrumental in advancing civil rights protections for all Americans in the 1950s and 1960s. As attorney general of California, Warren was the driving force behind the internment of Japanese civilians during World War II – putting a face on this shameful but important episode in our history. We believe he should be in Bonnie's history text.

3.3.4 WORLD LEADERS

It is startling to see a dozen foreign leaders featured in an elementary school–level American History textbook. But it makes sense. This editorial decision reflects a conscious and necessary effort to promote a more international view of the world.

World Leaders in Bonnie's Textbook

Sig	Person	Dates	C/G	Description
7	Adolf Hitler	(1889–1945)	C G	Fuehrer of Nazi Germany (WWII)
13	Elizabeth I	(1533–1603)	C G	Queen of England (The Virgin Queen)
18	Joseph Stalin	(1878–1953)	C G	Premier of USSR (WWII)
41	James I	(1566–1625)	C G	King of England (King James Bible)
58	George III	(1738–1820)	C G	King of England (American Revolution)
101	Benito Mussolini	(1883–1945)	C G	Italian fascist dictator (WWII)
236	Isabella I	(1451–1504)	C G	Queen of Castile (funded Columbus voyages)
346	Kublai Khan	(1215–1294)	C G	Fifth Great Khan of Mongol Empire
454	Ferdinand II	(1452–1516)	C G	King of Aragon (funded Columbus voyages)
637	Mikhail Gorbachev	(1931–)	C G	Last Soviet premier (fall of USSR)
928	Moctezuma II	(1480–1520)	C G	Ruler of Tenochtitlan (1502–1520)
1695	Musa I	(?–1337)	C G	Ruler of Africa's Malian Empire

Three of the world leaders are British rulers during the colonial era and revolution. But some of the rest have only a tenuous connection to the United States. *Kublai Khan (1215–1294)* [346] (China) and *Musa I (?–1337)* [1695] (Africa) led mighty pre-Columbian empires, meaning that they went to their graves completely unaware of the existence of the New World. Three others were foreign leaders during World War II, but it is shameful to include *Benito Mussolini (1883–1945)* [101] at the expense of *Winston Churchill (1874–1965)* [37]. Finally, it seems extravagant to devote two slots to *Ferdinand II (1452–1516)* [454] and *Isabella I (1451–1504)*

[236], who sponsored the voyages of *Christopher Columbus (1451–1506)* [20].

In addition to Churchill, we would add three other foreign leaders with stronger connections to American history:

- *Simón Bolívar (1783–1830)* [448] – Revolutionary leader of South America's struggle for independence from the Spanish Empire.
- *Toussaint Louverture (1743–1803)* [896] – Leader of the Haitian Revolution freeing the slaves of Haiti, thus challenging the institution of slavery throughout the Americas.
- *Mao Zedong (1893–1976)* [151] – Founder of the People's Republic of China, and leader during *Richard Nixon's (1913–1994)* [82] 1972 visit establishing relations between the two nations.

3.4 People by Time and Place

Interesting times make for interesting people. Four historic periods are particularly well represented in Bonnie's textbook: the Colonial era, the Revolutionary War, the winning of the West, and the Civil War.

3.4.1 THE COLONIAL ERA

Perhaps the most remarkable fact about America's Colonial era is just how long it lasted. If you start with Columbus in 1492 and end with the battles of Lexington and Concord in 1775, it ran 283 years – almost 50 years longer than the United States has been an independent nation. Even if you start with the settlement of Jamestown, Virginia in 1607, it still clocks in at 168 years. We credit 15 people from Bonnie's textbook as contributing to Colonial America, which does not appear extravagant given this vast expanse of time.

All these colonial figures achieve respectable significance rank, and several are known as the founder of their colonies: for example, *William Bradford* [1379] (Plymouth) and *John Smith* [529] (Virginia).

We are sympathetic to retaining *John Peter Zenger* [2022] as a symbol of the battle for freedom of the press. However, we recommend replacing several of the more expendable figures, such as *Pocahontas's* [428] husband

Colonial Figures in Bonnie's Textbook

Sig	Person	Dates	C/G	Description
35	Benjamin Franklin	(1706–1790)	C G	Founding father/scientist (captured lightning)
286	William Penn	(1644–1718)	C G	Founder of Pennsylvania colony
428	Pocahontas	(1595–1617)	C G	Indian princess in Jamestown colony
529	John Smith	(?–1631)	C G	English explorer (Jamestown)
678	Roger Williams	(1603–1683)	C G	Founder (Rhode Island colony)
687	James Wolfe	(1727–1759)	C G	British Army officer (Quebec)
894	Anne Hutchinson	(1591–1643)	C G	Colonial female religious leader
1113	John Winthrop	(1587–1649)	C G	Puritan founder (Massachusetts Bay colony)
1258	George Whitefield	(1714–1770)	C G	Anglican Protestant minister
1374	Bartolomé de las Casas	(?–1566)	C G	Spanish historian
1379	William Bradford	(1590–1657)	C G	Plymouth governor (Thanksgiving)
1682	James Oglethorpe	(1696–1785)	C G	Founder of the colony of Georgia
1873	John Rolfe	(1585–1622)	C G	Jamestown settler (tobacco)
2022	John Peter Zenger	(1697–1746)	C G	Publisher and journalist (press freedom)
2439	Thomas Hooker	(1586–1647)	C G	Puritan religious and colonial leader
5015	John White	(1540–1593)	C G	Failed English colonist (Roanoke)

John Rolfe [1873], with a group of more significant ministers:

- *John Wesley (1703–1791)* [359] – With a strong claim as the founder of Methodism, his Wesleyian name and teachings live on in several Christian denominations. He trumps the accomplishments of *George Whitefield* [1258]: an English Anglican priest who helped spread the Great Awakening in Britain.
- *Jonathan Edwards (1703–1758)* [913] – Reformed theologian and missionary to the Native Americans.
- *Cotton Mather (1663–1728)* [1676] – Fiery Puritan minister and protagonist of the Salem witch trials.

3.4.2 THE AMERICAN REVOLUTION

In contrast with the colonial figures, too many Founding Fathers appear in Bonnie's textbook. The following list includes 22 Founding Fathers, and is kept manageable only by ignoring presidents like *John Adams (1735–1826)* [61], American revolutionary military leaders like *Benedict Arnold (1741–1801)* [222], plus the Founding Mothers, to be discussed in Section 3.6.

Six of these names reside in the dead zone of significance above 5,000, a place where elementary school historical figures cannot naturally survive. Revere's riding mate *William Dawes* [7692] is adequately commemorated

Founding Fathers in Bonnie's Textbook

Sig	Person	Dates	C/G	Description
45	Alexander Hamilton	(1755–1804)	C G	U.S. Founding Father (National Bank)
113	Thomas Paine	(1737–1809)	C G	U.S. revolutionary ("Common Sense")
467	Samuel Adams	(1722–1803)	C G	U.S. Founding Father (Sons of Liberty)
472	Patrick Henry	(1736–1799)	C G	U.S. Founding Father ("Give me liberty")
481	Charles Cornwallis	(1738–1805)	C G	British Army officer (Revolutionary War)
616	John Hancock	(1737–1793)	C G	U.S. revolutionary (signer of Declaration)
627	Paul Revere	(1735–1818)	C G	U.S. Revolution ("British are coming!")
1050	Francis Scott Key	(1777–1843)	C G	Poet (lyrics of National Anthem)
1089	John Burgoyne	(1722–1792)	C G	British army officer, politician
1090	Nathan Hale	(1755–1776)	C G	Continental Army soldier/spy
1149	George Rogers Clark	(1752–1818)	C G	Revolutionary War officer
1175	Ethan Allen	(1738–1789)	C G	Revolutionary patriot (Green Mountain Boys)
1283	Benjamin Rush	(1746–1813)	C G	Physician and Declaration signer
1304	Francis Marion	(1732–1795)	C G	Revolutionary War officer (Swamp Fox)
1362	Richard Henry Lee	(1732–1794)	C G	Virginia statesman (Continental Congress)
1651	Thomas Gage	(1719–1787)	C G	British commander (Independence War)
2863	Mercy Otis Warren	(1728–1814)	C G	Political writer of the American Revolution
7692	William Dawes	(1745–1799)	C G	Patriot (American Revolution)
8476	Daniel Shays	(1741–1825)	C G	American soldier (Shay's Rebellion)
12010	William Prescott	(1726–1795)	C G	Colonel (Revolutionary War)
16052	Samuel Prescott	(1751–1777)	C G	Patriot (Battle of Bunker Hill)
23333	John Parker	(1729–1775)	C G	American soldier (Battle of Lexington)
25384	Peter Salem	(1750–1816)	C G	Soldier (Revolutionary War)

by a traffic island in Cambridge, Massachusetts. *Daniel Shays* [8476] was the instigator of the short-lived *Shays' Rebellion*, a memorable challenge to the authority of the new-born nation. But he was lucky to escape with his life, and died in obscurity. The causes of the rebellion (a weak federal government unable to pay its debts to its military veterans) are far more important than the name of the rabble-rouser who started it.

William Prescott [12010] and *Peter Salem* [25384] both fought at the Battle of Bunker Hill, though Prescott's fame is ultimately attributable to saying "Do not fire until you see the whites of their eyes." *John Parker* [23333] commanded at the Battle of Lexington, but then saw no further action. None of these people need to be in the textbook.

We are loath to add more figures from this era, but we must find room for the remarkable *Joseph Brant (1743–1807)* [1328], a Mohawk leader who fought ably on behalf of the British against the American revolutionaries. His story helps reveal many of the complexities of the Native American experience.

3.4.3 THE WINNING OF THE WEST

The westward expansion of the United States, growing from 13 coastal colonies to span the continent, is an exciting and important part of our history. We bought and fought and annexed our way into the mightiest nation on Earth. Most of this buying and fighting and annexing happened in the West.

Several Western figures in Bonnie's textbook are identifiable as founding figures of particular states, such as Father *Junípero Serra* [2835] (California) and *Brigham Young* [471] (Utah). *Joseph Smith* [55] is the founder of the Mormon faith, although he did not live to complete the migration West. Religion is a common denominator of many of these Western figures, including the husband-and-wife team of Marcus and *Narcissa Whitman* [13567]. Both were ultimately slain by the Native Americans they ministered to in the Oregon Territories.

Western Figures in Bonnie's Textbook

Sig	Person	Dates	C/G	Description
55	Joseph Smith	(1805–1844)	C G	American religious leader (Mormonism)
471	Brigham Young	(1801–1877)	C G	American Mormonism leader
498	Sam Houston	(1793–1863)	C G	First president of Texas Republic
501	Daniel Boone	(1734–1820)	C G	American frontiersman (coonskin cap)
606	Davy Crockett	(1786–1836)	C G	American folk hero (Alamo)
682	Antonio López de Santa Anna	(1794–1876)	C G	Mexican general, military dictator
934	Stephen F. Austin	(1793–1836)	C G	The father of Texas
1364	Levi Strauss	(1829–1902)	C G	Founder of Levi Strauss & Co.
2835	Junípero Serra	(1713–1784)	C G	Spanish missionary in California
3644	John Sutter	(1803–1880)	C G	Pioneer (California Gold Rush)
6599	Marcus Whitman	(1802–1847)	C G	Missionary to Indians (Oregon)
7740	James W. Marshall	(1810–1885)	C G	Discovered gold in California
10172	Charles Goodnight	(1836–1929)	C G	Western cattle rancher
10797	Juan Seguín	(1806–1890)	C G	Statesman (Texas Revolution)
13567	Narcissa Whitman	(1808–1847)	C G	Missionary to Indians (Oregon)

It is less obvious, however, who should replace the five figures with significance rank above 5,000. The editors wisely avoided the temptation to include the famous outlaws and lawmen of the West: *Billy the Kid (1859–1881)* [761], *Wyatt Earp (1848–1929)* [858], or *Jesse James (1847–1882)* [413]. But environmentalist *John Muir (1838–1914)* [569], the inspiration for our National Park System, clearly should make the cut. Certain explorers of the West, such as *Kit Carson (1809–1868)* [1215], *Zebulon Pike (1778–1813)* [2058], *John C. Frémont (1813–1890)* [1048], and *John Wesley Powell (1834–1902)* [1871], also seem appropriate choices.

3.4.4 THE CIVIL WAR ERA

The Civil War era contained an enormous cast of characters, including prominent abolitionists, political figures, and military leaders from both sides. All are well represented in Bonnie's textbook. The war cast a long shadow on American politics. All presidents from 1868 to 1893 served as Union officers during the war, except for *Grover Cleveland (1837–1908)* [98] – who hired a substitute to fight for him.

Bonnie's textbook generally captures the major figures of the period. The only unsupportable choices are those of Catherine and *Levi Coffin (1798–1877)* [10446], a husband-and-wife team of conductors on the Underground Railroad. *Harriet Tubman (1820–1913)* [1093], already in the text, much better fits the conductor slot.

Civil War Era Figures in Bonnie's Textbook

Sig	Person	Dates	C/G	Description
76	Robert E. Lee	(1807–1870)	C G	Confederate general (U.S. Civil War)
188	Jefferson Davis	(1808–1889)	C G	President of the Confederacy (U.S. Civil War)
233	William Tecumseh Sherman	(1820–1891)	C G	Union general (U.S. Civil War)
283	Stonewall Jackson	(1824–1863)	C G	Confederate general (U.S. Civil War)
304	John Brown	(1800–1859)	C G	American revolutionary abolitionist
379	George Armstrong Custer	(1839–1876)	C G	U.S. Army officer ("Custer's last stand")
735	William Lloyd Garrison	(1805–1879)	C G	Prominent American abolitionist
739	Stephen A. Douglas	(1813–1861)	C G	Senator (Lincoln-Douglas debates)
787	Winfield Scott	(1786–1866)	C G	U.S. Army general (Mexican/Civil Wars)
2014	Mathew Brady	(1822–1896)	C G	Civil War photographer
10446	Levi Coffin	(1798–1877)	C G	American Quaker, abolitionist

We would include *George B. McClellan (1826–1885)* [554], unsuccessful Union commander and Lincoln's opponent in the 1864 presidential election. Another significant omission is *Robert Gould Shaw (1837–1863)* [2834], the (white) colonel in command of the all-black 54th Regiment of the Union Army. For better or worse, he is the most famous figure celebrating African American military contributions to the Civil War. Two others are of particular interest, because of their revealing postwar activities:

- *Philip Sheridan (1831–1888)* [949] – An important Union general, who went on to lead the U.S. war efforts against the Native Americans in the Great Plains. Notorious for saying "the only good Indian is a dead Indian."
- *Nathan Bedford Forrest (1821–1877)* [975] – Innovative Confederate cavalry leader (the "Wizard of the Saddle"), who went on to found the Ku Klux Klan.

3.5 Cultural and Social Innovators

American ideas and culture remain our country's greatest export: the work of inventors, scientists, artists, business leaders, and explorers. They are properly and amply described in Bonnie's textbook.

3.5.1 INVENTORS AND BUSINESS LEADERS

"The business of America is business!" is the most famous saying of our otherwise silent president, *Calvin Coolidge (1872–1933)* [370]. Certainly no history of the United States would be complete without a focus on the building of our modern economy: the robber barons in the 1880s, the breakup of the trusts in the early twentieth century, the advent of the automobile, the Great Depression, the postwar boom, and the Information Age. We count 16 business leaders and inventors in Bonnie's textbook.

Inventors and Business Leaders in Bonnie's Textbook

Sig	Person	Dates	C/G	Description
40	Thomas Edison	(1847–1931)	C G	Inventor (lightbulb, phonograph)
104	Andrew Carnegie	(1835–1919)	C G	Steel magnate and philanthropist
106	Alexander Graham Bell	(1847–1922)	C G	Inventor (telephone)
148	Henry Ford	(1863–1947)	C G	American industrialist (Ford Motors)
172	John D. Rockefeller	(1839–1937)	C G	American oil magnate (Standard Oil)
328	Johannes Gutenberg	(1398–1468)	C G	German printer (movable type)
532	Samuel Morse	(1791–1872)	C G	Inventor (telegraph system)
758	Eli Whitney	(1765–1825)	C G	American inventor (cotton gin)
887	George Washington Carver	(1864–1943)	C G	Agricultural scientist (uses for peanuts)
954	Robert Fulton	(1765–1815)	C G	Engineer and inventor (steamboat)
2317	Cyrus McCormick	(1809–1884)	C G	American inventor (reaper)
2411	John Deere	(1804–1886)	C G	American manufacturer (Deere & Comp.)
2664	Samuel Slater	(1768–1835)	C G	Early American industrialist
3603	Lewis Howard Latimer	(1848–1928)	C G	African American inventor
8791	Francis Cabot Lowell	(1775–1817)	C G	American businessman (city of Lowell)
144560	George Shima	(1864–1926)	C G	First Japanese American millionaire

The practical arts of tinkering and invention are traits that Americans are quick to claim as our own. The industrial revolution coincided with the early years of an independent America, yielding inventions like the cotton gin (*Eli Whitney* [758]) and the steamboat (*Robert Fulton* [954]), both of which reshaped the economy of the fledgling nation. The modern world was built in the wake of technological developments in electricity and mechanics: electric lighting (*Thomas Edison* [40]), transportation (*Henry Ford* [148]), and communication (*Samuel Morse* [532] and *Alexander Graham Bell* [106]).

The line between businessmen and inventors can get blurry. *John Deere* [2411] invented "the plow that broke the plains," that made farming the Great Plains economically viable. He thus occupies a similar niche to *Cyrus McCormick* [2317], inventor of the harvesting machine or reaper. The companies they founded: John Deere and International Harvester (now Navistar), respectively, long dominated the manufacturing of agricultural machinery.

Who's missing? No figure associated with modern technologies such as computing, medicine, or radio/television appears in the book. This partially reflects the lack of visible inventors in our age of corporate research laboratories. But the advent of personal computing occurred in the 1970s, long enough ago to register on a historical time scale, with the key figures being *Bill Gates (1955–)* [904] and *Steve Jobs (1955–2011)* [2051]. Their contributions to our country exceed that of most military and political figures appearing in her text.

Bonnie's textbook does not contain anyone we would classify as a scientist. We would argue for the two leading atomic-era physicists: *Albert Einstein (1879–1955)* [19], who convinced *Franklin D. Roosevelt (1882–1945)* [43] to build the atomic bomb, and *J. Robert Oppenheimer (1904–1967)* [1296], the scientist most responsible for making it happen.

The business roster in Bonnie's textbook should also be significantly expanded. The entertainment industry traces a natural arc from *P. T. Barnum (1810–1891)* [767] to *Walt Disney (1901–1966)* [337]. The most significant figure in American financial history is *J. P. Morgan (1837–1913)* [322], ahead of *Warren Buffett (1930–)* [1858] and the robber baron *Jay Gould (1836–1892)* [1413]. The seminal figure in the growth of the railroads is *Cornelius Vanderbilt (1794–1877)* [925]. All belong in Bonnie's textbook.

3.5.2 LABOR LEADERS

The flip side of Capital is Labor. The battle for worker rights is a controversial but heroic story, with the winners being the ones who get to tell the tale.

Bonnie's history book includes a few seminal American labor leaders. *Samuel Gompers* [1575] founded the American Federation of Labor. Mother *Mary Harris Jones* [2398] fought against child labor, as an organizer for the United Mine Workers. *Dolores Huerta*'s [44244] accomplishments as labor leader to migrant farm workers are substantially (if perhaps unfairly) overshadowed by *César Chávez* [3948], rendering her redundant for the textbook.

Labor Leaders in Bonnie's Textbook

Sig	Person	Dates	C/G	Description
1575	Samuel Gompers	(1850–1924)	C G	American labor union leader (AFL)
2398	Mary Harris Jones	(1837–1930)	C G	American labor organizer
3948	César Chávez	(1927–1993)	C G	Cofounder of United Farm Workers
44244	Dolores Huerta	(1930–)	C G	Cofounder of United Farm Workers

We would be inclined to add one or two other figures to this list. *Eugene V. Debs (1855–1926)* [1340] was the major figure in the Pullman strike of 1894 and ran for president five times as a socialist, twice winning more than 900,000 votes. *A. Philip Randolph (1889–1979)* [6144] successfully organized the first predominantly black labor union, the Brotherhood of Sleeping Car Porters, and participated in the March on Washington with *Martin Luther King (1929–1968)* [221].

3.5.3 EXPLORERS AND ADVENTURERS

Explorers are people credited with discovering a significant landmass, region, or navigation route, while *adventurers* visit previously known places with new panache. By this definition, *Henry Hudson* [353] and *Ferdinand Magellan* [311] were explorers, while the first man on the moon (*Neil Armstrong* [1394]) or person to fly across the Atlantic (*Charles Lindbergh*

[736]) qualify as adventurers. We file 30 members of Bonnie's textbook under these categories.

Thirty names is an awful lot to get a handle on, and this group could do with substantial pruning. Several of the explorers never got close to the New World, including *Marco Polo* [147], *Zheng He* [464], and *Vasco da Gama* [281]. They seem unnecessary here.

Explorers and Adventurers in Bonnie's Textbook

Sig	Person	Dates	C/G	Description
20	Christopher Columbus	(1451–1506)	C G	Explorer, discoverer of the New World
147	Marco Polo	(1254–1324)	C G	Venetian merchant/traveler (China)
150	Francis Drake	(1540–1596)	C G	English captain, explorer, privateer
205	Samuel de Champlain	(1574–1635)	C G	French navigator (Quebec)
281	Vasco da Gama	(1469–1524)	C G	Portuguese explorer (India)
311	Ferdinand Magellan	(1480–1521)	C G	Explorer (first circumnavigation)
313	Walter Raleigh	(1552–1618)	C G	English aristocrat, writer, and explorer
353	Henry Hudson	(1570–1611)	C G	English sea explorer (Hudson River)
407	Amerigo Vespucci	(1454–1512)	C G	Italian explorer (named "America")
464	Zheng He	(1371–1433)	C G	Chinese explorer
473	Hernando de Soto	(1496–1542)	C G	Spanish explorer (Mississippi River)
528	Francisco Pizarro	(1471–1541)	C G	Spanish conquistador (Incas)
563	Leif Ericson	(970–1020)	C G	Viking discoverer of North America
631	Hernán Cortés	(1485–1547)	C G	Spanish conquistador (Aztecs)
736	Charles Lindbergh	(1902–1974)	C G	American aviator (New York to Paris)
817	Meriwether Lewis	(1774–1809)	C G	American explorer (Lewis & Clark)
910	Henry the Navigator	(1394–1460)	C G	Patron of early Portuguese exploration
929	Bartolomeu Dias	(1451–1500)	C G	Portuguese explorer
979	William Clark	(1770–1838)	C G	American explorer (Lewis & Clark)
1136	Vasco Núñez de Balboa	(1475–1519)	C G	Spanish explorer (Pacific Ocean)
1181	René-Robert Cavelier	(1643–1687)	C G	French explorer (Great Lakes)
1339	Jacques Marquette	(1637–1675)	C G	French Jesuit missionary (colonized Michigan)
1382	Francisco Vásquez de Coronado	(1510–1554)	C G	Spanish conquistador (7 cities of gold)
1394	Neil Armstrong	(1930–)	C G	Astronaut (first man to walk on the moon)
1407	Erik the Red	(950–1003)	C G	Founder of the Viking Greenland settlement
1450	Louis Jolliet	(1645–1700)	C G	French-Canadian explorer (Mississippi River)
2331	Álvar Núñez Cabeza de Vaca	(1490–1559)	C G	Spanish explorer
2977	John Glenn	(1921–)	C G	Astronaut (first American to orbit Earth)
3081	Juan Ponce de León	(1474–1521)	C G	Spanish explorer (Florida, fountain of youth)
5458	Pedro Menéndez de Avilés	(1519–1574)	C G	Spanish explorer (St. Augustine)

Adventurers naturally have a lesser claim on the history texts than explorers. But *Charles Lindbergh's* [736] flight helped define his time, and ushered in the age of international flight. Similarly, astronauts *John Glenn* [2977] and *Neil Armstrong* [1394] are the most emblematic figures of the still recent but somehow rapidly vanishing Space Age. Missions today involve larger crews and less dramatic accomplishments, making it unlikely that bigger spaceman-heros will emerge in our lifetimes. We think they got the right people here.

Pedro Menéndez de Avilés [5458] was the Spanish founder of St. Augustine (the oldest city in the United States), who drove the French out of Florida. He seems quite expendable from Bonnie's textbook in favor of *James Cook* [60], who discovered the Hawaiian Islands during his great voyages across the Pacific. *Álvar Núñez Cabeza de Vaca* [2331] was the first European to reach Texas, but *Hernando de Soto*'s [473] explorations in the region were more important. We would not hesitate to replace Vaca with *John Cabot* [314], the first to explore the landmass of North America.

3.6 Diversity

Diversity is an important goal in any history book, particularly one aimed at children. A dozen people telling a dozen different stories is far more interesting and informative than the same-sized crowd telling a single tale. America has been enriched by contributions from men and women of all races, colors, and creeds, and all these groups belong in Bonnie's textbook. But is each group telling its fair share of the story?

3.6.1 AFRICAN AMERICANS

The quest for multiculturalism was clearly a major factor in composing the roster of Bonnie's textbook. By our count, there were 11 Native Americans and 13 African Americans ranked among the 50 *least* significant figures in the text. This bias seems both heavy-handed and transparent. Worst of all, it is unnecessary. Nine of the 100 *most* significant figures in the book were African American, a strong showing that jibes with the racial demographic of the United States.

Several of the African Americans in Bonnie's textbook are footnote figures. Some fought (*James Armistead* [10241], *William Harvey Carney* [24221], *Peter Salem* [25384]) in the Revolution or Civil War. Others wrote largely forgotten books later rediscovered by modern historians, such as cowboy *Nat Love* [35100] and freed slave *Venture Smith* [97353]. Their stories are about being part of the story, faces in the crowd, as opposed to being leading actors of their times. Given the circumstances of slavery

Sig	Person	Dates	C/G	Description
221	Martin Luther King	(1929–1968)	C G	American civil rights leader
402	Frederick Douglass	(1818–1895)	C G	American social reformer, statesman
525	Booker T. Washington	(1856–1915)	C G	American educator (Tuskegee Institute)
545	W. E. B. Du Bois	(1868–1963)	C G	American civil rights activist
883	Dred Scott	(1799–1858)	C G	Fugitive slave (Dred Scott case)
887	George Washington Carver	(1864–1943)	C G	Agricultural scientist (uses for peanuts)
964	Rosa Parks	(1913–2005)	C G	African American civil rights activist
1093	Harriet Tubman	(1820–1913)	C G	African American abolitionist
1106	Malcolm X	(1925–1965)	C G	Black Muslim activist
1211	Langston Hughes	(1902–1967)	C G	Black poet (Harlem Renaissance)
1231	Sojourner Truth	(1797–1883)	C G	African American abolitionist
1320	Nat Turner	(1800–1831)	C G	Slave rebellion leader
1707	Thurgood Marshall	(1908–1993)	C G	First African American Supreme Court justice
1715	Olaudah Equiano	(1745–1797)	C G	Slave trade abolitionist
1797	Benjamin Banneker	(1731–1806)	C G	African American astronomer
1896	Phillis Wheatley	(1753–1784)	C G	Early African American poet
2686	Zora Neale Hurston	(1891–1960)	C G	American folklorist and anthropologist
3314	Colin Powell	(1937–)	C G	U.S. general / secretary of state (Gulf War)
3495	Crispus Attucks	(1723–1770)	C G	First victim of Revolutionary War
3603	Lewis Howard Latimer	(1848–1928)	C G	African American inventor
3783	Ida B. Wells	(1862–1931)	C G	African American journalist
7730	Madam C. J. Walker	(1867–1919)	C G	African American businesswoman
9500	Hiram Rhodes Revels	(1827–1901)	C G	First African American congressman
10241	James Armistead	(1760–1830)	C G	American revolution double agent
10625	David Walker	(1785–1830)	C G	African American activist (abolition)
12142	Prince Hall	(1738–1807)	C G	Abolitionist leader (Boston)
14221	Blanche Bruce	(1841–1898)	C G	First elected African American senator
17907	York	(1770–?)	C G	Slave and explorer with Lewis and Clark
24221	William Harvey Carney	(1840–1908)	C G	African American soldier (Civil War)
25384	Peter Salem	(1750–1816)	C G	Soldier (Revolutionary War)
35100	Nat Love	(1854–1921)	C G	African American cowboy (Civil War)
35465	Joseph Cinqué	(1814–1879)	C G	African slave leader (Amistad mutiny)
97353	Venture Smith	(1729–1805)	C G	African American child slave and author

FIGURE 3.4. African American figures in Bonnie's history book.

and institutionalized racism, which blocked opportunities to accumulate identifiable significance, no more could be humanly possible.

If we use our rule of thumb that a significance rank of 2,000 represents the boundary of general knowledge, half of the African Americans in Bonnie's textbook (Figure 3.4) probably should not be there. Even a more generous cutoff at rank 5,000 would eliminate more than one-third of these figures. Some we would replace with more prominent people: President *Barack Obama (1961–)* [111], baseball trailblazer *Jackie Robinson (1919–1972)* [841], ragtime composer *Scott Joplin (1867–1917)* [1508], jazz pioneer *Duke Ellington (1899–1974)* [495], singer and activist *Paul Robeson*

(1898–1976) [1278], and presidential mistress *Sally Hemings (1773–1835)* [1496]. All have more interesting and relevant stories to tell.

3.6.2 WOMEN

A similar story holds with women. By our rule of thumb standards, only 9 of the 25 women listed in Figure 3.5 pass the test of significance rank above 2,000. Seven more make the cut if we set it at 5,000.

Bonnie's textbook features four wives of U.S. presidents. In general, they have identified the most significant ones, although we would be inclined to drop *Martha Washington (1731–1802)* [1065] in place of Senator/Secretary of State *Hillary Rodham Clinton (1947–)* [575].

Part of the problem here is picking the wrong representative for a given historical niche. The textbook lists environmentalist *Marjory Stoneman Douglas (1890–1998)* [22679], savior of the Everglades. But it ignores

Sig	Person	Dates	C/G	Description
449	Harriet Beecher Stowe	(1811–1896)	C G	Abolitionist/author (*Uncle Tom's Cabin*)
517	Eleanor Roosevelt	(1884–1962)	C G	34th U.S. first lady and social activist
734	Elizabeth Cady Stanton	(1815–1902)	C G	Woman's suffrage leader
821	Abigail Adams	(1744–1818)	C G	2nd U.S. first lady
932	Amelia Earhart	(1897–1937)	C G	American aviation pioneer (disappeared)
1065	Martha Washington	(1731–1802)	C G	1st U.S. first lady
1138	Clara Barton	(1821–1912)	C G	Teacher, humanitarian (founded Red Cross)
1700	Dolley Madison	(1768–1849)	C G	4th U.S. first lady (saved papers)
1896	Phillis Wheatley	(1753–1784)	C G	Early African American poet
2208	Lucretia Mott	(1793–1880)	C G	American Quaker and abolitionist
2489	Ida M. Tarbell	(1857–1944)	C G	American investigative journalist
2686	Zora Neale Hurston	(1891–1960)	C G	American folklorist and anthropologist
2863	Mercy Otis Warren	(1728–1814)	C G	Political writer of the American Revolution
2926	Molly Pitcher	(1754–1832)	C G	Mythical female soldier of American Revolution
3783	Ida B. Wells	(1862–1931)	C G	African American journalist
3797	Deborah Sampson	(1760–1827)	C G	Female soldier in the American Revolution
5221	Sandra Day O'Connor	(1930–)	C G	First woman Supreme Court Justice
5857	Madeleine Albright	(1937–)	C G	Former secretary of state
6037	Carrie Chapman Catt	(1859–1947)	C G	American women's suffrage leader
14007	Eliza Lucas	(1722–1793)	C G	American female inventor (indigo)
16296	Belle Boyd	(1844–1900)	C G	Confederate spy (Civil War)
22679	Marjory Stoneman Douglas	(1890–1998)	C G	American journalist, feminist
23823	Maya Lin	(1959–)	C G	Landscape architect (Vietnam War Memorial)
101158	Mary Antin	(1881–1949)	C G	Author, immigration rights activist
206506	Luzena Wilson	(1821–1902)	C G	California Gold Rush entrepreneur

FIGURE 3.5. American women in Bonnie's history book.

Rachel Carson (1907–1964) [2322], who spearheaded the entire environmental movement through her book *Silent Spring. Emma Lazarus (1849–1887)* [8465], whose poem "The New Colossus,"[6] graces the Statue of Liberty, would be a stronger choice for a female author capturing the immigrant experience than *Mary Antin (1881–1949)* [101158].

3.6.3 NATIVE AMERICANS

The significance standard achieved by the Native Americans in Bonnie's textbook, shown in Figure 3.6, proves higher than those applied to women and African Americans. Only 6 out of 22 fall below our significance rank 5,000 barrier. We would replace these with two others of substantially greater significance: *Geronimo (1829–1909)* [941] and *Joseph Brant (1743–1807)* [1328].

Sig	Person	Dates	C/G	Description
428	Pocahontas	(1595–1617)	C G	Indian princess in Jamestown colony
613	Tecumseh	(1768–1813)	C G	Shawnee leader (Tippecanoe)
709	Sitting Bull	(1831–1890)	C G	Sioux leader (American-Indian wars)
1070	Sacagawea	(1788–1812)	C G	Shoshone woman (Lewis and Clark expedition)
1088	Liliuokalani	(1838–1917)	C G	Last monarch of Hawaii
1104	Crazy Horse	(1842–1877)	C G	Lakota war leader (Little Bighorn)
1612	Atahualpa	(1497–1533)	C G	Inca Emperor deposed by Pizarro
2409	Chief Joseph	(1840–1904)	C G	Nez Perce leader (Nez Perce War)
2731	Squanto	(1585–1622)	C G	Patuxet Indian (helped the Pilgrims)
3122	Sequoyah	(1767–1843)	C G	Creator of Cherokee written language
3325	Red Cloud	(1822–1909)	C G	Lakota leader (Red Cloud's War)
3586	Metacomet	(1639–1676)	C G	Indian war chief (King Phillip)
3589	Chief Powhatan	(1545–1618)	C G	Powhatan leader and father of Pocahontas
3744	Chief Pontiac	(1720–1769)	C G	Ottawa Indian leader (Pontiac's Rebellion)
3886	Osceola	(1804–1838)	C G	Seminole leader (2nd Seminole War)
4362	John Ross	(1790–1866)	C G	Cherokee chief (Trail of Tears)
6590	Massasoit	(1581–1661)	C G	Peaceful leader of the Wampanoag
8362	Hiawatha	(?–?)	C G	Founder of Iroquois Confederacy
20841	Samoset	(1590–1653)	C G	First Indian to contact the Pilgrims
23880	Tomochichi	(?–1739)	C G	Peaceful Creek leader
36510	Nampeyo	(1860–1942)	C G	Hopi Indian potter
48999	Hendrick Theyanoguin	(1692–1755)	C G	Mohawk Indian leader

FIGURE 3.6. Native Americans in Bonnie's history book.

6 "Give me your tired, your poor, Your huddled masses yearning to breathe free, The wretched refuse of your teeming shore. Send these, the homeless, tempest-tost to me, I lift my lamp beside the golden door!"

All of the Native Americans in Bonnie's textbook have been dead for at least 70 years. What our methods reveal is the startling lack of prominent contemporary Native Americans. We can identify several significant cultural figures belonging to the first generation after the Indian Wars, including Olympic athlete *Jim Thorpe (1888–1953)* [1356], humorist *Will Rogers (1879–1935)* [1921], and Vice President *Charles Curtis (1860–1936)* [4674]. The contributions of the Navajo code talkers of World War II and the Mohawk steelworkers building the tallest skyscrapers are well known, but we challenge the reader to name a single contemporary Native American. With a population of 2.5 million, according to the 2000 U.S. Census (half that of Jewish Americans), this *shouldn't* be so hard. That it is serves as an indictment against the insularity and terrible economic prospects of the reservation system. Recent Native American political leaders like Senator *Ben Nighthorse Campbell (1933–)* [22686] and Chief *Wilma Mankiller (1945–2010)* [64475] are rare, and rank far below their forefathers in stature.

Hawaiians are native Americans, too, as appropriately recognized in Bonnie's textbook by their last queen *Liliuokalani* [1088]. We might also add King *Kamehameha I (1758–1819)* [950], who formally established the Kingdom of Hawaii.

3.7 Educational Standards vs. Editorial Decision Making

State teaching standards clearly influence the contents of school textbooks. If a book does not satisfy the standard, it cannot be sold to schools within the state – thus providing a powerful incentive for publishers to get with the program.

We wondered to what extent state teaching standards influenced the choices in Bonnie's textbook. We collected all names listed in the teaching standards of 16 different states. Only 75 of the resulting 1,121 people were mentioned in at least half of the state standards. They prove to be a very illustrious bunch, with *George Marshall (1880–1959)* [1279] and *Harriet Tubman (1820–1913)* [1093] being the least significant members of the exceptionally strong team presented in Figure 3.7. The only name new to us here was the French philosopher *Montesquieu (1689–1755)* [268], but

#	Person	B	Sig	#	Person	B	Sig	#	Person	B	Sig
15	George Washington	*	6	10	Montesquieu		268	9	Augustus		30
14	Thomas Jefferson	*	10	10	Martin Luther		17	9	Aristotle		8
14	Abraham Lincoln	*	5	10	Malcolm X	*	1106	9	Andrew Jackson	*	66
13	William Clark	*	979	10	Leonardo da Vinci		29	9	Alexander the Great		9
13	Meriwether Lewis	*	817	10	John Smith	*	529	9	Alexander Hamilton	*	45
13	F. D. Roosevelt	*	43	10	Jefferson Davis	*	188	8	William Penn	*	286
13	Benjamin Franklin	*	35	10	James Madison	*	51	8	William L. Garrison	*	735
12	Ulysses S. Grant	*	28	10	Harriet Tubman	*	1093	8	Thomas Paine	*	113
12	Thomas Edison	*	40	10	Hammurabi		899	8	Roger Williams	*	678
12	Theodore Roosevelt	*	23	10	D. D. Eisenhower	*	110	8	Nicolaus Copernicus		74
12	Robert E. Lee	*	76	10	Bill Clinton	*	115	8	Niccolò Machiavelli		168
12	John Locke		100	9	Woodrow Wilson	*	47	8	Nelson Mandela		356
12	Harry S. Truman	*	94	9	Ronald Reagan	*	32	8	Lyndon B. Johnson	*	184
12	Adam Smith		56	9	Richard Nixon	*	82	8	Karl Marx		14
11	Susan B. Anthony	*	432	9	Plato		25	8	Julius Caesar		15
11	Michelangelo		86	9	Patrick Henry	*	472	8	John Marshall	*	401
11	Martin Luther King	*	221	9	Napoleon		2	8	John D. Rockefeller	*	172
11	Mao Zedong		151	9	Mohandas K. Gandhi		46	8	John Brown	*	304
11	Isaac Newton		21	9	Joseph Stalin	*	18	8	J. Rousseau		80
11	Frederick Douglass	*	402	9	John Calvin		99	8	George Marshall		1279
11	C. Columbus	*	20	9	John Adams	*	61	8	Eleanor Roosevelt	*	517
10	Winston Churchill		37	9	Henry Ford	*	148	8	Dred Scott	*	883
10	W. E. B. Du Bois	*	545	9	Galileo Galilei		49	8	Benito Mussolini	*	101
10	Socrates		68	9	E. C. Stanton	*	734	8	Andrew Carnegie	*	104
10	Rosa Parks	*	964	9	B. T. Washington	*	525	8	Adolf Hitler	*	7

FIGURE 3.7. Historical figures appearing in at least half of all state teaching standards. Starred (*) figures appear in Bonnie's textbook.

his theory of the separation of powers into executive, legislative, and judicial branches defined the structure of the government of the United States.

Several states have highly specific requirements. Nearly two-thirds (734 of 1,121) of the historical figures appearing in the standards are unique to one particular state. The Texas elementary school standards list the names of 138 different people, many sensible and worthy. But recent amendments have added many state, business, and religious figures – a good 42 of whom ranked in the very thin historical air above 30,000. The least significant people appearing in both the textbook and their standards were Texans *Charles Goodnight (1836–1929)* [10172] and *Juan Seguín (1806–1890)* [10797]. California state standards were comparably restrained, with only 6 of 67 figures ranked higher than 15,000. These outliers were primarily state figures.

Taken in aggregate, however, the state teaching standards prove pretty perceptive. Of the 74 people from Bonnie's textbook absent from all of the state teaching standards we studied, we would seriously regret seeing only three omitted from Bonnie's book: the traitor *Benedict Arnold (1741–1801)* [222], Utah founder *Brigham Young (1801–1877)* [471], and freedom of the press figure *John Peter Zenger (1697–1746)* [2022].

The absent figures also jibe nicely with the least significant people we identified computationally. Indeed, 16 of the 20 least significant people in Bonnie's book (and 29 of the bottom 40) went unrecognized by every state we studied. We cannot fairly blame state standards run amok for the contents of Bonnie's textbook.

3.8 A Nation of Fourth Graders?

With the aid of Stony Brook's Center for Survey Research, we asked 100 randomly selected Americans to rank the names appearing in Bonnie's text. Each question contained two of the historical figures from the text (say *Robert E. Lee* [76] and *Thurgood Marshall* [1707]), and asked the respondent to identify which was more historically significant.[7] Each respondent was asked 40 questions, and they were permitted to confess that they did not know either of the candidates.

This survey satisfies two objectives. First, it provides a rigorous test of how well our significance rankings jibe with general opinion. Second, it enables us to estimate public recognition as a function of significance ranking. In particular, what fraction of adult Americans know each of the figures in Bonnie's fifth grade textbook?

We estimated the difficulty of each question by the difference in the significance scores among the pair. Based on this difference, the questions were divided into ten equal-sized deciles. The questions in progressive deciles should get harder, because the significance difference between the pair being examined becomes smaller and thus is harder to discern. By the last decile, both figures are of essentially identical stature, enough so that the choice of who's bigger becomes fairly arbitrary. Should we really mark as incorrect anyone who insists that *George Washington* [6] is more significant than *Abraham Lincoln* [5]?

[7] In this regard, the questions are similar to those posed on our *Who's Bigger?* game App, presented in Appendix Section B.3.

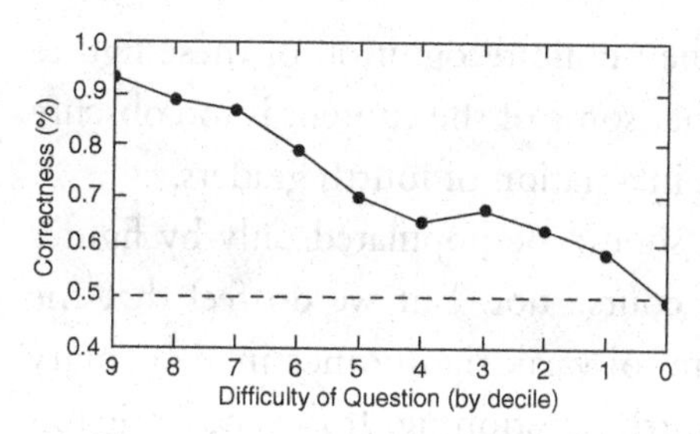

(a) Accuracy on "Who's Bigger?" questions by degree of difficulty (the difference in the significance scores of the pairs). For the easiest decile of questions, 92% of responses agree with our rankings. For the hardest decile of questions, where there is essentially no difference between the figure pair, respondents select both options equally.

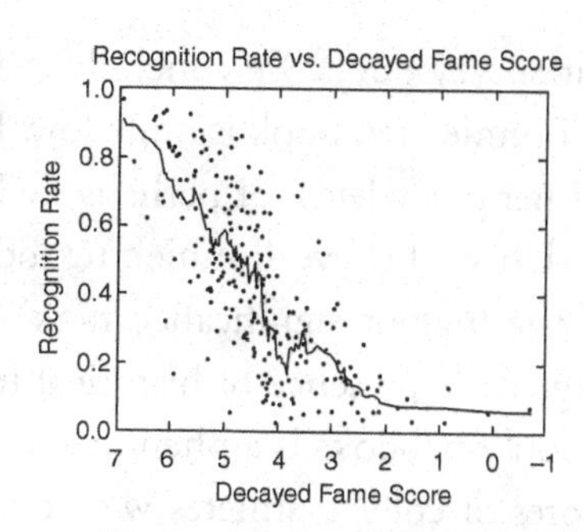

(b) Accuracy by historical figure. Most respondents recognize historical figures ranked in the top 1,000 (significance score about 4.5) but performance degrades rapidly on less significant people.

FIGURE 3.8. Who's Bigger? survey results.

The performance of our respondents on each tranche of questions is presented in Figure 3.8a, and is almost too good to believe. For the easiest group of questions, the surveyed respondents agreed with our ratings 92 percent of the time, clearly grade A performance. Their accuracy degrades progressively with each successive tranche. And for the last tranche, they score at the 50 percent mark, exactly as they should when forced to make arbitrary choices akin to picking heads or tails.

Our survey also enables us to estimate what fraction of respondents knew each historical figure in Bonnie's textbook, shown in Figure 3.8b. Almost all of the 45 figures we ranked among the top 200 in significance were known by a majority of our respondents. The only ones who failed this 50 percent standard were *James I* [41], *Joseph Smith* [55], and (by a narrow margin) *Benito Mussolini* [101]. Almost half of this cohort (22) were recognized by at least 80 percent of those surveyed. Only five of the 210 figures who ranked as less prominent achieved this level of recognition.

By contrast, of the 44 figures with significance rank above 5,000, only two were known by more than 50 percent of those surveyed: *Sandra Day O'Connor (1930–)* [5221] and *Madeleine Albright (1937–)* [5857]. Both are recent political figures. Only five others were known by even 20 percent of our survey pool. We conclude that the public recognition of these figures in Bonnie's textbook is quite low. Either some of the content is too obscure for her grade level, or perhaps we live in a nation of fourth graders.

Do we believe Bonnie's textbook should be populated only by figures of the highest significance rank? Of course not. But we do feel that the large-scale presence of historical figures of weak significance in elementary school textbooks is a phenomenon worth questioning. If the academic test scores of college athletes were observed to be substantially lower than the rest of the student body, should this result be dismissed because of biased testing? Or is it an opportunity to discuss the proper role of big-time sports at universities?

4

Reading through the Past

Reputations ebb and flow, leaving little trace of the course they followed in the historical record. *Plato (427–347 B.C.)* [25] and *Aristotle (384–322 B.C.)* [8] reign today as giants of Western civilization, but it was a very, very close call for both of them. With the descent of the Dark Ages after the fall of the Roman Empire, the writings of the ancient Greek philosophers were almost lost forever.

People's reputations often change posthumously. *Paul Revere (1735–1818)* [627] was a silversmith rescued from obscurity by *Henry Wadsworth Longfellow's (1807–1882)* [349] poem "The Midnight Ride of Paul Revere." Today, *Harriet Beecher Stowe (1811–1896)* [449] is far better known than her brother *Henry Ward Beecher (1813–1887)* [2905], but this wasn't the case during their lifetimes.

There is more to our computational study of history than significance rankings. Here we add another statistical arrow to our quiver. Google Ngrams is an amazing resource for monitoring changes in the cultural Zeitgeist, a dataset that lets us reconstruct how famous someone was 50, 100, or even 200 years ago. In this chapter, we introduce Google Ngrams and present the graphical representations we will use to make proper sense of this data. In particular, we will use it to study the phenomenon of posthumous changes in fame.

4.1 Google Ngrams

Scholarly works often include many pages of footnotes, with citations proving that the author has read hundreds of books. Ours doesn't, which might suggest Philistine standards of scholarship. Yet in a real sense, we have read far more broadly than any previous historian. We have garnered an understanding that can only come from reading millions of books.

Modern computational tools permit us to move backward in time and witness popular culture evolve to its current form. The key is analyzing the written legacy of books and newspapers that document the popular thinking of their time. Most documents are published shortly after they are written, and shortly before they are read. Thus, measuring the frequency with which people are mentioned in the written record, as a function of publication date, provides a good proxy for their significance over time.

Only recently have developments in scanning technology, computational/storage costs, and Internet economics collided so as to make large-scale historical trends analysis possible. As part of Google's mission to "organize the world's information and make it universally available and useful," it has engaged in a massive effort to scan the world's books. Massive indeed: they have digitized more than 15 million books, roughly 12 percent of all the books ever published.

Distributing the complete scanned texts to web surfers would invite the wrath of every copyright holder. The Google Ngrams project [Michel et al., 2011] cleverly avoids this by making the data available only in aggregate. Each Ngrams time series reports the annual frequency of a short (one- to five-word) phrase that occurs at least 40 times in their scanned book corpus. This eliminates obscure words and phrases, but leaves more than two billion time series available for analysis.[1]

Google graciously provides an Ngrams viewer, available at `http://books.google.com/ngrams`. Check it out: we promise you will enjoy playing with it. Figure 4.1 uses the Google Ngram data to display the prevalence of competing modes of transportation (*car* against *horse*) and communication (*telephone* supplanting *telegraph*), dominating national powers (the *United States* outgrowing *England*), and mealtimes (*breakfast* eating *lunch*'s lunch). Go ahead and play. Compare *hot dog* to *tofu*, *science* against *religion*, *freedom* to *justice*, and *sex* versus *marriage* to better understand this fantastic telescope for looking into the past.

1 Our analysis in this book is based on the data reported in Michel et al., 2011. A somewhat larger and improved corpus was released in August 2012. See Lin et al., 2012 for details.

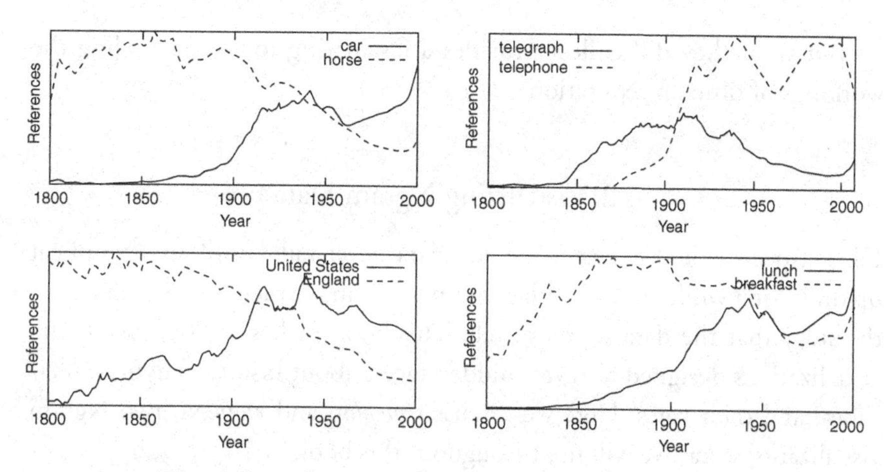

FIGURE 4.1. Historical trends in transportation, communication, national power, and mealtime habits revealed through Google Ngrams.

Google has even more graciously made the raw data available to researchers, enabling us to perform large-scale historical experiments that would otherwise be impossible. We will present our own graphic of representations of some of this Ngram data, which we think are pretty snappy. Still, the real credit for collecting it goes to Google.

There are certain biases inherent in using books to measure the cultural Zeitgeist. Books take a long time to write and publish (we can tell you something about that), so spikes naturally lag current events by a few years. We limit our attention here to English language books, which obviously underrepresents the significance of non-Anglo people and places. Biases in their sampling and scanning methodologies make data from before 1800 and after 2000 somewhat suspect: the former because of the relative dearth of early published matter, the latter because changes in Google's scanning criteria clog the counts with incomparable sets of books.

Not all aspects of popular culture are equally reported in books, particularly the books retained in the university library collections that served as the source of most scanned texts. The nineteenth-century Hall of Fame baseball player *Cap Anson (1852–1922)* [2342] ended his long career in 1897, yet does not appear in our book frequency data until 1917. We must excuse the university librarians of the time for shortsightedly focusing on science, literature, and history at the expense of sports, entertainment, and other vulgar aspects of popular culture.

But what they did collect provides a fascinating tool for revealing the workings of time on reputation.

4.2 Visualizing Ngram Data

Like suspects in old gangster movies, data never talks until after you beat up on it for a while. In particular, you need to make pictures to understand the story that the data wants to tell. This book makes use of several data visualizations designed to reveal hidden facets about historical figures, their times, and their roles. Here we discuss *timeplots* and *rugplots*, two Ngram visualizations that we will use throughout this book.

Timeplots

Google Ngram timeplots display the frequency that a given person is mentioned in books published each year. This provides a handle on how their reputation has ebbed and flowed over the years.

The Google Ngrams data consists of raw counts of the frequency with which a word or phrase is mentioned in books published in a given year. For example, *Paul Revere (1735–1818)* [627] was mentioned 254 times in books published in 1936. This by itself tells you nothing about his significance or popularity, because it only becomes meaningful when compared against other reference counts. That Revere got mentioned 587 times in 1997 does not mean he is becoming more popular with time, because the number of books published each year has grown continuously since Gutenberg. The right way to turn counts into trends independent of publishing and scanning variation is to normalize by the total number of scanned words published in each given year.

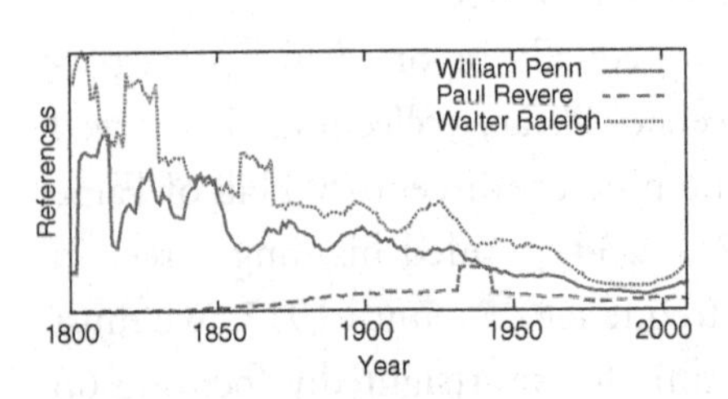

Let's look at a timeplot of the Google Ngrams data for the colonial/revolutionary era figures of *William Penn (1644–1718)* [286], *Walter Raleigh (1552–1618)* [313], and *Paul Revere (1735–1818)* [627]. The absolute counts for each figure each year are divided by the total number of references to *all* people in all books published that year, and smoothed by a moving average to reduce jitter. We see that Raleigh has consistently received more discussion than the other two over the past 200 years.

It is also clear that all three of these Founding Fathers are slowly being displaced, as new personalities take their place on the world stage. Their names remain recognizable to us because they each still account for about (1/1,000,000)th of all name references in newly published books. This is far below someone like *George Washington (1732–1799)* [6], but certainly nothing to sneeze at.

When properly normalized, even tiny timeplots can pack a punch. Figure 4.2 presents capsule timeplots for all U.S. presidents, ordered by term. The dark region represents a time course of each individual's reference frequency from 1800 to 2008, and so the peak generally shifts right with every successive presidency. We normalized each chart to its peak year, so it is impossible to distinguish the significance of an individual just by the amount of black ink in his time course. However, curves that swiftly decline from their peak generally reflect forgettable presidencies, like those of *Zachary Taylor (1784–1850)* [299], *Andrew Johnson (1808–1875)* [105], and *William McKinley (1843–1901)* [176].

The lightly shaded boxes in these capsule timeplots trace the life span of the individuals, clearly separating those who died in office from the few who lived long past their terms, like *Herbert Hoover (1874–1964)* [183] and *Gerald Ford (1913–2006)* [230]. These will be particularly revealing to illustrate the course of fleeting and posthumous fame.

Rugplots

It is difficult to compare the reference frequency of several historical figures simultaneously using time plots, because the lines tend to overlap and obscure each other. Instead, we prefer a graphical device we call a *rugplot*, so named because they suggest nice designs for rectangular throw rugs.[2]

Figure 4.3 is a rugplot of a half dozen early American figures. Each individual gets a band whose share of the area varies each year according to their relative reference frequency. Thus, the width of a person's band in the plot each year is a measure of their relative importance compared to the other figures in the plot.

[2] What we call rugplots sometimes go by the ugly moniker of "percent stacked area charts."

Term	Sig	Person	Ngrams Timeline
1	6	George Washington	
2	61	John Adams	
3	10	Thomas Jefferson	
4	51	James Madison	
5	220	James Monroe	
6	135	John Quincy Adams	
7	66	Andrew Jackson	
8	277	Martin Van Buren	
9	288	William Henry Harrison	
10	341	John Tyler	
11	240	James K. Polk	
12	299	Zachary Taylor	
13	446	Millard Fillmore	
14	427	Franklin Pierce	
15	237	James Buchanan	
16	5	Abraham Lincoln	
17	105	Andrew Johnson	
18	28	Ulysses S. Grant	
19	325	Rutherford B. Hayes	
20	285	James A. Garfield	
21	499	Chester A. Arthur	
22	98	Grover Cleveland	
23	339	Benjamin Harrison	
25	176	William McKinley	
26	23	Theodore Roosevelt	
27	153	William Howard Taft	
28	47	Woodrow Wilson	
29	242	Warren G. Harding	
30	370	Calvin Coolidge	
31	183	Herbert Hoover	
32	43	Franklin D. Roosevelt	
33	94	Harry S. Truman	
34	110	Dwight D. Eisenhower	
35	71	John F. Kennedy	
36	184	Lyndon B. Johnson	
37	82	Richard Nixon	
38	230	Gerald Ford	
39	462	Jimmy Carter	
40	32	Ronald Reagan	
41	363	George H. W. Bush	
42	115	Bill Clinton	
43	36	George W. Bush	
44	111	Barack Obama	

FIGURE 4.2. Capsule timeplots of all U.S. presidents, ordered by succession. Shaded areas denote the life span of each individual.

The people in Figure 4.3 all achieved renown by 1800, the leftmost year in the rugplot. Yet their historical fates have since diverged substantially. First note the relative mindshare decline of *Walter Raleigh (1552–1618)*

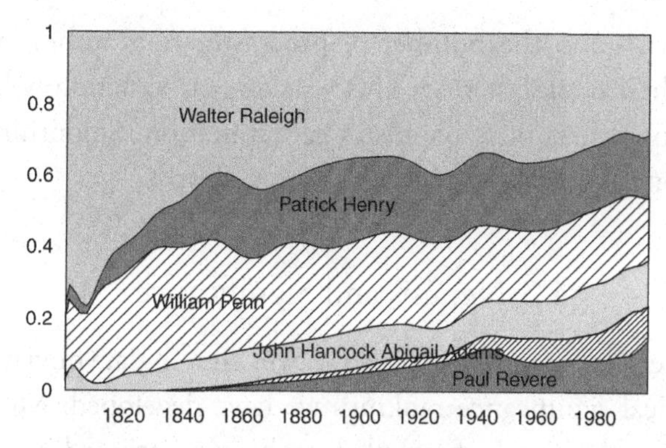

FIGURE 4.3. A rugplot of early American historical figures, showing posthumous growth for Patrick Henry, Abigail Adams, and Paul Revere.

[313] and *William Penn (1644–1718)* [286]. They were colonial figures who began to get crowded out from the record soon after the United States became an independent nation.

It is even more interesting to watch figures reemerge from the past. *Patrick Henry (1736–1799)* [472] was an important agitator of the Revolution, but did not take an active role in the national government that followed. He was rediscovered only after histories of the era began to be written. The mid-century rise of *Paul Revere (1735–1818)* [627] around 1850 was a result of poetry, not history. *Abigail Adams (1744–1818)* [821] has had a more recent resurrection. Interest in her has exploded since 1950, because she met a contemporary need for an intelligent and active female revolutionary-era protagonist.

Rugplots are informative and convincing representations of how popular conception of historical figures changes with time. However, be aware of certain quirks when interpreting them. Rugplots show how the pie slices every year, but does not represent how big the pie actually is. If all the people in a plot atrophy at the same rate, the bands will retain equal width. Far fewer books were published in 1800 than 2000, and thus the number of Ngram references to draw inference from were much smaller back then.

Much of the beauty of these plots comes from *smoothing*, which eliminates distracting variance in the time series data. The number of books to be published about *Theodore Roosevelt (1858–1919)* [23] this year depends, at least in part, upon the number of procrastinating authors who failed to complete their book last year. This year-to-year variation will affect the counts even for figures of stable historical reputation. Smoothing removes this fluctuation, to reveal the real trends more clearly.

4.3 Ngrams and Significance

The alert reader may wonder whether Google Ngram data obviates the need for the historical significance rankings we have developed. Might we just take the frequency with which people have been mentioned in books as the measure of their importance?

Several limitations of the Ngram data make this undesirable. First, Ngrams frequency cannot discriminate among multiple people who share the same name. Who is the real "Jim Clark": race car driver *Jim Clark (1936–1968)* [5363], entrepreneur *James H. Clark (1944–)* [44312], obscure baseball player *Jim Clark (1947–)* [829063] or any of the dozen other James and Jim Clarks appearing in Wikipedia? Such name collisions do not trouble our significance rankings because we separately analyze each Wikipedia page, but they play havoc in interpreting Ngram data.

A related issue is partial name matching. Any book reference to *Martin Luther King (1929–1968)* [221] also yields an Ngram for *Martin Luther (1483–1546)* [17], thus inflating his counts as well. These artifacts can be quite striking. The Ngram count for minor astronomer *Oliver Wendell (1845–1912)* [765197] is wildly inflated by the father-and-son pair *Oliver Wendell Holmes (1809–1894)* [945] and *Oliver Wendell Holmes (1841–1935)* [885]. More generally, it can be hard to establish exactly which Ngrams refer to a given person. Is *John F. Kennedy (1917–1963)* [71] best measured by *John Fitzgerald Kennedy*, *President Kennedy*, *John Kennedy*, *Jack Kennedy*, and/or *JFK*?

Nonetheless, it is valuable to compare our significance rankings against Google Ngrams. Showing a substantial correlation between these quantities helps validate our analysis. The discrepancies between these sources reveal phenomena worth noting, particularly the people who are overrepresented in books relative to Wikipedia.

Identifying the right measure of Ngram frequency proves somewhat tricky. To be fair to both recent and distant figures, we opted for a measure

Most Significant

Sig	Person	NGR	Ngrams Timeline
1	Jesus	2	
2	Napoleon	55	
3	Muhammad	583	
4	William Shakespeare	122	
5	Abraham Lincoln	1	
6	George Washington	7	
7	Adolf Hitler	16	
8	Aristotle	3284	
9	Alexander the Great	67	
10	Thomas Jefferson	13	
11	Henry VIII	29	
12	Charles Darwin	238	
13	Elizabeth I	228	
14	Karl Marx	22	
15	Julius Caesar	41	
16	Queen Victoria	11	
17	Martin Luther	27	
18	Joseph Stalin	64	
19	Albert Einstein	232	
20	Christopher Columbus	256	

Most References

NGR	Person	Sig	Ngrams Timeline
1	Abraham Lincoln	5	
2	Jesus	1	
3	Woodrow Wilson	47	
4	George W. Bush	36	
5	Charles de Gaulle	396	
6	Gautama Buddha	52	
7	George Washington	6	
8	Johns Hopkins	1281	
9	Theodore Roosevelt	23	
10	Richard Nixon	82	
11	Queen Victoria	16	
12	Louis XIV	26	
13	Thomas Jefferson	10	
14	Martin Luther King	221	
15	Winston Churchill	37	
16	Adolf Hitler	7	
17	George III	58	
18	Henry James	408	
19	Dwight D. Eisenhower	110	
20	Mark Twain	53	

FIGURE 4.4. Book Ngram frequency rankings (NGR) of the 20 most significant figures, with those of highest book reference frequency.

that used the peak 10-year period of reference frequency, normalized by the frequency of the most popular person during the window.

Figure 4.4 (left) presents the Ngram rank (NGR) for our twenty most historically significant figures. They generally have high Ngram frequencies, but with some interesting exceptions. Shakespeare, Darwin, Einstein, and Columbus have greater historical significance than full name Ngram references suggest precisely because they *are* Shakespeare, Darwin, Einstein, and Columbus. They are such recognizable figures that their last names suffice for unambiguous reference. Our Ngram frequencies count full name references, so these stars lose out to lesser figures with less distinguishable last names.

Aristotle (384–322 B.C.) [8] is the bigger mystery here. His distressingly low Ngram rank results from his unusually steady frequency in the books data. Our Ngram frequency statistic is based on each person's peak ten-year period, which works against those who were most active 2,500 years ago. Aristotle's mean Ngram frequency is actually quite high, and he would have ranked better by most other metrics.

By comparison, Figure 4.4 (right) shows the top Ngram frequency figures in our database. Nineteen of them also rank among the top 500 by significance. The lone exception is *Johns Hopkins (1795–1873)* [1281], who gets personal credit for every mention of his university.

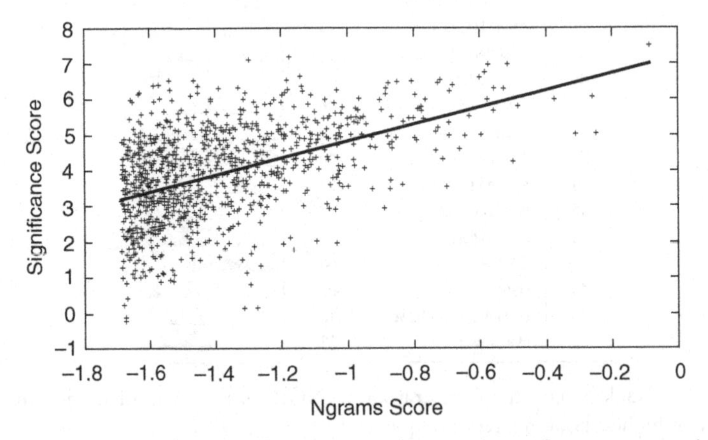

FIGURE 4.5. Dotplot of Ngram frequency vs. historical significance among the 1,000 most frequently mentioned people in books. The correlation is 0.504.

These are anecdotal examples of the relationship between Ngrams frequency and our significance rankings. Even more compelling evidence comes from correlation analysis. Figure 4.5 is a dotplot comparing significance and Ngrams scores of the 1,000 people most frequently mentioned in books. There is a nice linear relationship between them, with a correlation of 0.504. Of the top 1,000 people by Ngrams score, fully 362 of them rank in the top 1,000 by significance measure as well.

4.4 Posthumous Fame

Many of history's most significant figures never enjoyed the full fruits of public recognition during their lifetime. There is something particularly poignant about posthumous fame: a sense of regret and what-if, a rebuke of those who dismissed ideas that were simply too far ahead of their time. Which people achieved posthumous fame, and why did it happen to them?

We recognize two distinct forms of posthumous fame. The lesser variety involves the surge of grieving that a celebrity receives in the immediate aftermath of an unexpected death [Jones, 2005]. Important examples of this phenomenon include the musicians *John Lennon (1940–1980)* [162], *Elvis Presley (1935–1977)* [69], and *Kurt Cobain (1967–1994)* [1359]. Each of them arguably became more important and visible in death than they ever were in life. The death of an idol of one's youth exposes the fragility of our own lives, and so each generation's first passing occupies a particular niche forever.

The greater sense of posthumous fame concerns people who went unrecognized during their lifetimes but whose accomplishments first register after death. We used the Ngrams data to identify these figures, by computing the ratio of posthumous Ngram frequency to that of their lifetime. We define "living" to occur from age 30 to until 20 years after death, after which the "posthumous" period begins. These lags from actual life span capture the effects of publishing delays and living memory on the printed record.

This measure generally captures what we mean by posthumous fame, but misses figures who become famous in close proximity to their death.

The assassins *John Wilkes Booth (1838–1865)* [451] and *Lee Harvey Oswald (1939–1963)* [1435] both died within days of their infamous deeds. *Anne Frank (1929–1945)* [1821] died an unknown child in a concentration camp, but the publication of her diary two years after the war turned her into a symbol of the horrors of the Holocaust too rapidly for us to recognize as posthumous.

These posthumous figures are concentrated in relatively few domains. One doesn't see political leaders, because their work implies contemporary public recognition. Even when broader exposure accrues due to assassination (e.g., gay San Francisco supervisor *Harvey Milk (1930–1978)* [3680]), the rise is rapid enough to be instantaneous. One doesn't see athletes, because their achievements are immediately recognized on the field. A rare exception is the nineteenth-century baseball player *Roger Connor (1857–1931)* [10511], who was rediscovered after *Babe Ruth (1895–1948)* [434] popularized home runs, thus prompting a search to discover who had hit the most before Ruth.

Posthumous Figures

Sig	Person	Ngrams Timeline	PostR	Description
14	Karl Marx		22.69	Philosopher ("Communist Manifesto")
27	Ludwig van Beethoven		18.01	German composer ("Ode to Joy")
54	Edgar Allan Poe		33.27	American author ("The Raven")
73	Vincent van Gogh		49.16	Post-impressionist painter ("Starry Night")
102	William Blake		24.35	English poet and painter ("The Tyger")
131	Henry David Thoreau		34.57	American author ("Walden")
142	Georg Wilhelm Friedrich Hegel		10.51	German philosopher ("Historicism")
194	Emily Dickinson		33.79	American female poet (The Belle of Amherst)
198	Franz Schubert		9.13	Austrian composer (Romantic era)
223	Mary Shelley		17.40	English novelist (*Frankenstein*)
250	Gregor Mendel		8.75	Austrian monk and scientist (genetics)
350	Max Weber		14.58	German founder of modern sociology
384	Friedrich Engels		14.67	German Marxist (*The Communist Manifesto*)
601	Ramakrishna		45.80	Influential Indian Hindu mystic
724	Franz Kafka		10.43	German novelist (*The Metamorphosis*)
761	Billy the Kid		15.29	Old West gunman and outlaw
784	Émile Durkheim		9.42	French founder of modern sociology
978	Carl von Clausewitz		8.79	Prussian military theorist ("On War")
1038	Henri de Toulouse-Lautrec		9.56	French post-Impressionist painter
1281	Johns Hopkins		28.03	Entrepreneur, abolitionist (Baltimore)
1344	Wild Bill Hickok		9.00	Old West lawman and folk hero

Posthumous figures are generally individual artists/thinkers who produce works that can outlive them: painters, poets, philosophers, and composers. Very few figures in the table were completely unknown prior to their deaths, like *Emily Dickinson (1830–1886)* [194] or *Franz Kafka (1883–1924)* [724]. More typically their work has been previously recognized within a small community, positioned to promote it when conditions are ripe.

One corollary to this is that suicide is not the answer. Most of the leading posthumous figures in the accompanying table led relatively full lives, at least by the standards of their times. They needed this time to produce a body of work sturdy enough to survive in their stead.

Another way to achieve posthumous fame is through shifting agendas of later generations. Multiculturalism is a relatively recent idea, and its advocates scavenged history looking for figures from underrepresented groups to study and promote. The capsule timeplots given here demonstrate recent posthumous fame for several black and women figures from Bonnie's textbook.

Posthumously Recognized Figures from Underrepresented Groups in Bonnie's Textbook

Sig	Person	Ngrams Timeline	Sig	Person	Ngrams Timeline
1093	Harriet Tubman		3603	Lewis Howard Latimer	
1106	Malcolm X		9500	Hiram Rhodes Revels	
1231	Sojourner Truth		14221	Blanche Bruce	
1320	Nat Turner		35100	Nat Love	
1715	Olaudah Equiano		97353	Venture Smith	
2686	Zora Neale Hurston		206506	Luzena Wilson	

Finally, a less romantic but more reliable way to posthumous fame is by quietly collecting substantial resources and employing them imaginatively in one's will. *Johns Hopkins (1795–1873)* [1281] serves as the prototype here. We will push this point harder in Section 10.3.2, for those readers thinking about estate planning.

4.5 Faded Glory

It is equally interesting to identify the historical figures whose fame cooled the fastest: people who were well known in their times but have largely been ignored by subsequent generations.

Of course, the people who have been most thoroughly forgotten hold no cultural resonance for us today. Instead, we examine some of the most significant figures whose annual presence in the book Ngrams declined by at least 70 percent in the 50-year period following their passing from living memory (again taken as 20 years after their actual death).

Faded Figures

Sig	Person	Ngrams Timeline	PostR	Description
174	Arthur Wellesley		0.06	Duke of Wellington (Battle of Waterloo)
217	Gilbert du Motier		0.24	French military officer (Revolutionary War)
643	Humphry Davy		0.27	British chemist and inventor
903	Joseph Banks		0.12	English botanist (with Cook's expedition)
987	John Franklin		0.29	Explorer ("Northwest Passage")
1062	Joseph Bonaparte		0.21	Elder brother of Napoleon
1107	Charles Lyell		0.21	British lawyer, geologist
1608	Douglas Haig		0.22	Commanded British Expeditionary Force in WWI
1872	John Tyndall		0.16	Nineteenth-century physicist (Tyndall Effect)
1878	Robert Southey		0.20	English Romantic poet laureate
1923	John Herschel		0.12	English scientist (Cyanotype photography)
2057	Napoleon II		0.17	Son of Napoleon

Generally speaking, these forgotten figures fit into a small number of types. Some were military figures in battles that seemed of great importance at the time but were overtaken by later events. For example, General *Norman Schwarzkopf (1934–2012)* [10533] was the highly successful commander of U.S. forces during the first Gulf War with Iraq, but this conflict rapidly receded from memory after the larger scale involvement of the second Iraq War. Another group are writers and cultural figures whose work became less relevant, due to changing tastes and societal norms.

Other names have been forgotten for reasons of title. *Arthur Wellesley (1769–1852)* [174] defeated *Napoleon (1769–1821)* [2] at Waterloo, but history eventually settled on calling him the Duke of Wellington. Similarly, *Gilbert du Motier (1757–1834)* [217] became remembered as Lafayette.

On the flip side are those forgotten because of lost titles. Interest in *Napoleon's (1769–1821)* [2] brother *Joseph Bonaparte (1768–1844)* [1062] and son *Napoleon II (1811–1832)* [2057] plunged once the emperor was deposed and they were no longer in a position to succeed him. As *Wilson Mizner (1876–1933)* [58505] observed, "many a live wire would be a dead one except for his connections."

Several important figures in the making of the British Empire have faded from the record. Romantic era scientists like *Humphry Davy (1778–1829)* [643] and *Charles Lyell (1797–1875)* [1107] made important contributions to chemistry and geology, respectively, but these fields cut higher profiles in their times than they do today. Polar explorer *John Franklin (1786–1847)* [987] was a popular hero of the time, lost in Arctic ice seeking the fabled Northwest Passage. *Robert Southey's (1774–1843)* [1878] verse has long gone out of style, although he remains responsible for the original tale of Goldilocks and the three bears.

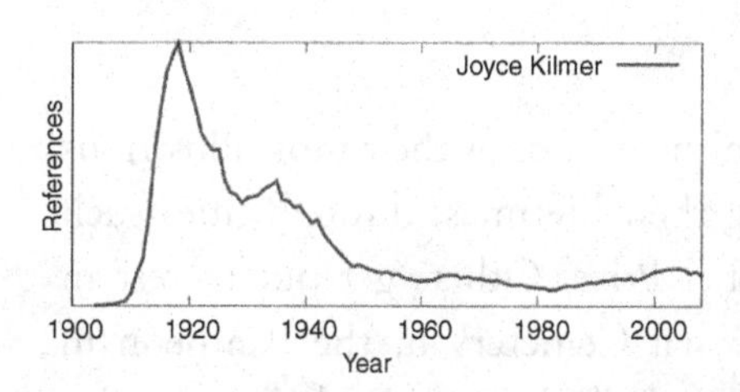

The poet *Joyce Kilmer (1886–1918)* [8092] ("I think that I shall never see a poem as lovely as a tree") represents an unusual combination of posthumous fame and rapidly faded glory. The combat death of this heroic young poet contributed to substantial fame in the years following World War I, but his style of sentimental lyric poetry then quickly fell out of fashion.

5

Great Americans and the Process of Canonization

Societies have developed many mechanisms to honor their most illustrious members. Some get recognized during their lifetimes, through titles such as knighthood or awards like the Nobel Prize. Others get put to rest in special places, such as Arlington National Cemetery or the Pantheon in Paris. Saints are canonized and so are baseball players – in their respective Halls of Fame.

Over the next two chapters, we will study the process of historical canonization by analyzing two long-standing New York institutions: the Hall of Fame for Great Americans in the Bronx and the Baseball Hall of Fame in the tiny village of Cooperstown. Both have held elections for more than 70 years, enough time to observe changes in each member's reputation in the years following selection. These institutions provide a natural laboratory to study how time erodes fame, and the limits to which knowledgeable observers can separate the gold from the dross of history.

5.1 The Hall of Fame for Great Americans

New York institutions rise and fall around real estate. The history of the city properly began when *Peter Minuit (1580–1638)* [3248] bought Manhattan from the Indians on May 24, 1626 for goods worth 60 Dutch guilders. Traditionally converted to $24, it was a steal: likely in more ways than one.

The Hall of Fame for Great Americans also began with a real estate transaction. Toward the end of the nineteenth century, New York University acquired land in the Bronx to serve as a new undergraduate campus.

One edge of the property faced a rocky hillside. The architect, *Stanford White (1853–1906)* [3303], took advantage of this setting to build an open air colonnade facing it. University Chancellor *Henry MacCracken (1840–1918)* [139484] seized this opportunity to create a shrine for renowned Americans, to fill the colonnade and drum up attention for his new campus.

The Hall of Fame was once a major American institution. Every five years, selections were made by a committee of one hundred or so prominent electors: elected officials, justices, university presidents, scientists, distinguished authors, and the like. Six electors eventually got elected to the Hall of Fame themselves: *Alexander Graham Bell (1847–1922)* [106], *Grover Cleveland (1837–1908)* [98], *Simon Newcomb (1835–1909)* [3666], *Alice Freeman Palmer (1855–1902)* [65220], *Theodore Roosevelt (1858–1919)* [23], and *Woodrow Wilson (1856–1924)* [47]. Special interest groups lobbied the electors, recruiting votes for their favorite candidates. Every five years, newspapers debated who should be admitted, and the announcement of each new class triggered another journalistic spasm. The Munchkins nominated Dorothy for the Hall of Fame in gratitude for killing the wicked witch, in the movie *The Wizard of Oz*. Committees were organized to raise the substantial funds necessary to sculpt a bronze bust worthy of each new immortal. This bust would be installed with grand ceremony – speeches and banquets culminating in a dramatic unveiling.

And then it suddenly withered away, a victim of changing times and places.

5.2 Who's In?

Why did this particular group of 102 historical figures receive their places in the Hall of Fame for Great Americans? Although they were the product of an electoral process by a distinguished body of academic, business, and political leaders, several selections prove baffling from our vantage point in history. The electoral results provide a unique laboratory to explore the shifting currents of fame.

The 10 most significant members all rank among the top 50 people in world history. Yet the bottom 10 are hopelessly obscure today. We will be impressed with any reader who can recognize more than one figure out of

Most Significant			Least Significant		
Sig	Person	Dates	Sig	Person	Dates
5	Abraham Lincoln	(1809–1865)	121367	Mark Hopkins	(1802–1887)
6	George Washington	(1732–1799)	65220	Alice F. Palmer	(1855–1902)
10	Thomas Jefferson	(1743–1826)	30412	C. S. Cushman	(1816–1876)
23	Theodore Roosevelt	(1858–1919)	26724	Lillian Wald	(1867–1940)
28	Ulysses S. Grant	(1822–1885)	24821	Rufus Choate	(1799–1859)
35	Benjamin Franklin	(1706–1790)	22095	John Lothrop Motley	(1814–1877)
40	Thomas Edison	(1847–1931)	21511	James Buchanan Eads	(1820–1887)
43	F. D. Roosevelt	(1882–1945)	20551	Sylvanus Thayer	(1785–1872)
45	Alexander Hamilton	(1755–1804)	16789	James Kent	(1763–1847)
47	Woodrow Wilson	(1856–1924)	16156	Emma Willard	(1787–1870)

these bottom 10, none of whom rank among the 15,000 most significant people in history.

The nature of the problem becomes even clearer in Figure 5.1, a cumulative distribution plot of the significance of the elected members of the Hall of Fame. The Hall of Fame has, as it should have, a very distinguished membership, with 60 percent of its members holding significance rank 2,000 or better. But the devil in any distribution lies at the tail. The remaining 40 percent prove nowhere near as elite a crowd. Indeed, a good 15 percent of them do not rank among the 10,000 most significant figures in history. This is not an impressive showing, when you are seeking to fill a Hall with only 100 seats in it.

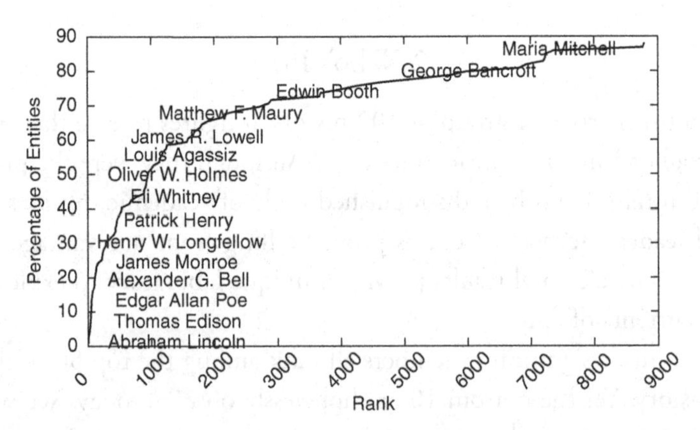

FIGURE 5.1. Cumulative distribution plot of Hall of Fame members by significance rank, with some representative names.

5.2.1 COUNTING THE VOTES

We can combine significance rankings with vote totals to gain a better sense of how accurately the electors were able to assess the greatness of their honorees. We plotted each honoree's margin of victory as a function of historical significance. To normalize for variation in the rules and the number of voters across different elections, we report the margin of victory in terms of the percentage of votes required to gain admission. Thus all members received at least 100 percent of the votes they needed, while the greatest (like *George Washington (1732–1799)* [6]) got almost twice as many votes as necessary.

Figure 5.2 shows that vote totals are correlated with significance. Generally speaking, the greater the immortal, the more votes they are likely to have received. This is shown by the upward-slanting regression line.

Certain cases are hard to understand from this temporal distance. It is difficult to see how *Frances Willard (1839–1898)* [5216] mustered more voter support than *Andrew Jackson (1767–1845)* [66], or how *William Crawford Gorgas (1854–1920)* [11592] beat *Theodore Roosevelt (1858–1919)* [23] in a head-to-head competition.

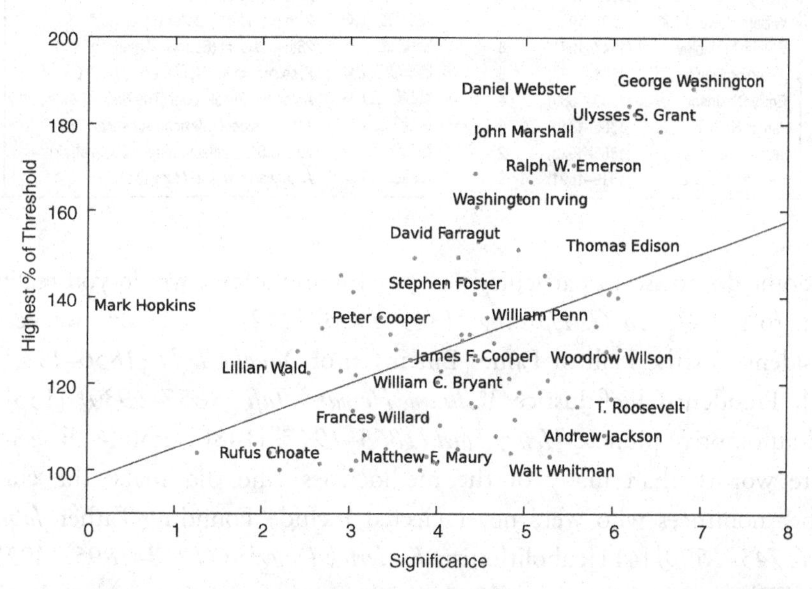

FIGURE 5.2. Margin of election victory vs. historical significance. More significant figures are generally elected by greater margins. Not all names appear on the chart.

However, the bigger outlier here is *Mark Hopkins (1802–1887)* [121367], who not only is the least significant figure in the Hall of Fame, but coasted in with almost 40 percent more votes than necessary. He served as president of Williams College for 36 years, and was apparently one hell of a teacher. President *James A. Garfield (1831–1881)* [285] said the best university was "Mark Hopkins on one end of a log and a student on the other." But I (Steve) am also one hell of a teacher, yet no one is going to make a statue out of me. Why did Hopkins make the Hall of Fame almost 30 years after his death?

5.2.2 WHO'S OUT?

The electors could have made other choices. Indeed, 10 available nominees rank among the 250 most significant figures in history (ok, 251), yet never passed muster with the voters.

Most Significant Non-Elected

Sig	Person	Dates	Noms	C/G	Description
93	Nikola Tesla	(1856–1943)	2	C G	Inventor (alternating current)
105	Andrew Johnson	(1808–1875)	3	C G	17th U.S. president (Reconstruction)
148	Henry Ford	(1863–1947)	1	C G	American industrialist (Ford Motors)
153	William Howard Taft	(1857–1930)	3	C G	27th president ("Dollar Diplomacy")
176	William McKinley	(1843–1901)	8	C G	25th president (Spanish-American War)
188	Jefferson Davis	(1808–1889)	8	C G	President of the Confederacy (U.S. Civil War)
194	Emily Dickinson	(1830–1886)	8	C G	American female poet (The Belle of Amherst)
240	James K. Polk	(1795–1849)	6	C G	11th president (Mexican-American War)
242	Warren G. Harding	(1865–1923)	2	C G	29th U.S. president (Teapot Dome affair)
251	Herman Melville	(1819–1891)	9	C G	American novelist (*Moby Dick*)

Some do not seem particularly inspired nominations: would you really want to see *Warren G. Harding (1865–1923)* [242], one of our weakest presidents, in the Hall of Fame? But inventor *Nikola Tesla (1856–1943)* [93], President/Chief Justice *William Howard Taft (1857–1930)* [153], and automotive pioneer *Henry Ford (1863–1947)* [148] certainly all seem more worthy than many of the mediocrities who did make the cut. Other nominees who were never elected include Founding Father *John Jay (1745–1829)* [411], abolitionists *Frederick Douglass (1818–1895)* [402] and *William Lloyd Garrison (1805–1879)* [735], and lexicographer *Noah Webster (1758–1843)* [1132].

Why did the electors reject these people in place of more obscure, apparently less deserving figures? We can identify several factors, some of which still plague selection processes today:

- *Cronyism* – Several of the marginal figures elected to the Hall of Fame were essentially contemporaries of prominent electors. The reputation of these figures generally declined rapidly after their selection.
- *Discomfort with gender and race* – It was only after electing 29 men that the absence of women as Great Americans was first noticed. Once the decision to admit women was made it was done clumsily, with several questionable choices and prominent omissions. The Hall has no Native Americans, and only two African Americans: *Booker T. Washington (1856–1915)* [525] and *George Washington Carver (?–1943)* [887].
- *Temporal biases* – The population of America has grown exponentially throughout its history. Thus, the number of famous people alive at any time has always been far more numerous than 50 years earlier, making it hard to compare the magnitude of figures over time.
- *Threshold comparisons* – Once a person has received a particular honor, they become a reference point to assess the worthiness of future candidates. Thus, one weak selection begets future weak selections, since that person has established the standard of membership.

We review each election, enabling you to play elector and see if you might have made different choices than they did.

5.3 Electoral History

The rules governing selection remained fairly consistent over the active history of the Hall of Fame, although certain details changed over time. The nominating committee generated a ballot of perhaps a hundred prominent Americans. Each candidate had to have been dead for a respectable period to encourage dispassionate thinking: initially at least 10 years, then extended to 25 years from 1925 on.

Each panel of electors, ranging from 93 in 1945 to a high of 141 in 1960, was charged with selecting new members from the nominees. Those picked by more than 50 percent of the electors (raised to 60 percent for elections between 1925 and 1940) became new members of the Hall of Fame, while those getting at least 20 votes were promised another appearance on the ballot next time.

Inaugural Election: 1900

This first election selected 29 people, more than a quarter of the ultimate composition of the Hall in one fell swoop. The nominators had free rein to choose anyone in American history who had died prior to 1890. The electors could define the hall on their terms, with no preconceptions, no prior models, only 51 votes out of 100 necessary for selection.

The 29 honorees fall into three tranches: the near unanimous (more than 90 votes), the comfortable (between 70 and 89 votes), and those on the bubble (who received between 50 and 69 votes). We first examine the greatest of the Great.

1900 Electees: Tier 1

Sig	Person	Dates	Votes	Type	Description
6	George Washington	(1732–1799)	97	Statesman	1st U.S. president (American Revolution)
453	Daniel Webster	(1782–1852)	96	Statesman	American statesman and Mass. U.S. senator
5	Abraham Lincoln	(1809–1865)	96	Statesman	16th U.S. president (U.S. Civil War)
35	Benjamin Franklin	(1706–1790)	94	Statesman	Founding Father/scientist (captured lightning)
28	Ulysses S. Grant	(1822–1885)	93	Military	18th U.S. president and Civil War general
10	Thomas Jefferson	(1743–1826)	91	Statesman	3rd U.S. president (Decl. of Independence)
401	John Marshall	(1755–1835)	91	Legal	U.S. chief justice (judicial review)

The top seven honorees include the preeminent representatives of all three branches of the U.S. government: the Executive (Washington and Lincoln), Judicial (Marshall), and Legislative (Webster) branches. It also includes Franklin and Jefferson, the two Founding Fathers whose intellect and versatility best capture the revolutionary spirit of American independence.

George Washington (1732–1799) [6] was the only nominee ever unanimously declared a Great American. The great esteem in which *Daniel Webster (1782–1852)* [453] was held may have dissipated somewhat over the past hundred years, but he tied for second in votes with *Abraham Lincoln (1809–1865)* [5]. To the electors, his historical distance and perhaps stature are analogous to the current perception of *Franklin D. Roosevelt (1882–1945)* [43], minus any association to a particular political party. Webster's party, the Whigs, collapsed shortly after his death.

The apparent odd man out is Grant, now more strongly perceived as an unsuccessful president than as the general who saved the Union. But the Civil War was as fresh in 1900 as the Vietnam War is today, and he was the primary military hero of this war. Grant's Tomb, the largest mausoleum in America, had been built by public subscription and opened in New York in 1897, just three years before this election.

1900 Electees: Tier 2

Sig	Person	Dates	Votes	Type	Description
122	Ralph Waldo Emerson	(1803–1882)	87	Author	American philosopher and lecturer
954	Robert Fulton	(1765–1815)	86	Inventor	Engineer and inventor (steamboat)
349	Henry W. Longfellow	(1807–1882)	85	Author	American poet ("Paul Revere's Ride")
433	Washington Irving	(1783–1859)	83	Author	Author ("Legend of Sleepy Hollow")
532	Samuel Morse	(1791–1872)	82	Inventor	Inventor (telegraph system)
913	Jonathan Edwards	(1703–1758)	82	Religious	Theologian (First Great Awakening)
1595	David Farragut	(1801–1870)	79	Military	U.S. Navy admiral ("Damn the torpedoes...")
252	Henry Clay	(1777–1852)	74	Statesman	Senator ("The Great Compromiser")
8785	George Peabody	(1795–1869)	74	Business	Entrepreneur, philanthropist (Peabody Trust)
227	Nathaniel Hawthorne	(1804–1864)	73	Author	Novelist (*The Scarlet Letter*)

The middle tier of honorees is dominated by authors and inventors. In general, the literary figures in the Hall of Fame have held up fairly well. During colonial times and the early days of the republic, the United States was an intellectual backwater, keenly aware that the cultural center of the world resided in Europe. In reaction, substantial glory accrued to our earliest literary figures.

George Peabody (1795–1869) [8785] was by far the weakest choice in this tier, a businessman and philanthropist who endowed several cultural and educational institutions. A stronger choice in this niche would have been *Johns Hopkins (1795–1873)* [1281], who was a regular nominee yet never mustered more than nine votes in any election.

Particularly noteworthy in the lowest tier of 1900 electees is the election of *Robert E. Lee (1807–1870)* [76], whose primary accomplishments were in opposition to the United States, as general of the Confederate Army.[1]

[1] I (Charles) believe that Lee unintentionally made a great contribution to the Union, through his strategy that saw Lee's home state of Virginia preserved before the vital Mississippi River.

1900 Electees: Tier 3

Sig	Person	Dates	Votes	Type	Description
4644	Peter Cooper	(1791–1883)	69	Business	Industrialist, philanthropist (Cooper Union)
758	Eli Whitney	(1765–1825)	69	Inventor	American inventor (cotton gin)
76	Robert E. Lee	(1807–1870)	68	Military	Confederate general (U.S. Civil War)
1037	John James Audubon	(1785–1851)	67	Scientist	Painter of birds (Audubon Society)
1193	Horace Mann	(1796–1859)	67	Educator	American education reformer
61	John Adams	(1735–1826)	65	Statesman	Founding Father and 2nd U.S. president
16789	James Kent	(1763–1847)	65	Legal	N.Y. Supreme Court justice and legal scholar
2862	Joseph Story	(1779–1845)	64	Legal	U.S. Supreme Court justice, lawyer
2905	Henry Ward Beecher	(1813–1887)	64	Religious	Clergyman, abolitionist, and social reformer
7203	William Ellery Channing	(1780–1842)	58	Religious	Unitarian preacher and theologian
1914	Gilbert Stuart	(1755–1828)	52	Artist	Painter ("George Washington" portrait)
2391	Asa Gray	(1810–1888)	51	Scientist	American botanist (*Gray's Manual*)

The Civil War ended just 35 years prior to this first election, but Lee's stature and sense of honor resonated in the North as well as the South. The controversy of how to treat Confederate figures would rage for most of the lifetime of the Hall of Fame.

Several honorees from this weakest tier, including *Peter Cooper (1791–1883)* [4644] and *Henry Ward Beecher (1813–1887)* [2905], were recently deceased New Yorkers who were likely known personally by some number of the electorate. The highest-ranking scientist to be elected was *John James Audubon (1785–1851)* [1037], the painter of birds better known today as an artist than a naturalist. With a large estate in the wilds of what is today upper Manhattan, Audubon was also a New Yorker. The other scientist, *Asa Gray (1810–1888)* [2391], was a botanist and correspondent of *Charles Darwin (1809–1882)* [12]. Their selections underscore the weakness of nineteenth-century American science in contrast with Europe.

Two honorees in this tranche are little known today. *William Ellery Channing (1780–1842)* [7203] was a founder of the Unitarian movement in the United States, which was more liberal and less dogmatic than the strict Calvinist doctrine of the founding colonists. *James Kent (1763–1847)* [16789] was an early American jurist and legal scholar. The nominee *John Jay (1745–1829)* [411], the first chief justice of the Supreme Court, would have been a much stronger selection in this space.

Who should have made the cut? We would happily replace Kent, Peabody, and Channing in the Hall with the three nominees who missed by the fewest votes. *Horace Greeley (1811–1872)* [986] (with 45 votes) was the dominant liberal figure of his era, an abolitionist and reformer whose influence came from the clout of his newspaper rather than political office. *Benjamin Rush (1746–1813)* [1283] (42 votes) was a signer of the Declaration of Independence, a founding father who played several roles in the new republic while making substantial contributions to American medicine. *Noah Webster (1758–1843)* [1132] was the pioneering lexicographer whose dictionary shaped and formalized the American language. The myriad variants of *Webster's Dictionary* published today have at best a tenuous connection to his original work, with his name having long fallen into the public domain. He got 36 votes in this election and would be nominated for the Hall of Fame a record 16 times during the next seventy years. Yet he never gained admission.

Gilbert Stuart (1755–1828) [1914] is now best known for his unfinished portrait of *George Washington* [6]. His rival and contemporary *John Singleton Copley (1738–1815)* [3609] mustered substantial support in every election from 1900 onward for decades, but was never elected, perhaps because he conveniently left the United States at the start of the American Revolution and painted his most important pictures in England. Another American painter of the time, *John Trumbull (1756–1843)* [1697], never received a vote.

1905

It did not go unnoticed that all 29 members of the inaugural class were male, as were the top ten runners-up. This put heat on the electors to come up with some women the second time around.

The three women they came up with are all relatively obscure today. *Mary Lyon (1797–1849)* [8765] was the founder and first president of Mt. Holyoke College, a pioneering women's liberal arts college, which served as a model for subsequent institutions of this type. *Emma Willard (1787–1870)* [16156] founded an earlier school for women, more a boarding school than today's notion of a college. *Maria Mitchell (1818–1889)* [7361] was an astronomer who discovered a comet, and went on to teach at Vasser.

1905 Electees

Sig	Person	Dates	Votes	Type	Description
135	John Quincy Adams	(1767–1848)	60	Statesman	6th U.S. president (Monroe Doctrine)
8765	Mary Lyon	(1797–1849)	59	Educator	American pioneer of women's education
1675	James R. Lowell	(1819–1891)	59	Author	American Romantic poet
233	William T. Sherman	(1820–1891)	58	Military	Union General (U.S. Civil War)
51	James Madison	(1751–1836)	56	Statesman	4th U.S. president (War of 1812)
1698	John G. Whittier	(1807–1892)	53	Author	American Quaker poet, abolitionist
16156	Emma Willard	(1787–1870)	50	Educator	American women's rights activist
7361	Maria Mitchell	(1818–1889)	48	Scientist	First American female professional astronomer

There were stronger candidates available, but the pickings slimmer than one might have supposed. Many fields were simply not open to women at that time. No female political figures existed; indeed, the Nineteenth Amendment granting women the right to vote would not be ratified until 1920. Many of the most historically prominent women of the era were still alive and thus not eligible for election, including *Susan B. Anthony (1820–1906)* [432] and *Clara Barton (1821–1912)* [1138].

But three nominated women were indeed vastly more significant than those who got elected. *Anne Hutchinson (1591–1643)* [894] was a minister expelled from the Massachusetts Bay Colony by Puritan clergy, and a major figure in the quest for religious freedom in the United States. She might have been too strong and controversial a figure for the conservative electorate. The 20 votes she received in 1905 proved her high-water mark; she rapidly lost support after the pressure to select women dissipated. *Louisa May Alcott (1832–1888)* [1024] was a pioneering children's author whose book *Little Women* remains popular today. Alcott fell just 11 votes short of election. *Dorothea Dix (1802–1887)* [2844], the superintendent of Army Nurses for the Union during the Civil War, and founder of the first modern American asylums for the mentally ill, would ultimately be nominated 15 times without ever being elected.

The biggest surprise among the also-rans was *Helen Hunt Jackson (1830–1885)* [6301]. An almost exact contemporary of Alcott who received almost as many votes, she was a writer of the West, whose popular novel *Ramona* drew attention to the federal government's mistreatment of Native Americans. Film versions of *Ramona* were produced in 1910, 1928, and 1936. She may be due for a revival, but even without it, she scores as more significant than any of the three women who were elected.

The men selected included two early presidents (*John Quincy Adams (1767–1848)* [135] and *James Madison (1751–1836)* [51]), plus two poets less well remembered today (*James Russell Lowell (1819–1891)* [1675] and *John Greenleaf Whittier (1807–1892)* [1698]): both victims of changing literary tastes and styles. *William Tecumseh Sherman (1820–1891)* [233], the other great Union general of the Civil War, was easily elected in his first year of eligibility. He would be the last military leader from the North to gain election, as the War between the States faded as a living memory from the minds of the electors.

1910

Harriet Beecher Stowe (1811–1896) [449] received the greatest number of votes in the 1910 election. The author of the book that arguably provoked the end of slavery (*Uncle Tom's Cabin*), she was of greater historical significance than seven of her nine co-inductees.

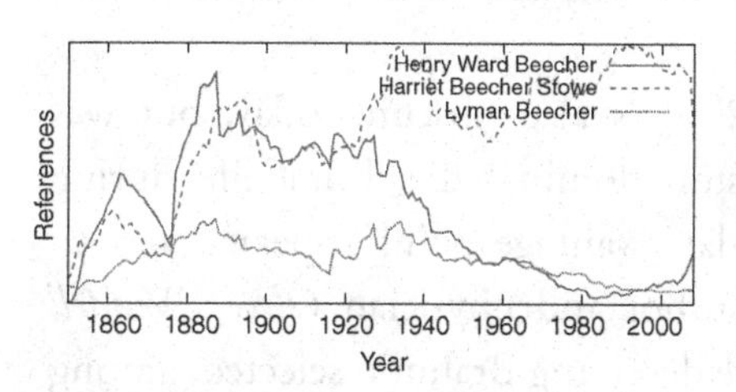

A timeplot shows how Stowe's stature has grown with the years, particularly in comparison with her father *Lyman Beecher (1775–1863)* [5203] and brother *Henry Ward Beecher (1813–1887)* [2905]. Both were famous preachers: Lyman the last significant Puritan minister, Henry arguably the most famous clergyman of his time [Applegate, 2006]. But literature outlives the spoken word, particularly from the era before recorded media.

All told, ten people were elected in that year. Authors dominate the honorees this time. The novelists and storytellers remain popular today, while the others have faded. *William Cullen Bryant (1794–1878)* [1779] was a prominent poet and journalist of the era. *George Bancroft (1800–1891)* [5977] was a great historian, whose *History of the United States* shaped much of the story we tell about the making of our nation. *John Lothrop Motley (1814–1877)* [22095] was a substantially less great historian, focusing on the Netherlands during the Age of Exploration and serving as an ambassador to Europe during the Civil War.

The other woman to gain inauguration was *Frances Willard (1839–1898)* [5216], a suffragist who founded the Woman's Christian Temperance

1910 Electees

Sig	Person	Dates	Votes	Type	Description
449	Harriet Beecher Stowe	(1811–1896)	74	Author	Abolitionist/author (*Uncle Tom's Cabin*)
945	Oliver Wendell Holmes	(1809–1894)	69	Author	American poet (Justice Holmes' father)
54	Edgar Allan Poe	(1809–1849)	69	Author	American author ("The Raven")
930	James Fenimore Cooper	(1789–1851)	62	Author	Writer (*Leatherstocking Tales*)
14279	Phillips Brooks	(1835–1893)	60	Religious	American Episcopal clergyman, author
1779	William Cullen Bryant	(1794–1878)	59	Author	American Romantic poet ("Thanatopsis")
5216	Frances Willard	(1839–1898)	55	Social	American educator and women's suffragist
5977	George Bancroft	(1800–1891)	53	Author	American historian, writer, and statesman
66	Andrew Jackson	(1767–1845)	53	Statesman	7th U.S. president ("Old Hickory")
22095	John Lothrop Motley	(1814–1877)	51	Author	American historian and diplomat

Union. Both causes were near the brink of success at the time of her election: the prohibition of alcohol and ratification of women's right to vote followed in 1920 from the Eighteenth and Nineteenth Amendments to the Constitution, respectively. She is more strongly linked to Prohibition than suffrage, however.

Phillips Brooks (1835–1893) [14279] is also obscure today, but was a prominent Boston clergyman. Boston dominated cultural life during the nineteenth century, and from today's vantage point appears overrepresented in the Hall of Fame. The author and physician *Oliver Wendell Holmes (1809–1894)* [945] was a more deserving Brahmin selected; among other accomplishments, he popularized that term for the well-bred elite of Boston.

Andrew Jackson (1767–1845) [66] just squeaked in the door with 53 votes. Jackson is a controversial figure; a founder of the modern Democratic party and the first populist president of the United States. He has always raised the hackles of conservatives. This ambivalence is reflected by his vote total, but historians generally rank him among our ten greatest presidents.

Martha Washington (1731–1802) [1065] was the top vote getter among the unselected. Her prominence is somewhat mystifying today, because most people cannot point to an individualized achievement or historical anecdote about her. Presumably she represented maternity in an era where there were few other female roles, although *George Washington (1732–1799)* [6] was never the father of more than our country.

1915

The top vote getter in the 1915 election was *Alexander Hamilton (1755–1804)* [45], who had never been previously nominated. Hamilton's stature among the Founding Fathers has only increased over time, as his vision of an industrial nation triumphed over our agrarian roots. Might the delay in honoring Hamilton pertain to the fact that he was born in the West Indies? Did Great Americans have to be native born?

1915 Electees

Sig	Person	Dates	Votes	Type	Description
45	Alexander Hamilton	(1755–1804)	70	Statesman	U.S. Founding Father (National Bank)
121367	Mark Hopkins	(1802–1887)	69	Educator	American educator and theologian
7235	Francis Parkman	(1823–1893)	68	Author	Historian ("The Oregon Trail")
1171	Louis Agassiz	(1807–1873)	65	Scientist	Paleontologist, etc. (hypothesized Ice Ages)
6927	Elias Howe	(1819–1867)	61	Inventor	Inventor (sewing machine)
2127	Joseph Henry	(1797–1878)	56	Scientist	American scientist (electromagnetics)
30412	C. S. Cushman	(1816–1876)	53	Artist	19th-century American stage actress
24821	Rufus Choate	(1799–1859)	52	Legal	Massachussets politician (U.S. senator, etc.)
501	Daniel Boone	(1734–1820)	52	Explorer	American frontiersman (coonskin cap)

This election picked two figures associated with the Western expansion of the United States. *Daniel Boone (1734–1820)* [501] was an early frontiersman and settler, back in the days when the West meant Kentucky. He filled a similar historical niche as *Davy Crockett (1786–1836)* [606], who never attracted much support from Hall of Fame voters but was a folk hero of comparable historical stature. *Francis Parkman (1823–1893)* [7235] was the first great historian of the American frontier, whose works reached a large popular audience.

Two are scientists. *Joseph Henry (1797–1878)* [2127] was a physicist, pioneering the study of electromagnetism and serving as the first secretary of the Smithsonian Institution. *Louis Agassiz (1807–1873)* [1171] was a naturalist and professor at Harvard. That he was born in Switzerland and did not arrive in the United States until he was 39 did not stop him from becoming a Great American. These concerns seem to have been relaxed for scientists: early admit *John James Audubon (1785–1851)* [1037] had been born on the French-Caribbean island of St. Domingue.

Elias Howe (1819–1867) [6927] was renowned as the inventor of the sewing machine, but the full history is complicated. Businessman

Isaac Singer (1811–1875) [4114] and mechanic *Walter Hunt (1796–1859)* [16684] have their own claims as the inventor of the sewing machine. Both were nominated to the Hall of Fame multiple times, but received a total of two votes between them.

Three of the six most obscure Hall of Fame members were selected in this, the fourth election. *Rufus Choate (1799–1859)* [24821] represents the dead white male marginalized today: a lawyer and orator from a prominent Boston family, he was an establishment figure in an establishment time. *Charlotte Saunders Cushman (1816–1876)* [30412] was a popular stage actress of her era. None of her work from this era before recorded media survives. That her reputation survived 40 years after her death to earn her election is testimony to the esteem her work must have held in its time.

The outlier among those who almost made it was *Horace Bushnell (1802–1876)* [54313], who peaked in second place this year with 44 votes, but drew significantly in all of the first seven elections. He was a Congregational clergyman and prolific author on Christian theology.

1920

Three members of this Hall of Fame class retain great historic stature today. *Mark Twain (1835–1910)* [53] authored perhaps the most American of novels. He is arguably the first modern celebrity, through his persona of the white-dressed river-boat captain of the Mississippi. *Roger Williams (1603–1683)* [678] founded the colony of Rhode Island as a refuge for freedom of worship. *Patrick Henry (1736–1799)* [472] was the orator whose "Give me liberty or give me death!" helped inspire the American Revolution, although he ultimately played a relatively small role in the building of the republic.

1920 Electees

Sig	Person	Dates	Votes	Type	Description
7233	William T. G. Morton	(1819–1868)	72	Medical	Dentist and pioneer of anesthesia
53	Mark Twain	(1835–1910)	72	Author	American author (*Huckleberry Finn*)
3993	Augustus Saint-Gaudens	(1848–1907)	67	Artist	Irish American sculptor
678	Roger Williams	(1603–1683)	66	Religious	Founder (Rhode Island colony)
472	Patrick Henry	(1736–1799)	57	Statesman	U.S. Founding Father ("Give me liberty")
65220	Alice Freeman Palmer	(1855–1902)	53	Educator	American educator (president of Wellesley)
21511	James Buchanan Eads	(1820–1887)	51	Engineer	American civil engineer (Eads Bridge)

The other four choices are substantially lower-wattage figures. *Augustus Saint-Gaudens (1848–1907)* [3993] was the preeminent American sculptor of his time. Dentist *William T. G. Morton (1819–1868)* [7233] was the primary driver behind surgical anesthesia for providing the first public demonstration of ether, even though priority of invention belongs to *Crawford Long (1815–1878)* [11578].

James Buchanan Eads (1820–1887) [21511] was an engineer: designer of the first substantial steel bridge, still in use in St. Louis, and several ironclad river ships for the Navy. He thrived in an era that produced tremendous feats of civil engineering. But he was the only engineer ever elected to the Hall of Fame, selected over others with more prominent monuments to their credit, such as *John A. Roebling (1806–1869)* [5444] (the Brooklyn Bridge) or *George Washington Goethals (1858–1928)* [11404] (the Panama Canal).

Two of the leading also-rans are particularly fascinating figures. *Benjamin Thompson (1753–1814)* [3435] was an early American-born scientist, who essentially spied for the British Army during the American Revolution and fled to Europe. He was eventually knighted by the British king and named Count Rumford of Bavaria. Thompson made substantial contributions to several areas of physics, but his credentials as a Great *American* appear scandalously thin. Yet he mustered 38 votes. *Adoniram Judson (1788–1850)* [7820] was a Baptist missionary to Burma, one of the first American missionaries to venture overseas. He was the choice of 29 electors. That he was categorized as an explorer instead of a religious figure makes a statement about how much smaller the world is today than it was back then.

This was the first election in which President *William McKinley (1843–1901)* [176] received votes, and he drew very little support. This seems surprising in light of his assassination. One might have imagined a greater sympathy vote for a recently martyred president – one who was elected to two terms by substantial margins. We attribute the fact that he didn't to two factors: the vigorous contrast with McKinley's successor, *Theodore Roosevelt (1858–1919)* [23], and the widespread horrors of World War I, which placed much greater demands on the public's capacity for mourning.

1925 to 1940

Admissions standards were toughened beginning in 1925, contributing to small enough classes that we will pool our discussion of several elections. The waiting period after death was extended from 10 to 25 years, which would depress the entry of fresh candidates for the next several elections. Also, the threshold for admission rose from a simple majority of electors to 60 percent. This meant that 63 of 105 electors were necessary for selection in 1925. As a consequence, only 10 people would be elected over the next four elections, from 1925 to 1940.

1925–1940 Electees

Sig	Person	Dates	Most Votes	Type	Description
1576	Stephen Foster	(1826–1864)	86	Artist	American songwriter ("Oh! Susanna!")
3686	Edwin Booth	(1833–1893)	85	Artist	Actor, brother of John Wilkes Booth
286	William Penn	(1644–1718)	83	Statesman	Founder of Pennsylvania colony
3666	Simon Newcomb	(1835–1909)	78	Scientist	American astronomer, mathematician
98	Grover Cleveland	(1837–1908)	77	Statesman	22nd and 24th U.S. president
1002	J. A. M. Whistler	(1834–1903)	74	Artist	Artist ("Whistler's Mother")
649	John Paul Jones	(1747–1792)	68	Military	American revolutionary naval hero
2170	Matthew F. Maury	(1806–1873)	66	Scientist	Navy scientist ("Scientist of the Seas")
220	James Monroe	(1758–1831)	66	Statesman	5th U.S. president (Panic of 1819)
160	Walt Whitman	(1819–1892)	64	Author	American poet ("Leaves of Grass")

John Paul Jones (1747–1792) [649] is a problematic figure. His reputation was built by attacking British ships off the coast of Europe using a merchant vessel, so he can be fairly viewed as a pirate instead of a naval officer. Further, he went on to serve the Russian Empress *Catherine the Great (1729–1796)* [108] as essentially a mercenary. Interest in Jones revived when his remains were discovered in France and re-interred in the Naval Academy Chapel in Annapolis in 1913. Indeed, he was never nominated for the Hall of Fame until 1915.

The leading vote getter of 1930 was artist *James Abbott McNeill Whistler (1834–1903)* [1002]. Whistler's selection seems peculiar. He lived his entire adult life in Europe after washing out of West Point, which would seem to undermine the case for him as a Great American, particularly in comparison with his contemporaries *Thomas Eakins (1844–1916)* [1341], *Winslow Homer (1836–1910)* [1466], and *John Singer Sargent (1856–1925)* [1727]. These others would not be eligible until later, however, by which time American art was much better established.

Matthew Fontaine Maury (1806–1873) [2170] is the least known of the 1930 honorees, but was a fascinating historical figure. A naval officer, his interests in charting wind and ocean currents made him a pioneering figure in oceanography. His years as superintendent of the U.S. Naval Observatory ended with the onset of the Civil War, when he sided with his home state of Virginia. He served the Confederacy as a diplomat, and played roles in founding both Virginia Tech and the Virginia Military Institute (VMI) after the war.

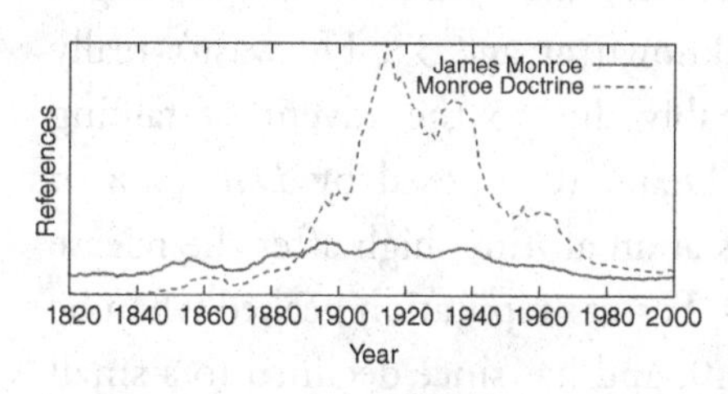

The selection of *James Monroe (1758–1831)* [220] may seem more dutiful than inspired: plugging the only Hall of Fame hole remaining among our first seven presidents. But 1930 was the height of Monroe's historical relevance. He is best remembered for the Monroe Doctrine, the statement that the United States would oppose any European efforts to meddle in North or South American nations. The chart reveals that the Doctrine was not visible in book Ngrams data before 1850, and didn't really take off until the United States began to assert itself in Latin America during the 1890s. *Theodore Roosevelt's (1858–1919)* [23] expansion of the Doctrine to justify further intervention made it an important instrument of American policy, peaking in the 1920s. Through all of this, Monroe's reputation has remained remarkably steady over time, a man overtaken by his Doctrine.

The election of 1935 added another president. *Grover Cleveland (1837–1908)* [98] was the eighth ranked president in *Arthur M. Schlesinger's (1888–1965)* [27072] 1948 historians poll. His reputation has faded somewhat since then, but he consistently ranks among the top half of American presidents.

Astronomer *Simon Newcomb (1835–1909)* [3666] was elected under the rules requiring a 25-year waiting period after death, which had kicked him off the ballot following his inaugural appearance in 1920. As an astronomer who combined legitimate scientific achievement with the ability to popularize science, he filled a role analogous to that of *Carl Sagan (1934–1996)* [2687] in more recent times. His historical significance ranks well below that of the philosopher and logician *Charles Sanders Peirce (1839–1914)*

[225] (nominated, but not elected, in 1950, 1955, and 1973), whose career Newcomb is credited with destroying.

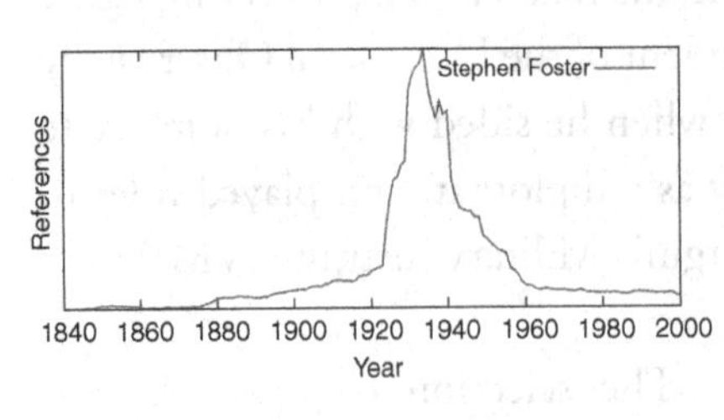

Songwriter *Stephen Foster (1826–1864)* [1576] worked at a time in which there was limited respect for copyright, so he managed to write several of America's most popular songs ("Oh! Susanna," "Suwanee River," and "Jeannie with the Light Brown Hair"), yet died poor and unknown at age 37. He wasn't really rediscovered until the 1920s, presumably due to the advent of talking pictures. Nostalgic sentiment about "Dixie" (composed by *Dan Emmett (1815–1904)* [14245], not Foster) was at an all-time high after the release of the film *Gone with the Wind* in 1939. Foster's reputation peaked when he was elected to the Hall of Fame in 1940, and has since declined to a small fraction of what it was during his heyday.

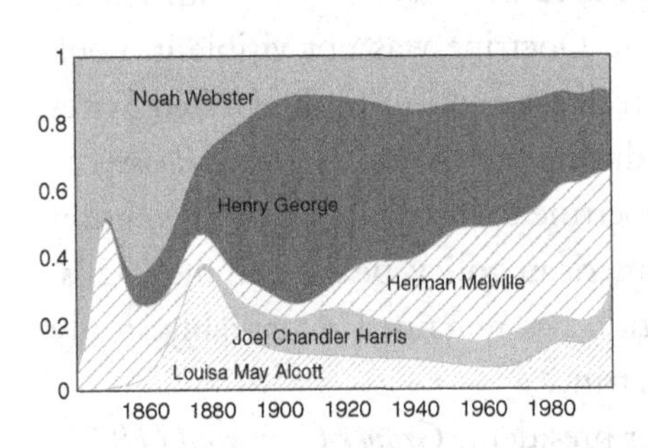

Five of the top also-rans in 1935 were authors. The Ngram data shows the wide fluctuations in their respective reputations over the years. *Herman Melville (1819–1891)* [251] had a particularly rocky ride, from his popular heyday around 1850 to a nadir where all his books were out of print by 1876. The critical revival of his work occurred in the 1920s, with the publications of his last manuscript *Billy Budd* and the declaration that *Moby-Dick* was the Great American Novel.

The reputation of *Joel Chandler Harris (1848–1908)* [6366] shadows that of his creation, the character Uncle Remus. Harris received 31 votes in 1935, but his popularity declined as his portrayal of black Americans fell out of favor. The Uncle Remus stories form the basis for Disney's popular 1946 film, *The Song of the South*, yet the film is so dated by changing attitudes on race that it has never been released on home video. Ngram data shows that *Louisa May Alcott (1832–1888)* [1024] overtook Harris around 1980, and the gap between them has continued to widen.

Henry George (1839–1897) [1538] missed election by only five votes. He was a political economist on the model of *Karl Marx (1818–1883)* [14], although George's philosophy was different. He advocated a land tax to redistribute wealth from nature, whereas people should own all that they create. George's influence peaked in the 1920s and declined rapidly after World War II.

1945 and 1950

In the 1945 election, *Booker T. Washington (1856–1915)* [525] became the first person of color installed in the Hall of Fame. An educator and the founder of the Tuskegee Institute, he was the major African American leader of his time. The electors had passed up good opportunities to break the color barrier earlier. In particular, *Frederick Douglass (1818–1895)* [402] was nominated repeatedly, yet amassed a total of one vote between the 1945 and 1950 elections.

1945–1950 Electees

Sig	Person	Dates	Most Votes	Type	Description
11592	William C. Gorgas	(1854–1920)	81	Medical	U.S. surgeon general and army physician
47	Woodrow Wilson	(1856–1924)	77	Statesman	28th U.S. president (World War I)
432	Susan B. Anthony	(1820–1906)	72	Social	American women's rights / civil rights leader
23	Theodore Roosevelt	(1858–1919)	70	Statesman	26th U.S. president (Progressive Movement)
106	Alexander G. Bell	(1847–1922)	70	Inventor	Inventor (telephone)
1244	Josiah W. Gibbs	(1839–1903)	64	Scientist	American physicist (Gibbs free energy)
525	B. T. Washington	(1856–1915)	57	Educator	American educator (Tuskegee Institute)
113	Thomas Paine	(1737–1809)	51	Author	U.S. revolutionary ("Common Sense")
4249	Walter Reed	(1851–1902)	49	Medical	U.S. Army physician (yellow fever)
6828	Sidney Lanier	(1842–1881)	48	Author	American musician, poet

Sidney Lanier (1842–1881) [6828] was a poet and musician along the *Stephen Foster (1826–1864)* [1576] model; that is, a Civil War–era Southern boy who died young. Luckily for him, he got his foot in the Hall by two votes when he did, because his reputation declined so rapidly after World War II that he never had a chance in any subsequent election.

Walter Reed (1851–1902) [4249] was the heroic Army doctor credited with defeating the tropical disease yellow fever, work that made completion of the Panama Canal possible. A deserving candidate who had received substantial support for several elections, it probably was the wartime prominence of the Walter Reed military hospital that pushed him over the threshold. Reed's clout also pulled in *William Crawford Gorgas (1854–1920)* [11592] during the subsequent election. Gorgas was the man of action to Reed's man of science in their quest to eliminate yellow fever. As the Army surgeon general, he drained swamps and eradicated mosquitoes after Reed identified them as the cause of the disease. We are personally more sympathetic to the case of *George Washington Goethals (1858–1928)* [11404], the army engineer who led the entire Panama Canal project.

The 1950 class also included two of our greatest presidents (*Theodore Roosevelt (1858–1919)* [23] and *Woodrow Wilson (1856–1924)* [47]) and the inventor of the telephone (*Alexander Graham Bell (1847–1922)* [106]), along with the Hall's most significant woman: *Susan B. Anthony (1820–1906)* [432], now generally recognized as the most important leader of the suffragettes. But this was not always the case. *Lucretia Mott (1793–1880)* [2208] organized the first women's rights convention in Seneca Falls, New York in 1848, but proved more interested in abolition than suffrage. *Lucy Stone (1818–1893)* [2738] was an early suffragist who retained similar stature in the movement until the turn of the century, according to the Ngram data.

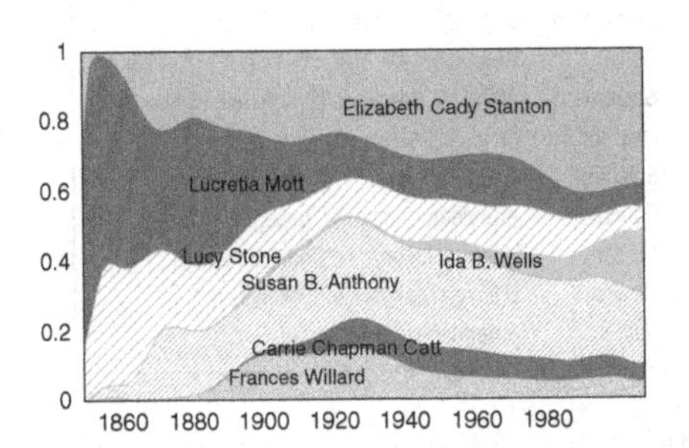

Anthony was arrested for voting in the 1872 presidential election; the subsequent trial brought greater attention to the suffragist cause. We will explore this issue in Section 15.4.1, but it does appears that Anthony is the single best representative in the battle for women's right to vote. Women gained the right to vote for the Hall of Fame well before the Nineteenth Amendment was ratified. *Alice Freeman Palmer (1855–1902)* [65220] served as an elector in 1900, which may have greased the wheels for her own election in 1920.

Josiah Willard Gibbs (1839–1903) [1244] was a pioneer in the study of thermodynamics, and the first theoretical physicist of note in the United States. He represents the professionalization of science, having received the first American doctorate in engineering. He served as the transitional figure between the natural science the Hall had previously honored, and the abstract technical discipline that better characterizes science today.

1955, 1960, and 1965

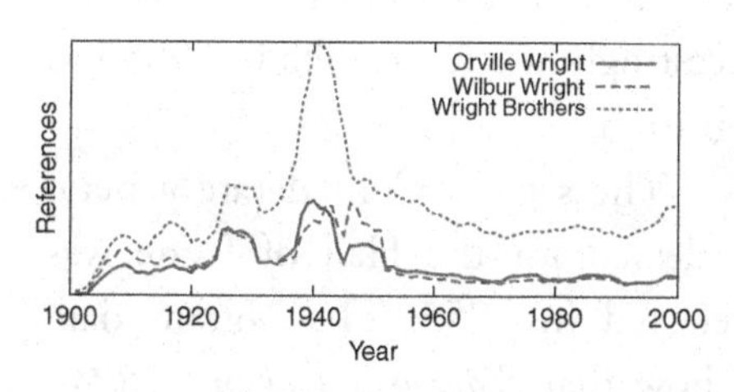

The leading vote getter in 1955, *Wilbur Wright*, presents a unique challenge to our methods, because his achievement as the inventor of powered flight is so completely shared with his brother Orville. They are so tightly intertwined that they share a single Wikipedia article. This renders us unable to directly compare the relative significance of the two brothers, but the Ngrams dataset enables us to track the evolution of their renown. Initially, Wilbur appears to have been the dominant figure. But starting around 1925, the brotherly meme asserts itself so strongly that they are now regarded as a single entity instead of a matched pair.

Their historical codependence is all the more remarkable considering that Wilbur died just nine years after the flight at Kitty Hawk and his younger brother outlived him by 36 years. Orville sold the Wright company shortly after Wilbur's death and essentially retired. Rules were changed to speed Orville's election in 1965, so that their two statues could enter the Hall of Fame together.

The accompanying table shows the rest of those chosen in the next three elections.

1955–1965 Electees

Sig	Person	Dates	Most Votes	Type	Description
40	Thomas Edison	(1847–1931)	108	Inventor	Inventor (lightbulb, phonograph)
1256	Jane Addams	(1860–1935)	94	Social	Nobel Peace Prize laureate (Hull House)
131	Henry David Thoreau	(1817–1862)	83	Author	American author (*Walden*)
885	Oliver W. Holmes	(1841–1935)	79	Legal	American jurist (Clear and present danger)
20551	Sylvanus Thayer	(1785–1872)	77	Educator	U.S. general ("The Father of West Point")
283	Stonewall Jackson	(1824–1863)	72	Military	Confederate general (U.S. Civil War)
12007	Edward MacDowell	(1860–1908)	72	Artist	American Romantic composer, pianist
1812	George Westinghouse	(1846–1914)	62	Inventor	American entrepreneur (Westinghouse Electric)

George Westinghouse (1846–1914) [1812] was honored for inventing a compressed air braking system that greatly improved railroad safety. But he would more accurately be described as an industrialist, whose Westinghouse Electric Corporation created the standards and infrastructure of the electrical power system. Like *Charles Goodyear (1800–1860)* [1984], who accidentally invented vulcanized rubber, his fame has been maintained by an eponymous company. The sale of Westinghouse to Toshiba in 2005 has already served to erode his public recognition.

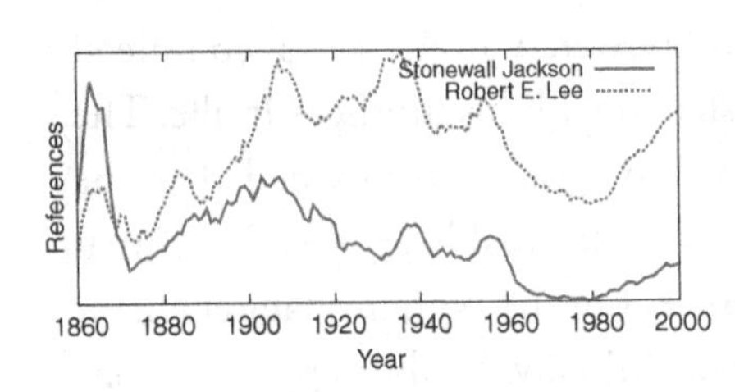

The second Confederate general selected for the Hall of Fame was elected in 1955. The Ngram data show that *Stonewall Jackson (1824–1863)* [283] dominated his commander, the redoubtable *Robert E. Lee (1807–1870)* [76], during the Civil War and its aftermath until about 1910. Stonewall Jackson's fame is partially a triumph of celebrity over gravitas. Slashing Confederate fighters like Jackson and *Nathan Bedford Forrest (1821–1877)* [975] are primarily remembered today like celebrities: recalled through stories and anecdotes of men in action. Higher-level commanders like Grant and Lee are primarily remembered for the great events they participated in, such as the surrender at Appomattox.

Immediately following Jackson's election in 1955, a lobbying effort emerged to support the candidacy of *Jefferson Davis (1808–1889)* [188], the president of the Confederacy. He jumped from one vote in 1955 to 44 in 1960, but never managed to cross the threshold of entry.

In 1960, *Thomas Edison (1847–1931)* [40], the father of our modern technological world, received the greatest number of votes in any Hall of Fame election. His "classmate" *Henry David Thoreau (1817–1862)* [131] has seen his reputation rise consistently since the 1920s, due to increasing interest in environmentalism. Indeed, his familiarity as measured by both Ngram frequency and significance ranking is beginning to eclipse that of his mentor *Ralph Waldo Emerson (1803–1882)* [122].

By contrast, it is difficult to explain the election of *Edward MacDowell (1860–1908)* [12007], a nineteenth-century classical composer, particularly in light of the later rejection of the still-popular jazz-age classical composer *George Gershwin (1898–1937)* [792]. MacDowell attracted substantial voter interest in five of the previous six elections, but Ngram data show that his reputation had peaked around 1935, and fallen into steep decline by 1960. He was a major American figure in a world of classical music dominated by Europeans, but more important proved the legacy of his MacDowell Colony, an artistic refuge in Petersboro, New Hampshire founded from his estate. More than 6,000 prominent artists have spent time in residency at the Colony. This living institution has preserved his memory, and the good will it generated no doubt made his election possible.

Jane Addams (1860–1935) [1256] received the most votes in 1965. Addams was a humanitarian most famous for founding the Hull House settlement in Chicago. Hull House outlived her as an organization, performing social welfare work for 122 years until its bankruptcy in 2012. She was the first American woman to be awarded the Nobel Peace Prize.

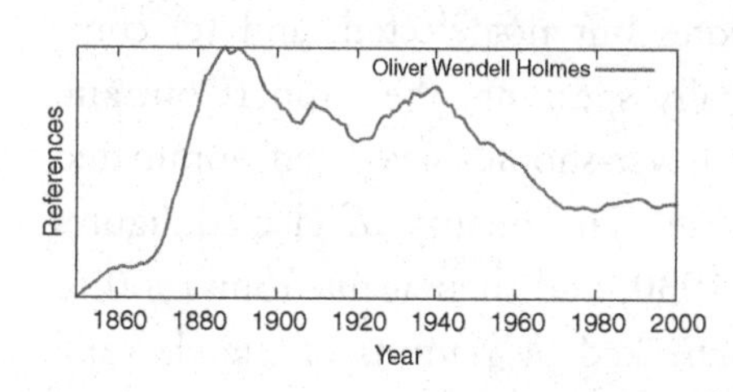

The election of *Oliver Wendell Holmes Jr. (1841–1935)* [885], a distinguished and long-serving justice of the Supreme Court, rests at least in part on the reputation of his father, *Oliver Wendell Holmes Sr. (1809–1894)* [945], who was elected to the Hall of Fame in 1910. Both were blessed with long and productive lives, which proves difficult to separate in the Ngram data. The name first became prominent around 1860 and soared to a peak during the 1880s. The son first joined the Supreme Court in 1902 and kept the name fresh through the 1950s.

Sylvanus Thayer (1785–1872) [20551] served as superintendent of the military academy at West Point and promoted engineering education. His name was linked to General/President *Dwight D. Eisenhower (1890–1969)* [110] as an upstanding representative of military education. Eisenhower served as president of Columbia University between his stints as Supreme Allied Commander and Commander in Chief.

1970 and 1973

The 1970 election marked a major milestone: for the first time the quality of failed candidates significantly trumped those who won. Figure 5.3

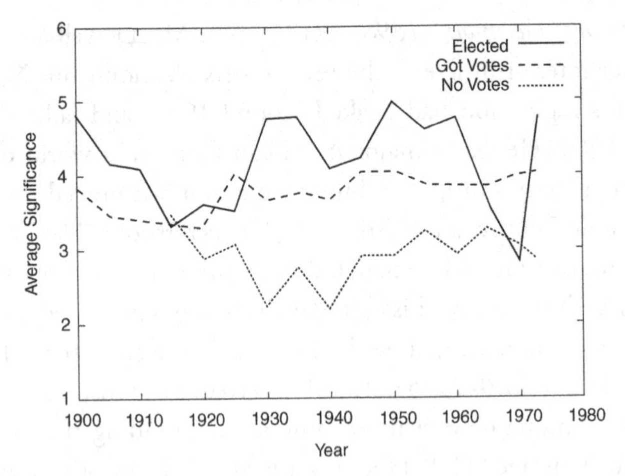

FIGURE 5.3. Significance of elected, losing, and unvoted-for Hall of Fame candidates.

plots (for each election) the average significance of the candidates (a) elected to the Hall of Fame, (b) receiving votes but not elected, and (c) completely shut out of the voting. Generally speaking, the winners outrank the losers, who ranked higher than the never-should-have-been-nominated class. But this was not always the case. The quality of elected figures sank steadily from a 1900 peak until 1930, and then again from 1960 to 1970. Both declines were eventually checked by reforms of the electoral procedures.

1970–1973 Electees

Sig	Person	Dates	Most Votes	Type	Description
887	George W. Carver	(1864–1943)	104	Scientist	Agricultural scientist (uses for peanuts)
1134	Louis Brandeis	(1856–1941)	98	Legal	Supreme Court justice ("People's Lawyer")
43	F. D. Roosevelt	(1882–1945)	87	Statesman	32nd U.S. president (New Deal, WW II)
2705	Albert A. Michelson	(1852–1931)	82	Scientist	American physicist (measured speed of light)
922	John Philip Sousa	(1854–1932)	78	Artist	Band music composer ("Stars and Stripes")
26724	Lillian Wald	(1867–1940)	68	Social	Humanitarian ("The House on Henry Street")
104	Andrew Carnegie	(1835–1919)	49	Business	Steel magnate and philanthropist
1138	Clara Barton	(1821–1912)	23	Social	Teacher, humanitarian (founded Red Cross)
5733	Luther Burbank	(1849–1926)	14	Scientist	American horticulturist (Burbank potato)

The stronger of the two 1970 inaugurates was *Albert Abraham Michelson (1852–1931)* [2705], a physicist who was awarded the Nobel Prize for his experiments measuring the speed of light. The Michaelson-Morely experiment disproved the existence of the ether as a medium for propagating light, and influenced *Albert Einstein's (1879–1955)* [19] work on relativity.

Far weaker was the choice of *Lillian Wald (1867–1940)* [26724], a respected nurse and social worker, whose Henry Street Settlement provided services to Jewish immigrants on the Lower East Side of New York. She was a praiseworthy humanitarian, but ultimately an important local/ethnic figure more than a national icon. We would have preferred clearing the backlog of long-standing nominees ahead of them, particularly *Noah Webster (1758–1843)* [1132] or *Frederick Douglass (1818–1895)* [402].

In 1973, the rules were changed to hold elections every three years instead of five, with the goal of ensuring an election during the U.S. bicentennial year. In contrast to the 1970 election, 1973 was a strong class. The real star was *Franklin D. Roosevelt (1882–1945)* [43], who led our country through the Great Depression and World War II.

Louis Brandeis (1856–1941) [1134] was a major figure on the Supreme Court, the primary advocate for the right to privacy. He was the first Jewish justice, joined later by *Benjamin N. Cardozo (1870–1938)* [3114] and *Felix Frankfurter (1882–1965)* [3252]. His legal reasoning continues to resonate today. His memory is preserved by Brandeis University, which was founded and named for him just seven years after his death.

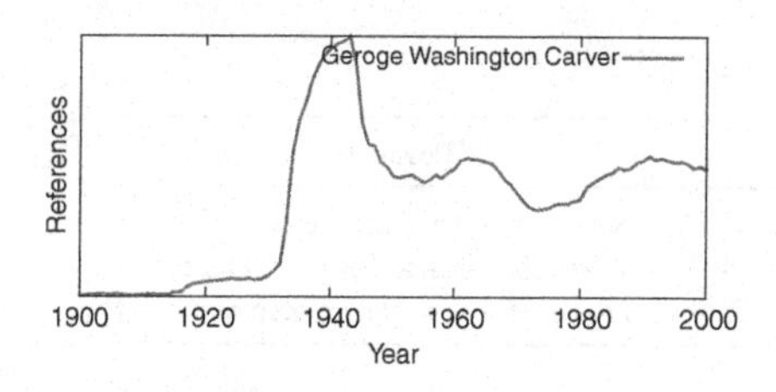

It is difficult to objectively assess the historical stature of top vote getter *George Washington Carver (1864–1943)* [887]. Carver was an important part of *Booker T. Washington's (1856–1915)* [525] Tuskegee Institute, and a respected and influential figure for more than 40 years. The importance of his scientific work and influence on the peanut industry is generally overstated. By no means did he invent peanut butter: *Marcellus Gilmore Edson (1849–1940)* [233015] received a patent for the stuff in 1884. Carver was primarily an educator: his research was published in agricultural circulars, not scientific journals.

As the Ngram dataset shows, real fame did not come until late in life, spiking immediately after his death in 1943. He has been justly recognized for his inspirational life story. He rose to public prominence during the 1920s: a living slave almost 60 years after Emancipation. His service to both his race and the greater Southern community made him irresistible at a time when the need for black heros was acute. This more than justifies his election, and will probably ensure his fame long into the future.

The scandal here was the rejection of *Henry Ford (1863–1947)* [148], whose innovations in transportation and manufacturing changed the world, yet yielded only 29 votes. At least 49 electors drove up in their cars and voted instead for military composer *John Philip Sousa (1854–1932)* [922], at a time when rock-and-roll music filled the air.

5.4 Decline and Fall

An election was indeed held during the 1976 bicentennial year. The rules were changed to create a complicated category-based point system, factoring in previous voting results and awarding more points to an elector's higher ranked selections. We have been unable to locate the full voting records from this election, but it is just as well. The new system proved to be rearranging deck chairs on the *Titanic.*

To its credit, the three people selected were all very sound choices: important Americans who had been overlooked for several elections.

1976 Electees

Sig	Person	Dates	C/G	Description
104	Andrew Carnegie	(1835–1919)	C G	Steel magnate and philanthropist
1138	Clara Barton	(1821–1912)	C G	Teacher, humanitarian (founded Red Cross)
5733	Luther Burbank	(1849–1926)	C G	American horticulturist (Burbank potato)

Others who did well this time around included lexicographer *Noah Webster (1758–1843)* [1132] (839 points), inventor *John Frank Stevens (1853–1943)* [37802] (740 points), Chief Justice *Charles Evans Hughes (1862–1948)* [1760] (935 points), and Native American *Chief Joseph (1840–1904)* [2409] (634 points). The winners Carnegie, Barton, and Burbank received 1,103, 942, and 740 points respectively. But none of them actually ever received their statues.

As we said at the beginning of the chapter, institutions in New York rise and fall around real estate. The Bronx fell into a morass of crime, poverty, and fear during the 1960s. Facing bankruptcy, New York University abandoned its Bronx campus, selling it off to the Bronx Community College in 1973.

The Hall of Fame was left in an ambiguous position. NYU had apparently made a formal declaration of trust when the Hall of Fame was established, implying a level of responsibility it did not want and could not afford. Tentative proposals to relocate the Hall met with opposition from local Bronx officials.

Robert Moses (1888–1981) [2779], the 90-year-old power broker responsible for building much of modern New York, had (as often) the soundest vision. The Hall of Fame, he declared, should move to the nation's capital, ideally to become part of the Smithsonian Institution. But the Smithsonian declined, as did the National Park Service, asserting that such a takeover would require an act of Congress.

On December 31, 1978, its entire board of trustees resigned, and the Hall of Fame has laid dormant ever since. Recent efforts have struggled to maintain the physical foundation of the colonnade and restore the existing statues. The busts of the last four honorees (Brandeis, Barton, Burbank, and Carnegie) were, well, busts – never sculpted, never installed.

The Hall of Fame may appeal to a somewhat childish notion of history: Great Men honored forever for doing Great Deeds. But we are all children once, and such a Pantheon still holds the power to inspire and educate. Perhaps the time has come to act on Moses' idea. Let's revive the Hall of Fame, either in the revived Bronx or else at a prominent location in Manhattan or Washington, DC. Time has dulled the passions and possessiveness of all the institutions involved, perhaps making politically viable what was impossible before.

In the Internet era, millions of people can be mustered to vote for just about anything. Imagine focusing the energy of distinct constituencies to get their champion – say *Ronald Reagan (1911–2004)* [32], *Martin Luther King (1929–1968)* [221], *César Chávez (1927–1993)* [3948], or *John F. Kennedy (1917–1963)* [71] – elected to the Hall of Fame for Great Americans. The institution could once again serve an important educational and cultural mission.

But until then, make it a point to visit the Hall of Fame on your next trip to New York. Spend a day in the Bronx. Take in a Yankees game, visit the Zoo or the Botanical Garden, and eat at Dominick's Restaurant on Arthur Avenue: New York's real Little Italy. And set aside an hour to walk the colonnade and visit the statues of some old friends. We know of no better place to channel the past and contemplate the future.

Notes

Several useful resources exist to document the rise and fall of the Hall of Fame for Great Americans. Voting records through 1973 have been drawn from the official guidebook *Great Americans: A Guide to the Hall of Fame for Great Americans* [Morello, 1977]. [Croce, 1978] gives a popular election-by-election review of the candidates and the reason for their selection. [Danilov, 1997] provides an overview of 274 different hall of fame museums, in more than 100 different fields.

[MacCracken, 1901] and [Johnson, 1935] are commemorative books on the founding of the Hall from 1901 and 1935, respectively. The archives for the Hall of Fame reside at the Bobst Library of New York University.

6

The Baseball Hall of Fame

A far more thriving institution resides in the tiny hamlet of Cooperstown, New York: baseball's ostensible birthplace. The Baseball Hall of Fame honors the game's greatest players, where its members are immortalized by bronze plaques. Like Mecca to the Muslims, a pilgrimage here once in a lifetime is essential for any baseball fan.

Baseball represents an essential part of the American character, making it worthy of historical attention. But our interest here is quite specific. Baseball is uniquely suited to quantitative analysis, a sport that has maintained a comprehensive statistical record spanning more than a century. The rules of play have remained relatively fixed over time, making it a meaningful question to ask whether *Babe Ruth (1895–1948)* [434] or *Barry Bonds (1964–)* [2217] was a better hitter, and *Walter Johnson (1887–1946)* [4303] or *Randy Johnson (1963–)* [6057] the better pitcher.

Statistical measures can be used to judge whether the best players have been recognized by the Baseball Hall of Fame. Objective measures of significance are rare outside of sports, so baseball's statistical record creates an excellent laboratory to study the forces of fame and canonization. Did the right players get selected for the Hall of Fame? Which deserving candidates have been forgotten? Who has been immortalized standing on feats of clay?

This statistical record makes baseball another test of our historical significance rankings. In principle, our Wikipedia-based significance rankings *should* reflect the statistical record left behind by each player's on-field performance. There seems no objective evidence one can bring to bear on whether, say, novelist *Stephen King (1947–)* [191] is *really* a more significant writer than *Theodore Dreiser (1871–1945)* [4885]. But career batting statistics make it absolutely certain that *Jimmy Dykes (1896–1976)* [28854] was a greater baseball player than *Roger Metzger (1947–)* [296263]. The extent to which this statistical performance record is reflected by our rankings provides an independent test of the veracity of our analysis.

Who's in, who's out, and why?

6.1 Performance and Posterity

Baseball is a game of numbers almost as much as of bat and ball. The most familiar statistics are useful for measuring particular skills. Home runs are the biggest hit a player can get, so power hitters who slug more home runs are more valuable than those who hit fewer. Batting average is the ratio of the hits a player gets to the number of chances he had to get them. A player with a higher batting average is generally better than one who hits for a lesser average.

But who was the more valuable player: banjo-hitter *Matty Alou (1938–2011)* [65370] with a batting average of .307 and 31 career home runs, or one-dimensional slugger *Rob Deer (1960–)* [78253], whose 230 home runs come with an anemic .220 average and massive strikeout total?

Wins above replacement (WAR) is a modern, *Moneyball*-era statistic designed to answer such questions, by encapsulating a player's total contribution on the baseball field. Games are won by teams that score more runs than their opponents. Each run is constructed from a series of hits, walks, and other atomic operations. The WAR statistic is based on how many runs are equivalent to a win, and then how much each hit or walk contributes, on average, to a run. Thus, a player's entire record can be compressed into a single number, reflecting how many runs they created. Similar WAR statistics can be calculated for pitchers from the components of their records: innings pitched, runs allowed, and strikeouts.

But the difficulty of scoring runs has varied with time. In 1968, a player with a .302 batting average could have won the batting title, yet that was less than the entire National League average of 1930. The "Above Replacement" part of the WAR statistic compares each player's achievements against what a marginal player would have accomplished given the same opportunities.

How many extra wins did the Yankees get by playing *Babe Ruth (1895–1948)* [434] than by playing some minor league outfielder in his stead?

The details of computing the WAR statistic are complicated, and not completely standardized between sources. We obtained the WAR values for our study from `www.baseball-reference.com`, the most complete source of baseball statistics on the open web. Matty Alou finished his career at 21.5 WAR, better than Rob Deer's WAR of 11.8. Both were good, but neither of these were Hall of Fame caliber players.

If we accept wins above replacement as a meaningful measure of a baseball player's career value, then Figure 6.1 should provide all the evidence you need about the general accuracy of our significance rankings. It shows a strong correlation (0.68) between our significance measure and the WAR achieved by Hall of Fame players. Increasing WAR generally implies greater historical significance.

More importantly, the outlier players make sense. *Jackie Robinson (1919–1972)* [841] was indeed more historically significant than his batting statistics show: he was the black man who integrated major league baseball. *John McGraw (1873–1934)* [4358] was a fine player, but one whose real fame came from 30 years managing the New York Giants – so yes, he was more significant than his playing statistics show. *Yogi Berra (1925–)* [4886] ("It ain't over 'til it's over!") and *Reggie Jackson (1946–)* [4603] ("The straw that stirs the drink") are cultural figures who rise above just being sluggers. On the other side, consensus choices for the weakest Hall of Fame selections, like *Chick Hafey (1903–1973)* [109931] and *Lloyd Waner (1906–1982)* [68198], receive our lowest significance scores.

Figure 6.1b presents the analogous plot for Hall of Fame pitchers. The greatest pitchers do not achieve quite the same significance as the greatest hitters, but there is a similarly strong correlation between fame and WAR. The outliers are again predictable: overpowering pitchers *Sandy Koufax (1935–)* [4855] and *Dizzy Dean (1910–1974)* [12369] had careers cut short by injury, limiting their ability to build up impressive WAR totals. *Jesse Haines (1893–1978)* [75009] and *Red Faber (1888–1976)* [58330] are considered weak Hall of Fame selections, with long careers of solid but unspectacular achievement.

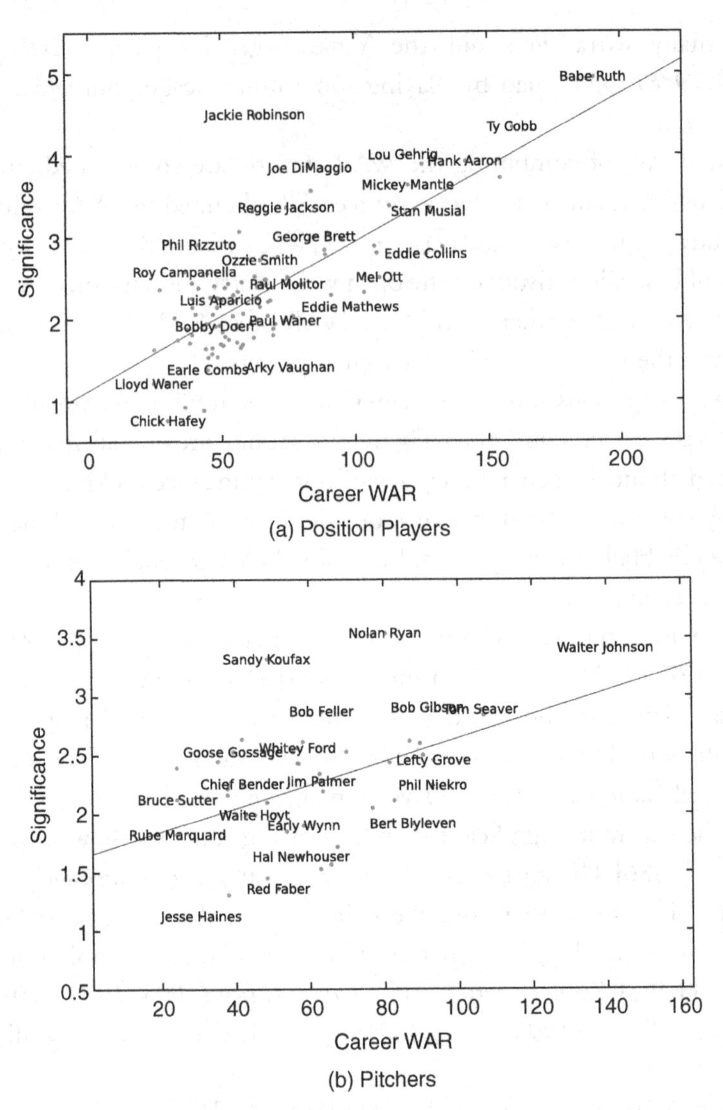

FIGURE 6.1. Dotplot of Hall of Fame player significance against Wins Above Replacement (WAR). Historical significance strongly correlates with performance.

6.2 The Electoral Record

The first election to the Baseball Hall of Fame was held in 1936. Our significance measure enables us to rigorously study the performance of the electorate. Did they make the best choices given the available candidates?

The Baseball Hall of Fame has always had a complicated electoral process. Usually two to four players get honored in a typical year. The rules have changed several times over the years in attempts to rectify problems, often making these troubles worse. For details on the policies and procedures, we refer the interested reader to [James, 1995], a comprehensive history of the Baseball Hall of Fame.

The root cause of these difficulties comes from having two distinct bodies voting on candidates, each with substantially different procedures. The better known group is the Baseball Writers' Association of America (BBWAA), consisting of roughly 300 newspaper sportswriters who regularly report on baseball. Each writer is allowed to vote for up to ten eligible candidates. Any player who receives votes on 75 percent of all submitted ballots is elected to the Hall of Fame, while any player who has been rejected for 15 years is dropped from further consideration. This proves to be a fairly stringent standard and, as we will see, the BBWAA generally selects qualified candidates.

The alternate path into the Baseball Hall of Fame is through the Veterans Committee, a group of about 15 former players, executives, and writers. Their mission is to identify worthy but neglected players who are no longer eligible for election by the BBWAA. Since the strongest candidates were presumably picked by the BBWAA, modesty dictates that the Veterans Committee should serve as an appeals process to bring forth a small number of overlooked individuals. However, the Veterans Committee has historically selected about as many people as the BBWAA. Only repeated rule changes have restrained them from admitting even more from this weaker pool.

6.2.1 THE WEAKNESS OF THE TAIL

We can use a cumulative frequency distribution to get a good sense of the quality spectrum among the members of a given group. In particular, Figure 6.2 shows the fraction of the Baseball Hall of Fame that has achieved any given significance rank. The top 70 percent or so are distributed fairly uniformly among the 30,000 or so most significant figures in history. Above this, the quality deteriorates rapidly, with fully 10 percent of the "immortals" living outside the 100,000 most significant people

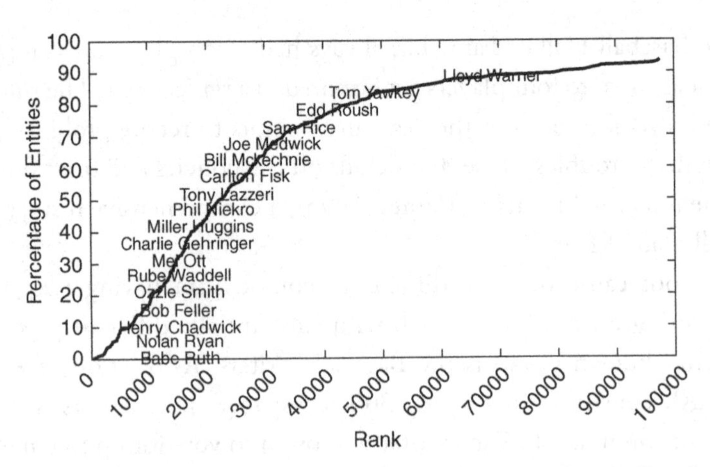

FIGURE 6.2. The weakness of the tail: cumulative distribution of Baseball Hall of Fame members by significance rank.

in Wikipedia. Hundreds of former baseball players of greater historical significance can get into the Hall only by paying admission.

We saw this phenomenon before, in Figure 5.1, the cumulative distribution significance plot for the Hall of Fame for Great Americans. It had the same basic shape as Figure 6.2: a linear phase capturing the 65 percent most qualified candidates, followed by a long tail that accumulates members on a fairly arbitrary basis. Literally thousands of people with equal or greater qualifications were ignored. We also saw this in Bonnie's fifth grade history textbook, in Figure 3.2. All the selection processes we have seen function effectively to identify the best 65–70 percent of a group, but break down badly in filling out the remainder of the class.

Having seen the same effect several times in substantially different contexts, we posit this to be an important truism about *any* human selection process, be it college admissions, employee interviewing, sports drafts, or prison sentencing. Look around at the people you study with, work with, or serve time with. We suspect you will agree that roughly a third are noticeably less qualified than the rest of the group.

Why is this? The extreme outliers in any cohort are quite easy to identify: the greatest presidents, the undisputed geniuses, those who walk on water. But things get fuzzier around the boundary, particularly because any select group sits at the tail of a talent distribution. In a perfect world, a university like Stony Brook would identify the objectively best 35 percent of its applicant pool. But there are far more candidates whose real abilities sit just outside this threshold than just inside. Suppose each reviewer's estimate of a candidate's "true" significance was subject to some random noise. Candidate quality is difficult to accurately differentiate at this resolution, and noisy selection will more likely favor one of the unwashed masses over the guy who really should have made the cut.

This isn't just theory. We have performed computational experiments where we add a modest amount of random noise to our evaluation of historical significance, and simulated the resulting halls of fame. The cumulative distribution charts of these simulations fit extremely well with what we have seen here. Our results for four different Halls of Fame (Baseball, Football, Basketball, and Great Americans) are presented in Figure 6.3. In all cases,

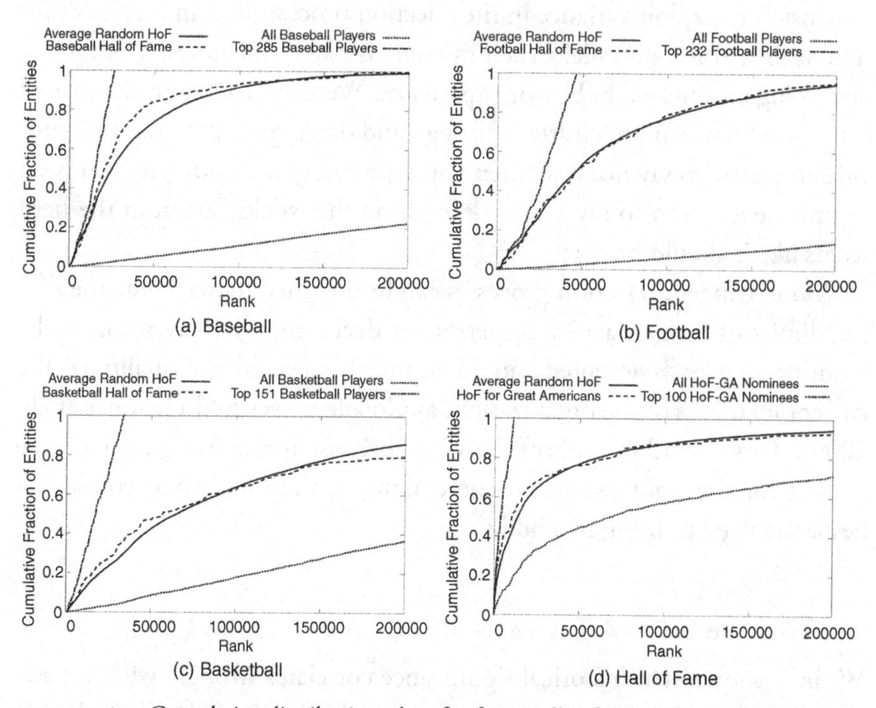

FIGURE 6.3. Cumulative distribution plots for four Halls of Fame. The solid line represents a simulated selection process with noise added to the significance estimation. This provides an excellent fit to the observed quality of the actual selections (center-dashed line). These can be contrasted with the profiles of two other models: the electorate always picks the best candidates (upper-dashed line) or selects members at random (lower-dashed line).

our simulated quality distributions look startlingly like what the voters did in their respective halls of fame. We conclude that this random noise model explains much of the mystery of human selection processes.

Making Selection Processes Work for You

Now that we understand some of the dynamics of human selection processes, what can you do to rise above a sea of similarly qualified people?

The key is variance and repetition. Since each college has an imperfect but independent admissions process, apply to a bunch of places, and maybe you will be lucky enough to be misevaluated above the threshold. Apply for every possible job opening: you will find employment unless rejected in *all* of these independent evaluations. Repeatedly appeal your conviction to higher courts, and maybe you will be the rare one who will catch a break.

The other key, we believe, is striving to stand out in some arbitrary dimension to exploit variance in the selection process. Resume experts caution against using gimmicks such as colored paper, including a photo, or describing an unusual hobby or experience. We would agree in the case of jobs for which you are clearly a strong candidate: you never want to provide an excuse to say no. But when you are a marginal candidate, you need to provide a reason to say yes, and anything that sticks out from the herd seems like it should be a win.

More stringent selection processes can be designed to overcome the natural limits of human acuity in personnel decisions. The key seems to be ensuring that each accepted candidate meet or exceed the quality of the current members of the organization, as Google strives to do [Broder et al., 2009]. This is hard to pull off, however, because it requires great patience to wait for the right person to come along at the time when you are in desperate need to bring in a body.

6.2.2 WAR AGAINST THE ELECTORATE

We have shown that historical significance correlates strongly with the statistical evidence (wins above replacement) for Hall of Fame baseball players. But even though height correlates strongly with weight, one does a much better job of measuring how tall you are. Here we use our significance measure to study the behavior of the electorate in greater detail.

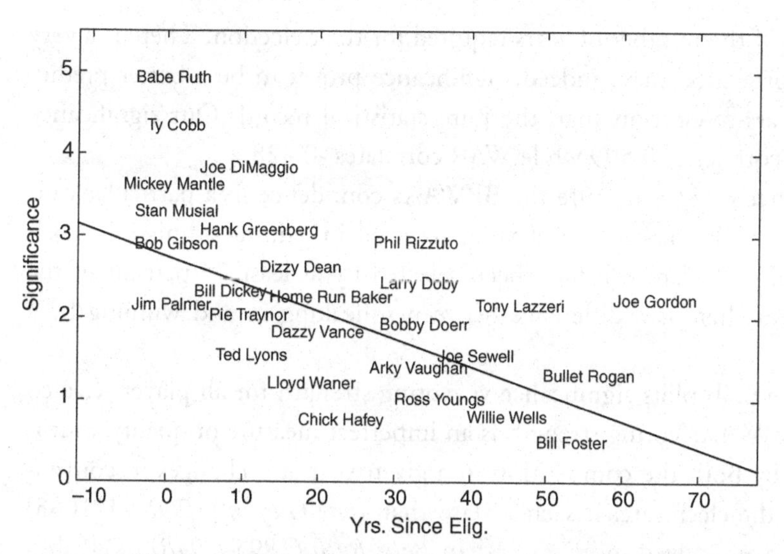

(a) Significance of Hall of Fame baseball players vs. years of electoral eligibility.

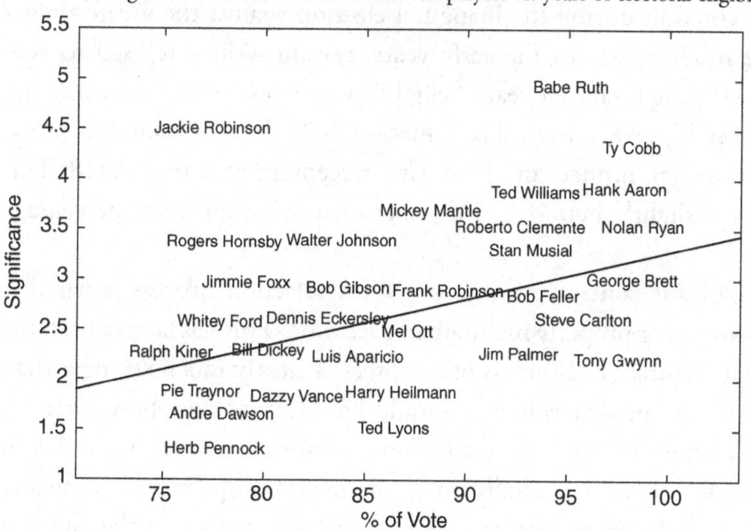

(b) Significance of Hall of Fame baseball players vs. voting strength. More significant players generally get elected with higher percentages of the vote.

FIGURE 6.4. Significance of Hall of Fame baseball players vs. electoral criteria.

Baseball players generally become eligible for selection five years after their retirement. The more deserving the honoree, the quicker their merits should be recognized by the electorate. This indeed proves to be the case. Figure 6.4a plots the historical significance of baseball players as a function of the number of years required for their election. There is a very strong correlation here: indeed, significance proves to be a better predictor of years-to-election than the pure statistical record. Our significance measure correlates -0.599 while WAR correlates -0.428.

Another way to measure the BBWAA's confidence in a particular candidate is by the percentage of votes received in gaining admission. Each successful candidate will have been selected by at least 75 percent of the voters, but there is a difference between squeaking in and winning by a landslide.

Figure 6.4b plots significance vs. voting strength for all players elected by the BBWAA. Voting strength is an imperfect measure of quality, as it is affected by both the competition in a given year and changes in conventions by the electorate. It seems weird that *Tony Gwynn (1960–)* [21088] should have received more votes than *Babe Ruth (1895–1948)* [434], but Ruth was voted in during the inaugural election against the greatest players of the modern era. In the early years, certain writers refused to vote for anyone during their first year of eligibility, but yesterday's orneriness has been replaced by a trend toward acclamation. Still, the regression line shows that historical significance correlates with percentage of vote at 0.418. This correlation is slightly better than that between WAR and vote percentage (0.415).

An important concern of any long-term selection process is whether standards are being properly maintained over time. This has largely been the case. The inaugural (1936) class of honorees is clearly much stronger than any other class, but that is how it should be. Our analysis shows that the weakest selections of the Veterans Committee were bunched in the 1960s and 1970s, but there is generally no strong relationship between time and strength. Baseball creates worthy new Hall of Famers at about the rate that it elects them.

6.3 The Glory of Their Times

We now attempt to trace the history of major league baseball through its Hall of Fame players. Cooperstown is not an impossibly elite address, for it contains almost 2 percent of all the players who ever reached the Major Leagues. Hall of Famers represent roughly 10 percent of all plate appearances (at bats) by major league hitters [James, 1995] because they tend to play much more regularly, during much longer playing careers, than their lesser peers. This makes them a large enough sample to represent the state of the game.

Professional baseball dates back to 1869, and has since passed through several distinct eras. The pre-1900 game was marked by continuous experimentation with rules, playing tactics, and financial models. The modern era begins in 1900, with the founding of the American League. Babe Ruth's popularity marked the end of the dead-ball era in 1920, with equipment changes that greatly increased the frequency of home runs. World War II caused a marked deterioration in the quality of play: the leagues were kept running for morale purposes, but all able-bodied players served in the military. Amputees *Pete Gray (1915–2002)* [68081] (arm) and *Bert Shepard (1920–2008)* [226629] (leg) were pressed into baseball service during the war years.

All this history plays itself out in Figure 6.5, which measures the median significance of the Hall of Famers active in each year, broken down by position (infielders, outfielders, and pitchers). The earliest stars were primarily infielders, although general parity was realized before the modern era.

The dip in significance from 1920 to 1945 has two distinct causes. The new lively ball inflated the statistics of its players relative to their earlier peers. Respect for pitching declined during this offense-minded

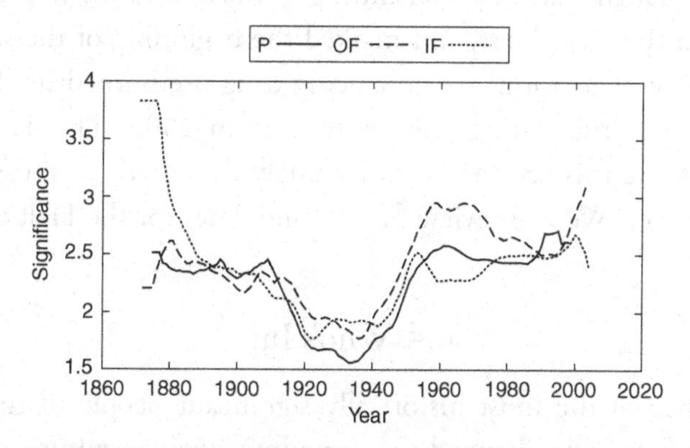

FIGURE 6.5. Temporal significance of Hall of Fame baseball players, broken down by position: Pitchers (P), Infielders (IF), and Outfielders (OF). The relative prominence of outfielders has increased in the post–WWII era.

era, with the significance of pitchers dropping well below position players. Lax work by the Veterans Committee admitted too many players from this time period. A record 55 Hall of Famers played in 1928. Such extravagance caused the average significance of Hall of Fame players to plummet.

The other factor was the professional price young men paid during World War II. Players active in 1941 typically had their career interrupted for three years, and often never returned to previous form. A generation of younger players never got the opportunity to develop skills or lost their prime athletic years to the war.

The postwar era was marked by racial integration. Baseball rebounded to its greatest popularity ever, aided by the end of this color ban. Great hitters dominated great pitchers in the 1950s and 1960s, and many of baseball's greatest stars of this period were black. Of the ten most significant Hall of Fame outfielders, five were black and seven played during this time period. These stars clearly outclassed those of other positions.

Power pitchers grew to dominance in the 1980s. Only three pitchers (*Lefty Grove (1900–1975)* [15761], *Warren Spahn (1921–2003)* [13010], and *Early Wynn (1920–1999)* [31295], won 300 games between 1925 and 1981. But they were joined by six more between 1982 and 1990, plus a final four in the interval from 2003 to 2009. Expansion from 20 teams in 1968 to 30 teams by 1998 diluted the quality of play, causing the significance of the Hall of Famers to dip.

Unprecedented feats of power hitting in the late 1990s swung momentum back to the outfielders, and marked the beginning of the steroid era. This epidemic of performance-enhancing drug use burned itself out with more stringent drug testing policies starting in 2006. The data on Hall of Famers active this century is misleadingly thin, because most have not been elected yet. We will review future candidates for the Hall of Fame in Section 6.6.

6.4 Who's In?

The influence of the most historically significant people in the Baseball Hall of Fame extended beyond the game into popular culture, so they are worth knowing just to understand the frequent literary allusions to them. However, we are aware that many readers will find the names in this section as meaningless as a roster of the greatest corporate accountants. If so, we give you our blessing to pass on to the next chapter.

A second subtext here will involve the relative effectiveness of the BBWAA and the Veterans Committee in identifying candidates for the Baseball Hall of Fame. Each BBWAA electee is presented along with his percentage of the vote.

6.4.1 INFIELDERS

The most interesting infielders in the Baseball Hall of Fame are those whose historical significance rank seems out of proportion to their statistical accomplishments (WAR). *Jackie Robinson (1919–1972)* [841] was the first black player allowed to play in the major leagues, a true historical figure whose significance transcends his skills in baseball. What made him important was his character more than his skills: he played with a fierceness and dignity that made him the most exciting player of his time.

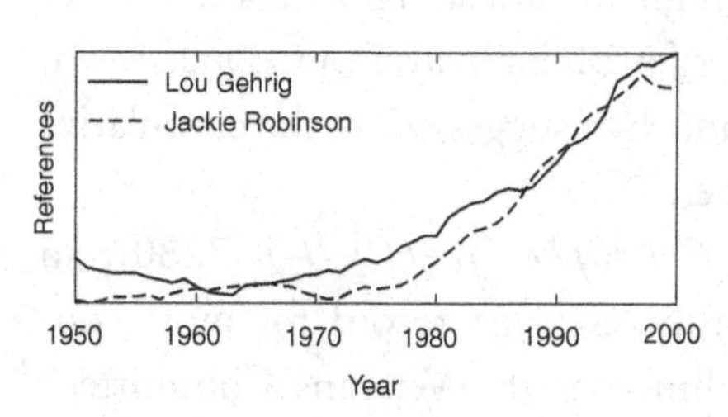

The Ngrams timeplot tracks the relative frequency trajectory of the two most significant infielders. Robinson's has exploded since 1980, when the historical record of the Civil Rights movement began to reflect on the centrality of his experience. *Lou Gehrig*'s *(1903–1941)* [1820] exceptional significance rank is

Top Hall of Fame Infielders

Person	Sig	Pos	WAR	Career	Years	Celeb/Grav	Team	Elected	Pct
Jackie Robinson	841	2B	63.20	(1945–1956)	11	C G	Dodgers	1962	77.50
Lou Gehrig	1820	1B	118.40	(1923–1939)	17	C G	Yankees	1939	
Honus Wagner	2513	SS	134.50	(1897–1917)	21	C G	Pirates	1936	95.13
Rogers Hornsby	4623	2B	127.80	(1915–1937)	23	C G	Cardinals	1942	78.11
Yogi Berra	4886	C	61.90	(1946–1965)	19	C G	Yankees	1972	85.61
Nap Lajoie	6604	2B	104.20	(1896–1916)	21	C G	Naps	1937	83.58
Hank Greenberg	6922	1B	56.80	(1930–1947)	13	C G	Tigers	1956	84.97
Cal Ripken	7280	SS	89.90	(1981–2001)	21	C G	Orioles	2007	98.53
George Brett	7559	3B	85.00	(1973–1993)	21	C G	Royals	1999	98.19
Jimmie Foxx	7901	1B	95.20	(1925–1945)	20	C G	Athletics	1951	79.20
Phil Rizzuto	8570	SS	41.80	(1941–1956)	13	C G	Yankees	1994	
Mike Schmidt	9930	3B	108.30	(1972–1989)	18	C G	Phillies	1995	96.52

partially due to his on-field performance: he totaled more WAR than any other first baseman, and owned the glamorous record of playing 2,130 consecutive games. But the tragedy of his premature death at age 37 from amyotrophic lateral sclerosis (now called Lou Gehrig's Disease) created a persona that transcended the sport. Ngrams data shows that references to this disease have surged since 1975, and now account for perhaps half of Gehrig's references.

Catcher *Yogi Berra (1925–)* [4886] was a three-time Most Valuable Player and the leader of Yankee teams that went to the World Series 14 times from 1947 to 1963. Yet his most enduring legacy is as the apocryphal source of a particular brand of twisted wisdom, such as "Nobody goes to that restaurant anymore. It's too crowded." He also served as the inspiration for the cartoon character Yogi Bear.

Hank Greenberg's (1911–1986) [6922] short but brilliant career makes his WAR score misleading: his high vote total and early election mark him as a first-class Hall of Famer. He was the outstanding Jewish ballplayer of his era, which contributes to his high historical significance. Pitcher *Sandy Koufax (1935–)* [4855], the other Jewish ballplayer in Cooperstown, also ranks as more historically significant than suggested by the cumulative statistics of his short, but brilliant, career.

The most significant shortstop is *Cal Ripken Jr. (1960–)* [7280], an outstanding hitter and fielder who broke Gehrig's record for most consecutive games played. Close behind him was the Veterans Committee's most inspired choice, *Phil Rizzuto (1917–2007)* [8570], an excellent shortshop who went on to even greater fame during his 40-year career as a radio/television announcer for the Yankees.

The BBWAA selected 14 of the 15 most significant infielders in the Baseball Hall of Fame. Phil Rizzuto was the lone exception, and has the lowest WAR of this elite cohort. By contrast, eight of the ten least significant infielders in the Hall were selected by the Veterans Committee.

6.4.2 OUTFIELDERS

Outfield is where the strongest hitters traditionally live, and has generally been the glamour position since *Babe Ruth (1895–1948)* [434]. Four top centerfielders had personas that reached beyond baseball. *Ty*

Cobb (1886–1961) [1108] was arguably the greatest player and the worst human being ever to play baseball, as portrayed by *Tommy Lee Jones (1946–)* [8938] in the movie *Cobb. Joe DiMaggio (1914–1999)* [2293] serves as a symbol of grace in the books of *Ernest Hemingway (1899–1961)* [248] and the songs of *Paul Simon (1941–)* [6156]. *Willie Mays (1931–)* [2803] and *Mickey Mantle (1931–1995)* [3173] were the color and style poles of New York baseball during its glory days in the 1950s and 1960s.

Top Hall of Fame Outfielders

Person	Sig	Pos	WAR	Career	Years	Celeb/Grav	Team	Elected	Pct
Babe Ruth	434	RF	190.00	(1914–1935)	22	C G	Yankees	1936	95.13
Ty Cobb	1108	CF	159.50	(1905–1928)	24	C G	Tigers	1936	98.23
Hank Aaron	2066	RF	141.60	(1952–1976)	24	C G	Braves	1982	97.83
Ted Williams	2185	LF	125.30	(1939–1960)	19	C G	Red Sox	1966	93.38
Joe DiMaggio	2293	CF	83.60	(1936–1951)	13	C G	Yankees	1955	88.84
Willie Mays	2803	CF	154.69	(1948–1973)	26	C G	Giants	1979	94.68
Mickey Mantle	3173	CF	120.20	(1951–1968)	18	C G	Yankees	1974	88.22
Roberto Clemente	3487	RF	83.80	(1955–1972)	18	C G	Pirates	1973	92.69
Rickey Henderson	4499	LF	113.10	(1979–2003)	25	C G	Athletics	2009	94.81
Reggie Jackson	4603	RF	74.59	(1967–1987)	21	C G	Athletics	1993	93.62
Stan Musial	4975	LF	127.80	(1941–1963)	22	C G	Cardinals	1969	93.24
Frank Robinson	8874	RF	107.40	(1956–1976)	21	C G	Reds	1982	89.16

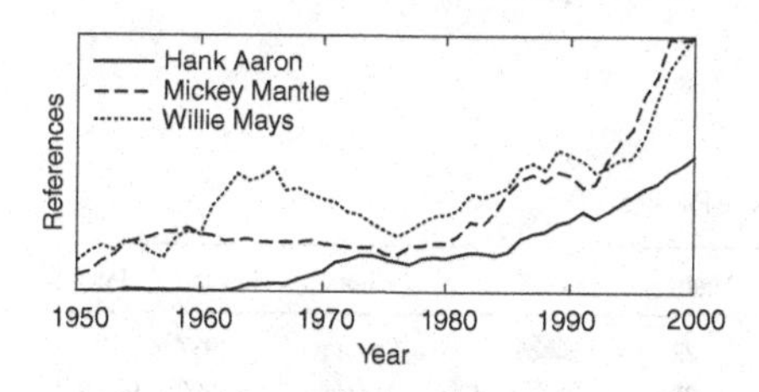

An Ngrams timeplot reveals the frequency trajectory of the greatest outfielders of the 1950s and '60s. *Willie Mays (1931–)* [2803] has a noticeable bulge in the middle of his playing career. For most of this period, he, rather than *Hank Aaron (1934–)* [2066], was the popular candidate to break Ruth's home run record.

Right field is also a slugger's position, populated by the first great home run hitter (Ruth) as well as the man who broke his home run record (Aaron). *Roberto Clemente (1934–1972)* [3487] was the greatest Latin American ballplayer. He died a hero: lost in a plane crash delivering supplies to earthquake-stricken Nicaragua.

Among left fielders, *Ted Williams (1918–2002)* [2185] was broadly regarded as the greatest hitter of all time, and *Rickey Henderson (1958–)* [4499], the greatest baserunner. *Stan Musial (1920–2013)* [4975] was the most respected player of the 1950s.

The dichotomy between the choices of the BBWAA and the Veterans Committee here are particularly stark. The 15 most significant outfielders were all selected by the BBWAA. The 10 least significant outfielders were all selected by the Veterans Committee.

6.4.3 PITCHERS

Pitching is supposedly 90 percent of baseball, yet pitchers themselves do not register as strong individually as position players. The most significant pitcher is *Cy Young (1867–1955)* [2461], who holds the record for games won by a huge margin, and is the namesake of the award given to the outstanding pitcher each season. Still, seven position players rank as more significant than he does.

Only four other pitchers rank among the 5,000 most significant people in history. Two are among the greatest stars of early baseball: *Walter Johnson (1887–1946)* [4303] and *Christy Mathewson (1880–1925)* [4348]. The other two are record-setting strikeout pitchers: *Nolan Ryan (1947–)* [3599] and *Sandy Koufax (1935–)* [4855]. The four modern pitchers who finished with 100 WAR all rank among the top ten in historical significance.

Top Hall of Fame Pitchers

Person	Sig	Pos	WAR	Career	Years	Celeb/Grav	Team	Elected	Pct
Cy Young	2461	P	143.19	(1890–1911)	22	C G	Spiders	1937	76.12
Nolan Ryan	3599	P	80.50	(1966–1993)	27	C G	Angels	1999	98.79
Walter Johnson	4303	P	139.80	(1907–1927)	21	C G	Senators	1936	83.63
Christy Mathewson	4348	P	90.70	(1900–1916)	17	C G	Giants	1936	90.71
Sandy Koufax	4855	P	48.70	(1955–1966)	12	C G	Dodgers	1972	86.87
Bob Gibson	8708	P	91.80	(1959–1975)	17	C G	Cardinals	1981	84.04
Tom Seaver	8918	P	106.10	(1967–1986)	20	C G	Mets	1992	98.84
Bob Feller	9030	P	63.30	(1936–1956)	18	C G	Indians	1962	93.75
Grover C. Alexander	9525	P	106.80	(1911–1929)	19	C G	Phillies	1938	80.92
Dizzy Dean	12369	P	41.80	(1930–1947)	12	C G	Cardinals	1953	79.17
Steve Carlton	12706	P	86.90	(1965–1988)	24	C G	Phillies	1994	95.82
Warren Spahn	13010	P	89.70	(1942–1965)	21	C G	Braves	1973	82.89

Our rankings of the top Hall of Fame pitchers table favor power pitchers, which helps explain the presence of *Nolan Ryan* [3599] in second place. He barely broke even during his long career, winning just 52.6 percent of his games, yet set spectacular strikeout records and pitched effectively into his mid-40s.

The three modern-era pitchers who pitched at least three no-hit games appear among the top eight in our ranking: Ryan (7), Koufax (4), and *Bob Feller (1918–2010)* [9030] (3). No-hit games are chance events, curiosities that receive outsized attention despite having no particular significance in the standings. Terrible pitchers have thrown no-hitters: the one *Bobo Holloman (1923–1987)* [248284] threw in his first start represents 33.3 percent of his career victory total. *Don Larsen (1929–)* [21037] achieved great fame for throwing the only perfect game in World Series history, yet ended his career with a losing record.

By contrast, 300-game winner *Lefty Grove (1900–1975)* [15761], proclaimed the greatest pitcher ever in [James, 1985], never pitched a no-hitter. Not by coincidence, he also ranks as the least significant modern pitcher with more than 90 wins above replacement.

Relief pitching is a relatively recent innovation in baseball, with the first identifiable stars dating back only to the 1950s. It is telling that the most significant relief pitcher (*Dennis Eckersley (1954–)* [12768]) also had considerable success as a starter during the first half of his career.

Hall of Fame Relief Pitchers

Person	Sig	Pos	WAR	Career	Years	Celeb/Grav	Team	Elected	Pct
Dennis Eckersley	12768	P	58.30	(1975–1998)	24	C G	Athletics	2004	83.20
Goose Gossage	14305	P	39.50	(1972–1994)	22	C G	Yankees	2008	85.82
Rollie Fingers	17061	P	24.30	(1968–1985)	17	C G	Athletics	1992	81.16
Hoyt Wilhelm	21335	P	37.90	(1952–1972)	21	C G	WhiteSox	1985	83.80
Bruce Sutter	23879	P	24.30	(1976–1988)	13	C G	Cubs	2006	76.90

None of the Hall's pitchers have reputations that significantly transcend baseball. Notable perhaps is *Jim Bunning (1931–)* [16270], who served in the U.S. Senate after a successful career as a starting pitcher. Senator Bunning was an excellent selection of the Veterans Committee, who otherwise amassed 5 of the 10 least significant Hall of Fame pitchers, while 14 of the top 15 were selected by the BBWAA.

6.4.4 THE EXECUTIVE SUITE

The Hall of Fame also honors nonplaying personnel who contributed significantly to the development of the game. These fall into three categories: managers, executives, and umpires.

Managers

As of this writing, 21 people have been inaugurated into the Hall of Fame primarily as managers. They are generally a distinguished group, with only recent Veterans Committee inductees *Billy Southworth (1893–1969)* [40300] and *Frank Selee (1859–1909)* [46077] ranking in the thin air above 35,000.

Top Hall of Fame Managers

Person	Sig	Years	Dates	Celeb/Grav	Elected
Connie Mack	3538	57	(1862–1956)	C G	1937
John McGraw	4358	33	(1873–1934)	C G	1937
Casey Stengel	5929	28	(1890–1975)	C G	1966
Harry Wright	10202	23	(1835–1895)	C G	1953
Tommy Lasorda	10366	21	(1927–)	C G	1997
Leo Durocher	10475	24	(1905–1991)	C G	1994

The top half dozen managers include several whose reputations transcend baseball. *Connie Mack* [3538] was the grand old man of baseball, managing for 50 years until he retired at age 87. That he owned the team and made all the hiring decisions made this possible. His grandson *Connie Mack III (1940–)* [60367] and great-grandson *Connie Mack IV (1967–)* [99703] served in the U.S. Senate and House of Representatives, respectively.

Casey Stengel [5929] managed the New York Yankees at the height of their success, but is perhaps most remembered for his colorful, disjointed way of speaking. *Leo Durocher* [10475] was a notorious competitor, famous for saying that "Nice guys finish last." Both Stengel and Durocher managed to advanced ages, giving them time to shape their historical legacies.

Executives

A collection of 23 management figures have been elected to the Hall of Fame. The weakest (*Pat Gillick (1937–)* [85359]) were elected by the

Veterans Committee in 2011. The top five include several important people.

Top Hall of Fame Executives

Person	Sig	Years	Dates	Celeb/Grav	Elected
Kenesaw Mountain Landis	3269	25	(1866–1944)	C G	1944
Branch Rickey	4660	31	(1881–1965)	C G	1967
Albert Spalding	6487	8	(1850–1915)	C G	1939
Charles Comiskey	7500	32	(1859–1931)	C G	1939
Ban Johnson	7777	28	(1865–1931)	C G	1937

Kenesaw Mountain Landis (1866–1944) [3269] was a federal judge appointed as the first Commissioner of Baseball in the wake of the Black Sox scandal, where eight members of the American League champion White Sox conspired with gambler *Arnold Rothstein (1882–1928)* [10599] to throw the 1919 World Series. His leadership is credited with restoring integrity to baseball. His counterpart is *Charles Comiskey (1859–1931)* [7500], the longtime owner of the White Sox, credited with provoking the scandal by shabby treatment of his players.

It was *Branch Rickey (1881–1965)* [4660] who hired *Jackie Robinson (1919–1972)* [841] to break baseball's color line. He was an innovative and successful executive, who also created the modern minor league farm system. *Albert Spalding (1850–1915)* [6487] was a star pitcher in the early era of professional baseball, but is today best known as the co-founder of the Spalding sporting goods company.

Umpires

Umpires are the on-field arbiters of play. With faces hidden by protective masks and crouched behind the plate, they generally go unrecognized even by serious fans.

Nine major league umpires have been selected for the Hall of Fame. The most outstanding, *Billy Evans (1884–1956)* [28820] and *Bill Klem (1874–1951)* [41013], have significance ranks comparable to low-end Hall of Fame players. The rest are far weaker selections, culminating with *Bill McGowan (1896–1954)* [172464].

6.4.5 EARLY FIGURES IN BASEBALL

The claim that Civil War general *Abner Doubleday (1819–1893)* [2722] invented baseball in 1839 is a romantic myth, although it had enough legs to get the Hall of Fame established in Cooperstown. Doubleday died 12 years before the Mills Commission relied on hearsay evidence to declare him the sport's inventor. Doubleday never claimed this himself and left no mention of baseball in his papers.

Two men have stronger claims as founding figures. *Alexander Cartwright (1820–1892)* [7164] wrote the rules governing the first game played by the Knickerbocker Base Ball Club in 1846. Writer *Henry Chadwick (1824–1908)* [6753] provided the first coverage of the fledgling sport and shaped its rules and statistics.

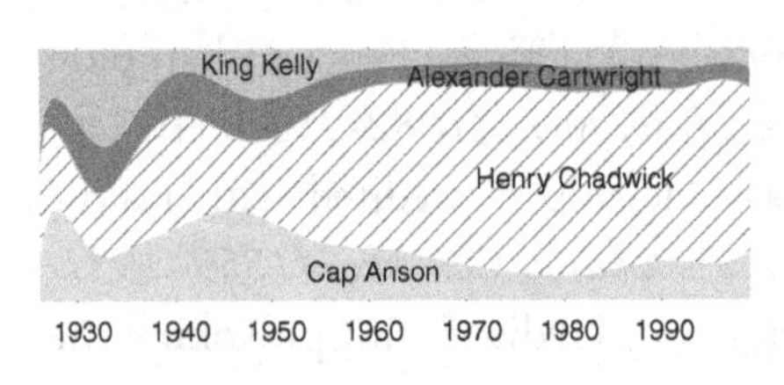

Examining a rugplot of the relative frequency of major nineteenth-century baseball figures since 1925 makes several trends obvious. The founding figures have held their ground or even expanded market share, particularly Chadwick, the journalist who left behind a written legacy. Of the superstar players of the time, the interest in *Cap Anson (1852–1922)* [2342] has held up better than that in *King Kelly (1857–1894)* [7378].

Anson had a long career, producing a statistical record that *looks* like a Hall of Famer of today: he was the first to reach the important milestones of 3,000 career hits and 2,000 runs batted in. By contrast, Kelly's reputation is based largely on defense and baserunning. The website `www.baseball-reference.com` uses a similarity metric to compare the statistics of different players. The ten most similar players to Anson are all Hall of Famers. By contrast, only three of the ten most comparable to Kelly are members, and none of them are recent. But during their time, Anson and Kelly were stars of similar magnitude.

Certain early players reemerged because of post-career changes in the game. Long after *Babe Ruth (1895–1948)* [434] popularized home run hitting came the question of who had hit the most before him. The answer turned out to be *Roger Connor (1857–1931)* [10511], whose prominence boomed again after *Hank Aaron (1934–)* [2066] broke Ruth's home run record in 1974. Connor was elected to the Baseball Hall of Fame in 1976.

Similarly, the fall of baseball's color line in 1947 sparked interest in its history. A black catcher, *Moses Fleetwood Walker (1857–1924)* [10549], played in the American Association starting in 1884, but was banned at *Cap Anson's (1852–1922)* [2342] instigation in 1888. *Frank Grant (1865–1937)* [49145], who began his career in 1889, is recognized as the first Negro League star.

Top Nineteenth-Century Baseball Hall of Famers

Person	Sig	Pos	WAR	Career	Years	Celeb/Grav	Team	Elected
Cap Anson	2342	1B	99.30	(1871–1897)	27	C G	White Stockings	1939
Henry Chadwick	6753	Pioneer				C G	Official	1938
A. Cartwright	7164	Pioneer				C G	Official	1938
King Kelly	7378	RF	47.50	(1878–1893)	16	C G	White Stockings	1945
Dan Brouthers	7710	1B	81.70	(1879–1904)	19	C G	Bisons	1945
Charles Radbourn	10306	P	73.59	(1881–1891)	11	C G	Grays	1939
Roger Connor	10511	1B	87.20	(1880–1897)	18	C G	Giants	1976
Hughie Jennings	11855	SS	46.40	(1891–1918)	28	C G	Orioles	1945
Ed Delahanty	11881	LF	74.70	(1888–1903)	16	C G	Phillies	1945
George Wright	11909	Pioneer	28.60	(1867–1882)	16	C G	Red Stockings	1937

In total, we count 30 figures from nineteenth-century baseball in the Hall of Fame. All have reasonable historical significance: they lived long enough ago that reputational decay has been completed. The least significant nineteenth-century member of the Hall of Fame appears to be *Tommy McCarthy (1863–1922)* [33661].

6.5 The Negro Leagues

The Negro Leagues provided the only refuge for black professional baseball players from 1889 until 1947. The economic instability of these leagues, coupled with a limited statistical record, makes it difficult to assess the general quality of play. Still, it is clear that many players were of Hall of Fame caliber.

Negro League players were admitted to the Hall of Fame in three waves. In the 1970s, a special committee selected nine players, with two others later picked by the Veterans Committee. The second wave selected one Negro League player in each of the elections from 1995 to 2001. It ended with a tsunami in 2006, when a special committee selected a final 17 figures, doubling the Negro League presence in the Hall of Fame. In total, 33 Negro League figures were admitted in these special elections, of which we consider the ten most significant.

Top Hall of Fame Negro Leaguers

Person	Sig	Years	Dates	Celeb/Grav	Elected
Satchel Paige	4831	29	(1906–1982)	C G	1971
Josh Gibson	12535	17	(1911–1947)	C G	1972
Rube Foster	22160	25	(1879–1930)	C G	1981
Cool Papa Bell	27999	22	(1903–1991)	C G	1974
John Henry Lloyd	30110	27	(1884–1964)	C G	1977
Monte Irvin	44334	18	(1919–)	C G	1973
Frank Grant	49145	18	(1865–1937)	C G	2006
Bullet Rogan	59453	20	(1889–1967)	C G	1998
Judy Johnson	62431	20	(1899–1989)	C G	1975
Buck Leonard	67297	18	(1907–1997)	C G	1972

Satchel Paige (1906–1982) [4831] and *Josh Gibson (1911–1947)* [12535] stand as consensus choices for the greatest pitcher and hitter in the Negro Leagues. Several other figures from the first wave of admission (1971–1981) have significance ranks similar to white Hall of Famers. But those in the second batch (1995–2006) are much weaker, and the third wave much weaker still. Indeed, Negro Leaguer *Alex Pompez (1890–1974)* [240953] ranks as the least significant person ever inaugurated into the Baseball Hall of Fame.

There is some academic research to suggest that black players may receive fewer BBWAA votes than equally skilled white counterparts [Findlay and Reid, 1997], although the effect appears small [Desser et al., 1999; Jewell et al., 2002]. But it seems clear that the Hall of Fame has bent over backwards in an attempt to atone for the sins of baseball's color line.

How many Negro Leaguers *should* be in the Hall of Fame? A pioneering history of the Negro Leagues, *"Only the Ball Was White"* [Peterson, 1992], argued that blacks made up 10 percent of the population of the United States in the first half of the twentieth century, so 10 percent of the Hall of Famers during this period should be black. This worked out to eight players in 1970, a total that should be slightly increased to compensate for Hall of Fame admissions after Peterson performed his calculations.

We reason toward a number similar to Peterson, but computed in a different way. Seven graduates of the Negro Leagues earned admission to the Hall of Fame based on their later play in Major League Baseball: *Hank Aaron (1934–)* [2066], *Ernie Banks (1931–)* [23154], *Roy Campanella (1921–1993)* [13397], *Larry Doby (1923–2003)* [16845], *Willie Mays (1931–)* [2803], and *Jackie Robinson (1919–1972)* [841]. Two others who had meaningful major league careers were elected on the strength of their earlier Negro League play: *Satchel Paige (1906–1982)* [4831] and *Monte Irvin (1919–)* [44334].

Taken together, this suggests that about ten Hall of Fame caliber players debuted during the last 15 years of the color line. Although several barnstorming teams existed earlier, *Rube Foster (1879–1930)* [22160] is credited with organizing the first major Negro League in 1920. Assuming a similar quality of play from 1920 to 1935, this would suggest about ten additional worthy Negro League players debuted in this period, plus a handful more from earlier days. We conclude that the players displayed in the table listing Top Hall of Fame Negro Leaguers represent the Negro League members who should be in the Baseball Hall of Fame.

A bigger mystery is how the 2006 selection committee missed two of the most historically significant absent members in black baseball. As discussed in the last section, *Moses Fleetwood Walker (1857–1924)* [10549] was the first player excluded by the color line. *Buck O'Neil (1911–2006)* [30703] was a star Negro League player and manager, who became the first black coach in the major leagues. He did much to renew interest in the Negro Leagues, and received the Presidential Medal of Freedom for his work. Yet both were ignored in favor of 17 vastly more obscure figures.

6.6 Who's Out?: Missing Players

The flip side of the coin concerns historically significant players who have not yet received admission to the Hall of Fame. Figure 6.6 identifies the more prominent omissions in two categories: those old enough to have received several chances at election and those still active or freshly eligible for consideration.

Several of the most prominent missing old-timers (Figure 6.6a) made their names outside baseball. *Jim Thorpe* [1356] was primarily an Olympic

Person	Sig	Dates	Celeb/Grav
Jim Thorpe	1356	(1888–1953)	C G
Pete Rose	2817	(1941–)	C G
S. J. Jackson	3445	(1887–1951)	C G
Billy Sunday	3625	(1862–1935)	C G
George Halas	5358	(1895–1983)	C G
Roger Maris	6964	(1934–1985)	C G
Don Mattingly	8628	(1961–)	C G
Billy Martin	8986	(1928–1989)	C G
Joe Torre	9913	(1940–)	C G
Moses F. Walker	10549	(1857–1924)	C G
Jimmy McAleer	11501	(1864–1931)	C G
Keith Hernandez	12012	(1953–)	C G
F. Valenzuela	12814	(1960–)	C G
Chuck Connors	13253	(1921–1992)	C G
Bob Ferguson	13993	(1845–1894)	C G
Thurman Munson	15161	(1947–1979)	C G
Eddie Cicotte	15219	(1884–1969)	C G
Denny McLain	15771	(1944–)	C G

(a) Eligible for selection

Person	Sig	Dates	Celeb/Grav
Barry Bonds	2217	(1964–)	C G
Alex Rodriguez	3304	(1975–)	C G
Derek Jeter	4076	(1974–)	C G
Albert Pujols	4607	(1980–)	C G
Roger Clemens	5205	(1962–)	C G
Bo Jackson	5454	(1962–)	C G
Ichiro Suzuki	5481	(1973–)	C G
Sammy Sosa	5494	(1968–)	C G
Randy Johnson	6057	(1963–)	C G
Mariano Rivera	6666	(1969–)	C G
Mark McGwire	6716	(1963–)	C G
David Ortiz	7769	(1975–)	C G
Pedro Martínez	8662	(1971–)	C G
Tom Glavine	9386	(1966–)	C G
Greg Maddux	9587	(1966–)	C G
Johnny Damon	11054	(1973–)	C G
Curt Schilling	11877	(1966–)	C G
Gary Sheffield	12201	(1968–)	C G

(b) Ineligible for selection

FIGURE 6.6. The most significant players missing from the Baseball Hall of Fame.

star, *Billy Sunday* [3625] was an evangelist, and *Chuck Connors* [13253] was an actor. Great football stars who also played some baseball include *George Halas* [5358] and *Bo Jackson* [5454].

Another group is recognized as great players, but who have been denied admission because of serious gambling infractions. *Shoeless Joe Jackson (1887–1951)* [3445] and *Eddie Cicotte (1884–1969)* [15219] were both expelled from baseball in the Black Sox scandal. Similarly, *Pete Rose (1941–)* [2817] and *Denny McLain (1944–)* [15771] have unsavory pasts that render them *non grata* from the Hall of Fame.

Once these names are eliminated, we are left with a list of borderline Hall of Fame candidates: all very good, but maybe less than great. Among the modern omissions, the top four all have substantial associations with the New York Yankees. We believe that the strongest candidates here are catcher/manager *Joe Torre (1940–)* [9913] and slugger *Don Mattingly (1961–)* [8628]. We conclude that relatively few worthies have been left on the table.

Figure 6.6b lists the most significant active or recently retired players. The majority of the names on this list should, in principle, easily win admission to the Hall of Fame when their time comes. The wild card is how voters will eventually sort out the issue of performance-enhancing drugs. Five of the top 11 names on this list have been linked to steroid use (*Barry Bonds (1964–)* [2217], *Alex Rodriguez (1975–)* [3304], *Roger Clemens (1962–)* [5205], *Sammy Sosa (1968–)* [5494], and *Mark McGwire (1963–)* [6716]. McGwire has appeared on only 20 percent of ballots through his first five years of eligibility, well short of the 75 percent necessary for admission. All of the eligible candidates here were rejected by the voters during their first election.

6.7 The Market for Collectibles

I (Steve) collected baseball cards during my rapidly receding youth, and in particular owned a beautiful 1969 card of *Mickey Mantle (1931–1995)* [3173]. I was convinced that it would be worth something some day, and indeed, as I write this, one is selling on Ebay for $175.[1] Further, I was convinced that, like other collectibles, it would only increase in value with age. Just think what it will be worth 100 years from now!

[1] My younger brother Rob damaged it beyond repair, a crime I have never completely forgiven him for.

Probably much less than today, in dollar-adjusted terms. Our fame decay model, presented back in Section 2.4, suggests that historical reputations generally peak about 70 years after birth. Mantle was born in 1931, meaning that his fame began declining about ten years ago, and should only shine about one-tenth as brightly by 2113.

The market for collectibles literally puts a price on the head of famous people, which is in part a function of their historical significance. We study the phenomenon in two different domains: baseball cards and autograph prices.

6.7.1 BASEBALL CARDS

We analyze published prices for all Topps baseball cards since 1951, from `freebaseballcardspriceguide.com`. Ten-year old kids from 1951 are wealthy and nostalgic 75-year-olds today, so this is not quite long enough to observe the predicted effects of price decay to kick in. But it is sufficient to study the extent to which historical significance reflects collector interest.

Current-issue baseball cards are a commodity bought by children, using their spare nickels and dimes. Thus, we would not expect the collector prices to substantially increase until these kids have grown, experienced the pain of Mother discarding their cards, and then reached an economic position where they can afford to restore these collections.

The price data in Figure 6.7a suggests that collector pressure kicks in at around forty years of age, when the average price per card rises beyond a few pennies and the value of individual cards starts to appreciably diverge. Figure 6.7b shows that prices correlate approximately 0.4 with historical significance during the years of collector interest. This is quite substantial, since we ignore such important pricing effects as the higher values for cards from each player's rookie season, and the systematic scarcity of higher-numbered cards compared with those appearing earlier in each annual series.

We limited our analysis to the Topps brand for a reason. Rising prices for older cards ruined a kid's game in the early 1990s by viewing baseball cards as collector's items certain to appreciate in value. New lines of cards sprang up to accommodate demand, charging higher prices to create artificial

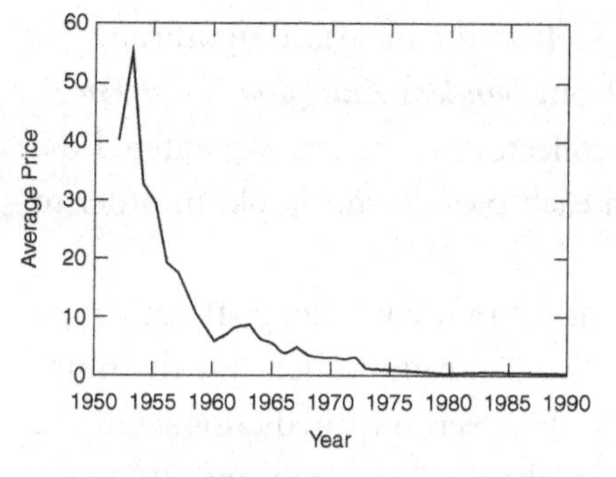

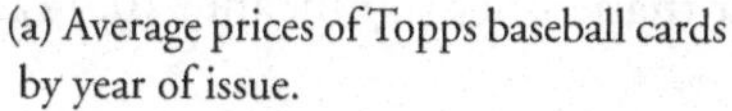
(a) Average prices of Topps baseball cards by year of issue.

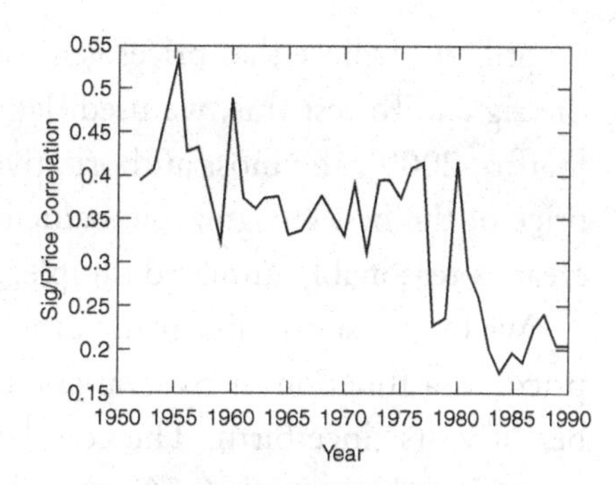

(b) Correlation of baseball card prices with historical significance.

FIGURE 6.7. Analysis of baseball card price data. There is high correlation for older, actively collected series.

"scarcity." I (Charles) remember being quite swept up in this baseball card bubble. My 1994 *Chipper Jones (1972–)* [21772], in some especially rare subseries now forgotten, peaked in the neighborhood of $80. Alas, this bubble quickly burst, and my cards are worth no more than a few dollars today. Our pricing model assumes a rational market with no such brand effects. Let the buyer beware.

6.7.2 AUTOGRAPHS

Autograph collectors also put a price on the head (or hand) of historical figures. We hypothesize that the price such collectors are willing to pay depends upon the historical significance of the person who signed it. Therefore, the correlation between autograph prices and our historical significance measure provides another way to validate our ranking methodology.

Several factors go into determining the value autograph collectors place on particular signatures. Supply and demand trumps everything. For example, *Button Gwinnett (1735–1777)* [7357], an otherwise forgotten signer of the Declaration, has commanded prices as high as $150,000 for his signature. Gwinnett managed to get himself killed in a duel less than a year after signing the Declaration, and left behind only 51 known signatures. This, coupled with the many collectors who covet a complete set of the Signers, has made his autograph one of the most valuable in America.

Still, we believe that prices generally reflect the historical significance of the signer. To test this, we used data from *Sander's Autograph Price Guide* [Saffro, 2009], the most authoritative collector's reference. We entered the price of the first recognizable name on each page of this book, in order to create a reasonably unbiased sample.

We built a simple linear regression model to predict (logged) autograph prices as a function of two variables: historical significance and the number of years since birth. The correlation between our predictions and the (logged) real prices was 0.56, much better than either significance (0.44) or birth year (-0.47) acting alone.

Figure 6.8 demonstrates the correlation between predicted and actual prices, and identifies outliers. Those above the line are priced higher than our model predicts, including several whose short lives (e.g., *John Wilkes Booth (1838–1865)* [451], *Edgar Allan Poe (1809–1849)* [54], and *Wolfgang Amadeus Mozart (1756–1791)* [24]) precluded leaving a sufficient paper trail for future generations. Our model overvalues contemporary film/TV stars (*Oprah Winfrey (1954–)* [1844] and *Chow Yun-fat (1955–)* [7694]) and politicians (*Dick Cheney (1941–)* [1059]) who are all regularly asked to sign autographs by their admirers, so supply is plentiful.

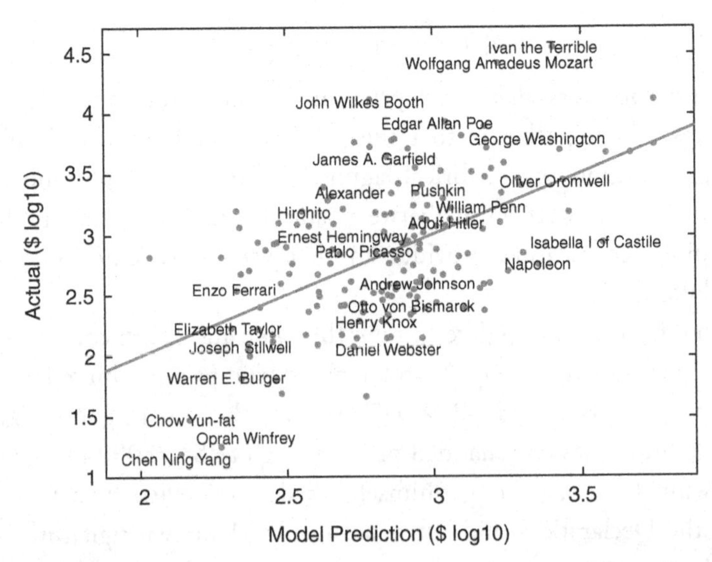

FIGURE 6.8. Autograph prices correlate well (0.56) with our model's predicted prices. People whose prices are higher (above the line) or lower (below) than the model predicts.

At the low end, our model values the autograph of the least significant person in Wikipedia at only $4.07. By contrast, we project that *Jesus' (7 B.C.–A.D. 30)* [1] autograph would sell for $5,780,960 were one to be found, which may well be a low-ball estimate. However, there aren't enough data points of ancient autographs to provide a good idea of how to fit that end of the scale.

Notes

[James, 1995] is the best general study on the history and selection process of the Baseball Hall of Fame. Recent popular books on these subjects include [Chafets, 2009; McConnell, 2010]. Academic studies of Baseball Hall of Fame voting patterns include [Desser et al., 1999; Findlay and Reid, 2002, 1997; Jewell et al., 2002]. Studies on baseball card prices include [Findlay and Santos, 2012; Matheson and Baade, 2003; Mullin and Dunn, 2002]. Price models for paintings and autographs are explored in [Collins et al., 2006; Renneboog and Spaenjers, 2012].

7

Historical Timescales

I (Steve) recently turned the big 5-0, a lifetime long enough to measure up against historical time scales in a nontrivial way. The grand old game of baseball's American League was founded in 1901, exactly 60 years and two days prior to my own founding. We have shared almost half of its history together. I recall the United States Bicentennial in 1976, amazed that our country was really 200 years old. Now I have personally witnessed more than 20 percent of American history since the Declaration of Independence. The Jewish calendar marks this year as 5774, meaning that my life will soon cover 1 percent of the span since the world's biblical creation. Young readers may smirk, but if you are lucky someday this will happen to you.

Time is the basic organization of history, measured across many scales: the calendar year, the human life span, and the historical age. But the flow of time seems deceptively slow moving, and most of it occurred off of our watch. We need calendars, timelines, and other tools to chart the course of history, and help us understand it better. In this chapter, we will analyze historical figures across different scales of time: ranging from the days of the year to vast historical epochs.

7.1 Life-Span Analysis

All men (and women) are created equal, in that they have but one life dedicated to the pursuit of happiness and achievement. But life spans are not equal, differing greatly in length and the period in which they are lived. There are several revealing ways to look at history by life span.

7.1.1 LIFE-SPAN DEMOGRAPHICS

Demographically, we can consider how the life spans of significant people match up to the population at large. The exact amount of time we will get on earth remains unknown to us until our dying day, yet statistical analysis provides an idea of what we should feel entitled to. Accurate demographic records go back only to the nineteenth century, but we can compute a longer life expectancy time series by analyzing the people in Wikipedia.

Figure 7.1 integrates two sources of data on life span: demographic data on adult life expectancy in the United States since 1850 [Infoplease, 2007], and the median life spans of all the people in Wikipedia, as a function of the year they died.

Life expectancy has grown continually since the fifteenth century, at least among the elite represented in Wikipedia. Since 1900, life spans have sharply increased – a development attributable to modern medical knowledge, economic growth, and the improvements in nutrition and hygiene that go along with them. This increase in life span has been an egalitarian affair, as shown by the associated demographic data. From 1850 until 1920, the people who are in Wikipedia lived longer than the typical American by up to five years (in the median). This is partially a result of selection bias: people appearing in Wikipedia *had* to live long-enough lives to accomplish something noteworthy. But class issues are also at work: people in a

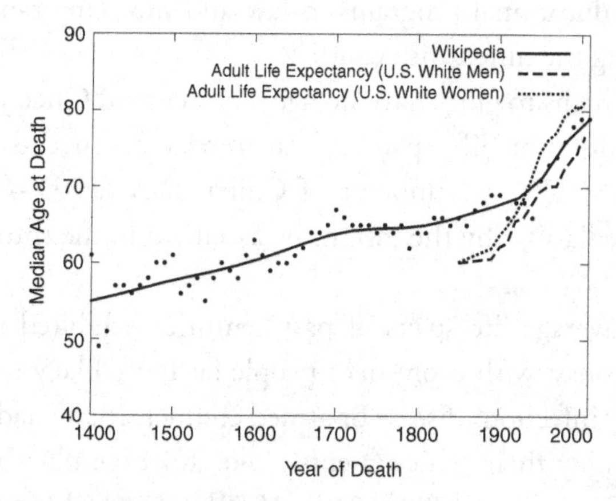

FIGURE 7.1. Median life span of figures in Wikipedia compared to U.S. adult life expectancy.

position to receive advancement opportunities were also likely to have been in a position to receive adequate food and shelter.

But this gap has disappeared over the past century. The life span of recent Wikipedia elites now sits squarely between that of the average white American man and woman. Why? Most advantages of modern technology quickly trickle down to the general population. The wealthy could not independently fend off the effects of plague or pillaging bands of thugs: such risks must be broadly eliminated in the population before anyone can reap their own advantage [Pinker, 2011].

We would expect a longer life to allow one more time for accomplishment. Thus, the historical significance of people should increase with their age, presuming that they use this extra time to polish their resume for posterity. We compared the average significance of people by life span for two cohorts: those who died before 1900 and those who have died since.

The significance distribution of the pre-1900 cohort proved surprisingly flat with respect to age at death. People used to become independent adults at much younger ages. A childhood like that of *Thomas Edison (1847–1931)* [40], out of school selling newspapers on the railroad from the age of 12, seems unimaginable today. The shorter life spans of this era did not encourage the notion of working one's way up the ranks. Long periods of training and apprenticeship did not exist. *Walter Reed (1851–1902)* [4249] received his medical degree from the University of Virginia at age 18, testimony that even the best educational programs in law and medicine ran only one to two years during the nineteenth century.

Traditional paths to glory have no age requirement. Once you have outlived your predecessor, life span has no impact on success in achieving hereditary office. The last emperor of China, *Puyi (1906–1967)* [3006], became Wikipedia-worthy the instant he ascended to the throne at the age of two.

The short average life spans of past centuries exhibited much greater variance than today, with prominent people far more likely to be suddenly struck down by infectious disease or armed conflict. Many had distressingly little time to enjoy their achievements. Take, for example, the case of the Indian guide *Sacagawea (1788–1812)* [1070], who died from fever at age 25, just six years after her service to Lewis and Clark.

We live in a different world. Historical significance now generally increases with life span. The biggest jump occurs between the ages of 30 to 35. Before 30, the main opportunities for fame are limited to fields like sports and entertainment. But the entire world of achievement is open by age 35. You can even become president. The peak of many careers comes early: as *F. Scott Fitzgerald (1896–1940)* [747] said, "there are no second acts in American lives."

7.1.2 DOMINATING LIFE SPANS

We have only limited control on the lengths of our lives, measured by what we make of the time allotted to us. Figure 7.2 highlights the most significant individuals of every possible life span from 17 to 106 years. There are several interesting things to observe here:

- The greatest of the great accomplished their missions during relatively short life spans. *Muhammad (570–632)* [3] was the longest lived of the five most significant figures in history, and he only reached 62.
  Of our top 20 historical figures, only 10 of them made it past 60. Several started young, and hence managed amazingly eventful lives. *Henry VIII (1491–1547)* [11] died at 55, yet reigned for 38 years and ran through six wives. *Napoleon (1769–1821)* [2] took just 51 years to become emperor of France, conquer most of Europe, suffer defeat and exile, then be restored to emperor and repeat the cycle of conquer, defeat, and exile.
- Others associated with long periods of activity did not have particularly lengthy life spans by contemporary standards. *Queen Victoria (1819–1901)* [16] reigned as queen for more than 63 years. Her popular historical persona is that of a stern, aged woman, yet she only lived age to 81. Victoria went into seclusion on becoming a widow at age 42, which probably contributed to her appearing older than she actually was.
  Thomas Jefferson (1743–1826) [10] is famous for dying 50 years to the day after the signing of his Declaration of Independence, yet he only made it to 83.
- Two pairs of quickly slain assassins and their presidential victims dominate their respective life spans: both *Abraham Lincoln (1809–1865)* [5] and *John Wilkes Booth (1838–1865)* [451], and *John F. Kennedy (1917–1963)* [71] and *Lee Harvey Oswald (1939–1963)* [1435]. An arguable third pair is *Charles I (1600–1649)* [39] and *Oliver Cromwell*

Life	Person	Dates	Sig
17	Lady Jane Grey	(1537–1554)	675
18	Tutankhamun	(1341–1323 B.C.)	255
19	Joan of Arc	(1412–1431)	95
20	Évariste Galois	(1811–1832)	2262
21	Billy the Kid	(1859–1881)	761
22	Pocahontas	(1595–1617)	428
23	Saint George	(280–303)	260
24	Sacagawea	(1788–1812)	1070
25	John Keats	(1795–1821)	305
26	John Wilkes Booth	(1838–1865)	451
27	Jimi Hendrix	(1942–1970)	338
28	Caligula	(A.D. 12–A.D. 41)	210
29	Percy B. Shelley	(1792–1822)	329
30	Akhenaten	(1379–1349 B.C.)	388
31	Franz Schubert	(1797–1828)	198
32	Richard III	(1452–1485)	344
33	Jesus	(7 B.C.–A.D. 30)	1
34	Jesse James	(1847–1882)	413
35	Wolfgang A. Mozart	(1756–1791)	24
36	Anne Boleyn	(1500–1536)	154
37	Vincent van Gogh	(1853–1890)	73
38	Louis XVI	(1754–1793)	83
39	Joseph Smith	(1805–1844)	55
40	Edgar Allan Poe	(1809–1849)	54
41	Richard I	(1157–1199)	120
42	Elvis Presley	(1935–1977)	69
43	Edward II	(1284–1327)	375
44	Mary	(1542–1587)	127
45	Cnut the Great	(990–1035)	355
46	John F. Kennedy	(1917–1963)	71
47	Attila	(406–453)	193
48	Charles I	(1600–1649)	39
49	Thomas Aquinas	(1225–1274)	90
50	James Cook	(1728–1779)	60
51	Napoleon	(1769–1821)	2
52	William Shakespeare	(1564–1616)	4
53	P. I. Tchaikovsky	(1840–1893)	63
54	Charles II	(1630–1685)	78
55	Henry VIII	(1491–1547)	11
56	Abraham Lincoln	(1809–1865)	5
57	Socrates	(469–399 B.C.)	68
58	Charles Dickens	(1812–1870)	33
59	Oliver Cromwell	(1599–1658)	50
60	Theodore Roosevelt	(1858–1919)	23
61	Benito Mussolini	(1883–1945)	101
62	Muhammad	(570–632)	3
63	Ulysses S. Grant	(1822–1885)	28
64	Karl Marx	(1818–1883)	14
65	Genghis Khan	(1162–1227)	38
66	J. Rousseau	(1712–1778)	80
67	George Washington	(1732–1799)	6
68	Edward I	(1239–1307)	155
69	Elizabeth I	(1533–1603)	13
70	Carl Linnaeus	(1707–1778)	31
71	Charlemagne	(742–814)	22
72	John Locke	(1632–1704)	100
73	Charles Darwin	(1809–1882)	12
74	Joseph Stalin	(1878–1953)	18
75	Augustine of Hippo	(354–430)	72
76	Albert Einstein	(1879–1955)	19
77	Augustus	(63 B.C.–A.D. 14)	30
78	Mohandas K. Gandhi	(1869–1948)	46
79	Immanuel Kant	(1724–1804)	59
80	Plato	(427–347 B.C.)	25
81	Queen Victoria	(1819–1901)	16
82	J. W. von Goethe	(1749–1832)	88
83	Thomas Jefferson	(1743–1826)	10
84	Isaac Newton	(1643–1727)	21
85	James Madison	(1751–1836)	51
86	Nikola Tesla	(1856–1943)	93
87	Giuseppe Verdi	(1813–1901)	228
88	Michelangelo	(1475–1564)	86
89	Ramesses II	(1302–1213 B.C.)	293
90	Winston Churchill	(1874–1965)	37
91	Thomas Hobbes	(1588–1679)	167
92	John Dewey	(1859–1952)	417
93	Ronald Reagan	(1911–2004)	32
94	George Bernard Shaw	(1856–1950)	213
95	W. E. B. Du Bois	(1868–1963)	545
96	Katharine Hepburn	(1907–2003)	1396
97	John D. Rockefeller	(1839–1937)	172
98	Georgia O'Keeffe	(1887–1986)	1178
99	Mimar Sinan	(1489–1588)	1706
100	Bob Hope	(1903–2003)	1098
101	E. Bowes-Lyon	(1900–2002)	832
102	Amos Alonzo Stagg	(1862–1965)	3839
103	Edward Bernays	(1891–1995)	7664
104	Rose Kennedy	(1890–1995)	7394
105	Anthony the Great	(251–356)	1066
106	Soong May-ling	(1897–2003)	7688

FIGURE 7.2. Most significant individuals by age at death.

(1599–1658) [50], who signed Charles's death warrant. *Guy Fawkes (1570–1606)* [392] achieved his notoriety by attempting to kill *James I (1566–1625)* [41].

- The ranks of lives fully lived before 40 include several artists who died tragically young: romantic poets *John Keats (1795–1821)* [305], *Percy Bysshe Shelley (1792–1822)* [329], and *Edgar Allan Poe (1809–1849)* [54], classical composers *Franz Schubert (1797–1828)* [198] and *Frédéric Chopin (1810–1849)* [195], and painter *Vincent van Gogh (1853–1890)* [73].
  A surprise in this category is mathematician *Évariste Galois (1811–1832)* [2262], who purportedly wrote his manuscript outlining the foundations of group theory the night before dying in a duel at the age of 20.
- There are other paths open to achieving great fame with a short lifetime. Notorious outlaws here include *Billy the Kid (1859–1881)* [1411] and *Jesse James (1847–1882)* [413]. Like guitarist *Jimi Hendrix (1942–1970)* [338] and pharaoh *Tutankhamun (1341–1323 B.C.)* [255], they lived fast, died young, and left a good-looking corpse.

Women appear most prominently at extreme life spans: the very oldest and the very youngest. Survival effects are clearly at work at the senior end of the spectrum. Women outlive men, and the effects become more apparent the longer you wait. On July 1, 2005, the U.S. Census officially counted only 14,013 male centenarians (those over 100) against 56,091 women.

Something different seems to happen among historical women fated to die young. A distressing percentage of them were executed for marital (*Anne Boleyn (1500–1536)* [154], *Marie Antoinette (1755–1793)* [125]) or political reasons: *Joan of Arc (1412–1431)* [95], *Lady Jane Grey (1537–1554)* [675], *Mary (1542–1587)* [127]. Clearly martyrdom is part of their historical appeal, but one cannot escape thinking that gender factored into why they lost their heads in the first place.

7.1.3 EXTREME LIFE SPANS

Particularly fascinating are the historical figures at the very extremes of life span, either tragically short or unimaginably long. The most significant figures at each end are presented in Figure 7.3.

The shortest life in Wikipedia belongs to *John I (1316–1316)* [9137], who reigned as king during his entire five days of existence. His experience is shared by several others here: born into royalty but fated to die achingly

Extreme Young				Extreme Old			
Age	Person	Dates	Sig	Age	Person	Dates	Sig
0	John I	(1316–1316)	9137	107	George Abbott	(1887–1995)	13250
1	James	(1507–1508)	33580	108	Gorgias	(483–375 B.C.)	3704
2	William IX	(1153–1156)	16333	109	Boris Yefimov	(1899–2008)	25254
3	Simon of Trent	(1472–1475)	19734	110	Allauddin Khan	(1862–1972)	12357
4	Albert Kamehameha	(1858–1862)	34087	111	Harry Patch	(1898–2009)	10998
5	Allegra Byron	(1817–1822)	16011	112	Thomas Peters	(1745–1857)	38303
6	Margaret	(1283–1290)	4531	113	Henry Allingham	(1896–2009)	7454
7	Guru Har Krishan	(1656–1664)	14975	114	Alivardi Khan	(1641–1756)	23607
8	Emperor B. of Song	(1271–1279)	11191	115	Edna Parker	(1893–2008)	18179
9	R. of Shrewsbury	(1473–1483)	5950	116	Ermanaric	(260–376)	14378
10	Louis XVII	(1785–1795)	1907	117	M. Meilleur	(1880–1998)	49862
11	Hamnet Shakespeare	(1585–1596)	6572	118	Hsu Yun	(1840–1959)	16439
12	Edward V	(1470–1483)	2204	119	Ramanuja	(1017–1137)	1363
13	Agnes of Rome	(291–304)	3409	120	K. of Glendalough	(498–618)	8379
14	Emmett Till	(1941–1955)	4597	121	Thomas Gilbert	(1752–1873)	350556
15	Edward VI	(1537–1553)	206	122	Jeanne Calment	(1875–1997)	3329
16	Edward the Martyr	(962–978)	4475	123	Hava Rexha	(1880–2003)	248650

FIGURE 7.3. Most significant individuals by age at death: lives extremely short and extremely long.

young, often at the hands of an ambitious relative. Others were young children of prominent writers: *Allegra Byron (1817–1822)* [16011] and *Hamnet Shakespeare (1585–1596)* [6572], both with enigmatic relationships with their famous fathers.

Martyrdom is a major theme here. Christian martyrs include St. *Agnes of Rome (291–304)* [3409] and *Simon of Trent (1472–1475)* [19734], whose death provoked a blood libel against the Jews and was on the calendar of Saints until removed in 1965. *Emmett Till (1941–1955)* [4597] became a martyr for the American Civil Rights movement: an African American boy murdered for flirting with a white woman. Several young royals fill historical roles as martyrs, such as *Emperor Bing of Song (1271–1279)* [11191] who bravely accepted death instead of surrender.

The extreme old are famous more for the length of their lives than the content, a trait they share with *Methuselah* [3767], who the Bible says lived 969 years, during which his only mission was to be begat and then beget others.

The longest verified life span belongs to *Jeanne Calment (1875–1997)* [3329], who at 90 sold her Paris apartment to a speculator, in exchange for a stream of monthly payments until her death. The speculator died 30 years later, out $180,000 and still sans the apartment. Wikipedia contains several people whose long, unverified life spans rest at least partially on the paucity of accurate birth records. Wikipedia's extreme old fall into two categories: those who died recently, and much older figures whose life spans are essentially legendary. Nine of the 17 people in the old side of Figure 7.3 have died since 1995, while four others date to ancient or medieval times.

7.2 Birth Date Effects

The discipline of astrology asserts that the date and place of one's birth have important consequences on life's course, through the influence of celestial objects. We are skeptical of the proposed mechanism, but agree that the specifics of birth date and birthplace can have lifelong influence. Birthplace often determines national citizenship, and we feel safe in asserting that the fates of millions of people have been profoundly shaped by which side of the border they happened to be born on.

But birth date seems a much more arbitrary phenomenon. Everybody's birthday gets celebrated once per year. In the long run, can it really matter if someone is born today as opposed to a few days earlier or later? We can analyze the Wikipedia population and find out.

Figure 7.4 presents frequency by birth date in three roughly equal-length eras: those born in 1900–1940, 1940–1980, and 1980–2010, whom we will call the "Greatest Generation," the "Baby Boomers," and the "Hipsters," respectively. The frequency distribution by birth date is noticeably different for each group. For the Greatest Generation, it was basically uniform, with no time of the year proving either particularly fertile or barren. Things look somewhat different in the post–World War II era, with a clear "baby bump" emerging in September/October. During the last 30 years the distribution has changed again, with a strong trend emerging that Wikipedia hipsters have birthdays in the early part of the year. What could cause such changes in the natural rhythm of life?

7.2.1 DEMOGRAPHIC TRENDS AND ADVANTAGES OF BIRTH

Explaining these effects requires decoupling two distinct phenomena: (1) that birth frequencies are not uniform across the calendar year, and (2) the

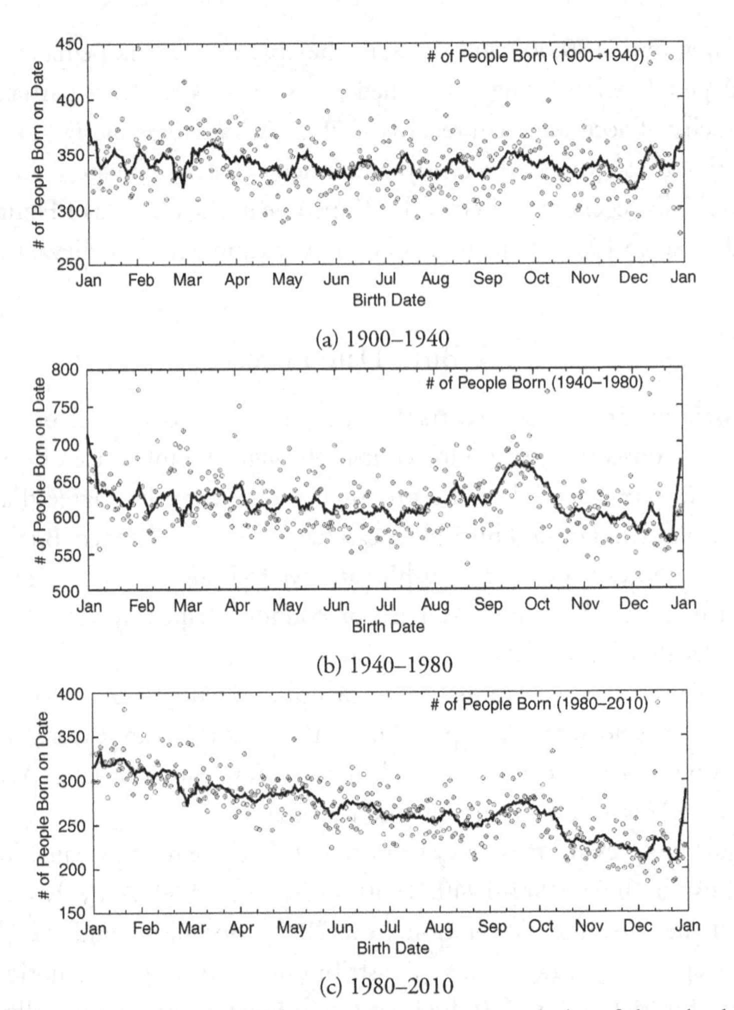

FIGURE 7.4. The number of Wikipedia people born on each day of the calendar year, during the beginning (top), middle (center), and end (bottom) of the twentieth century. A September birth bulge emerges after World War II. Recent people with early-year birthdays are strongly overrepresented in Wikipedia.

future prospects for any newborn are at least partially dependent on their date of birth.

Birth date frequencies in a population depend upon two factors: the rate at which babies are made and the dates on which they are delivered. Not much could be done to affect either of these until quite recently, which resulted in a uniform distribution of birthdays. But technological improvements have given parents greater control. On the production front, effective birth control is now widely available, enabling parents to decide when they want to have children. Junior's date of arrival has also become subject to manipulation, with an increasing percentage of births either induced or Caesarian deliveries. As a result, fewer kids are born on weekends and holidays, because people (particularly doctors) seek to avoid procedures on these dates. There is also evidence that people time deliveries to minimize taxes [Dickert-Conlin and Chandra, 1999], since the dependent child deduction applies only to those born before the end of the tax year.

September is officially the month with the highest frequency of births in the United States [CDC, 2012], with September 16 reported as the single most popular birthday of the year [*Times*, 2006]. This peak follows the Christmas/New Year's Holiday season by nine months, which presumably explains the September bulge in Figure 7.4b. But why didn't this trend begin until after 1940? We speculate that this is the result of the widespread distribution of condoms during World War II, which introduced effective family planning throughout the general population.

But birth statistics cannot explain the biggest curiosity here. Why have people with early-year birthdays become so strongly overrepresented in Wikipedia since 1980? One hypothesis might be that since January births in any given year are (by definition) older than that of December births, a few months more effort might substantially increase the odds of qualifying for Wikipedia before you turn 30. However, this idea can safely be rejected, because the birth data show an annual cycle: more Wikipedia hipsters were born in January/February of year 198x than November/December of 198x – 1.

We conclude that the specific birth dates of this younger cohort actually play a part in their chances for future success. Arbitrary calendar dates are often granted special status by tradition or statute. Particularly critical is the cutoff date for admission to kindergarten. Each state sets a fixed calendar date (typically September 1) such that five-year-olds with birthdays before the cutoff go to kindergarten, while the rest get to stay home another year.

There is a general belief that children born just after the cutoff have a significant advantage in school. They will be the strongest and most

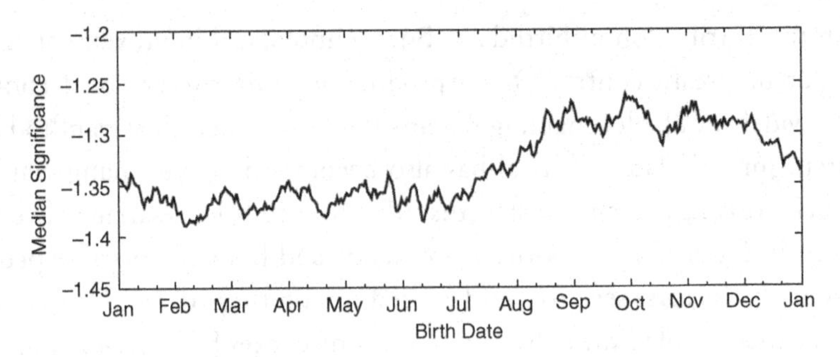

FIGURE 7.5. Median significance of Wikipedia people born since 1980, by birth date. Those born in the later half of the year appear to be a more accomplished cohort.

mature members of their kindergarten class, leading to academic success and greater self-confidence. In principle, this should also contribute to the post–September 1 Wikipedia bulge.

Another clue appears in Figure 7.5. It plots the median significance of this cohort by birthday, and shows that those born in the second half of the year are noticeably more significant than their more frequent early-year brethren. Thus, the surplus of early-year people must consist of a more historically marginal set of characters.

It has been observed that early-year births are strikingly overrepresented in sports such as hockey and soccer [Gladwell, 2008; Musch and Grondin, 2001]. These sports are characterized by active youth leagues, which stratify players by birth year. Thus, the January children are almost a full year bigger and stronger than the December kids, advantages that lead to greater athletic success. The stars of the six-year-old leagues get better coaching and more playing time, advantages that accrue to a better chance of stardom in the seven-year-old league and beyond: even after all players have reached their ultimate size.

The hipster cohort consists of people who reached Wikipedia prior to their thirtieth birthday. Sports is a domain unique in the extent to which it favors youth over experience. Very few professional hockey players are destined to enter Wikipedia after they turn 30. We believe the birthday distribution of this hipster cohort in Wikipedia will even out over time, once succeeding waves of authors, businesspeople, politicians, and centenarians stake their claims to fame and start to outrun the athletes.

7.2.2 BIRTHDAY GREETINGS AND DEATH NOTICES

Each of us shares our birthday with approximately (1/365)th of the human race. It is nice to know which important people celebrate along with you. Indeed, news outlets often broadcast "Happy Birthday" greetings to celebrities each day, more for our sake than theirs.

For fun, we computed the most significant people born on each date of the year, and placed the full tables in the Appendix, Section B.2. Take a peek to find the historical figure born under your star. Steve is proud to share his birth date with *Franklin D. Roosevelt (1882–1945)* [43], while Charles has something nice in common with *Bill Gates (1955–)* [904].

Similar statistics can be computed with respect to death date instead of birth date. Figure 7.6 plots the frequency of Wikipedia mortality by death date, revealing a strong seasonal effect. People are much more likely to die in winter than summer, an effect attributable to the greater prevalence of infectious disease in the colder months of the year. The trend here remains unchanged when restricted to recent (post-1960) deaths, showing the limitations of modern antibiotics and the advantages of retiring to Florida.

We have constructed tables of the most significant historical figures by death date, which also appear in Section B.2. Perhaps significantly, *Charles I (1600–1649)* [39] was beheaded on Steve's birthday. The most significant person to die on our Charles's birthday was *John Locke (1632–1704)* [100]. You can check out who died on yours in Figures B.3–B.4.

Particularly poignant is the thought of dying on your birthday. The most prominent people to pull this off include the Italian painter *Raphael*

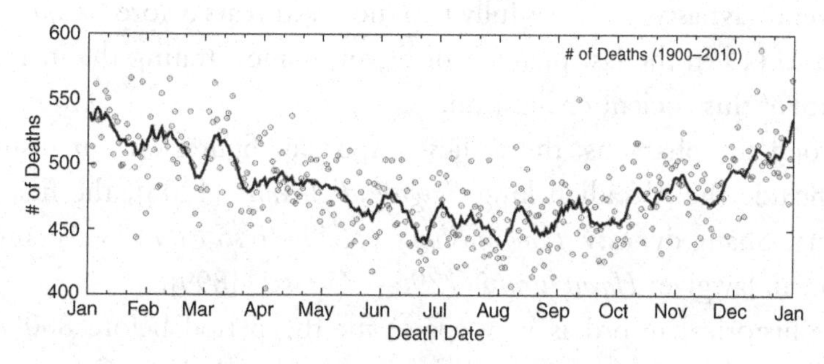

FIGURE 7.6. The number of Wikipedia people who died on each day of the calendar year, since 1900. Winter is the hardest season.

(1483–1520) [140] and *William Shakespeare (1564–1616)* [4], although there is evidence that Shakespeare's traditional birth date was in fact an eighteenth-century scholar's mistake. *John Harrison (1693–1776)* [1400], the clockmaker whose invention of the chronograph provided the first way to accurately measure longitude during long sea voyages, was also punctual enough to die on his birth date. Assuming a uniform distribution, we would expect one out of 365 people to have their lights go out when they blew out their candles, consistent with the people appearing in Wikipedia.

But the relationship between death date and birth date is not quite uniform, either. An analysis of more than 2.7 million death certificates showed that mortality changes right around birthdays [Phillips et al., 1992]. For men, mortality peaks right before their birthday, while women tend to hang on until just after their birthday.

Wikipedia offers us a dataset of about 200,000 people with complete birth and death records to test this against. We see elevated mortality on the actual birth date for both men and women. Further, both genders appear to survive beyond their birth date in greater than expected numbers, somewhat at odds with the published findings.

7.3 Ancient History

The earliest historical figure for whom Wikipedia has a precise date of death is *Khasekhemwy (?–2686 B.C.)* [8062], the final king of Egypt's second dynasty. Indeed, the 24 earliest dated figures in Wikipedia are all Egyptian pharaohs. The most notable of them was *Khufu (?–2566 B.C.)* [1119], builder of the Great Pyramid of Giza. The earliest figure for whom we know both birth and death dates is *Mentuhotep II (2061–2010 B.C.)* [12382], from the eleventh dynasty. This was fully two thousand years before *Cleopatra VII (?–30 B.C.)* [144], the last pharaoh of Egypt, demonstrating the incredible resilience of this ancient civilization.

Beyond the pharaohs, the earliest important figures on our historical stage include the Akkadian king *Sargon of Akkad* [1203], the first king of China's Shang dynasty *Tang of Shang (1675–1646 B.C.)* [9832], and the Babylonian lawgiver *Hammurabi (1796–1750 B.C.)* [899].

The historical record is very sparse for the period before 800 *B.C.E.*, and it is difficult to separate real figures from mythology. Our standard is whether Wikipedia has reported birth or death dates, which eliminates figures like *Moses* [203] and *Abraham* [382]. With the exception of *Hammurabi (1796–1750 B.C.)* [899], the most significant of these "pre-Ancient" figures are either biblical or Egyptian pharaohs.

Pre-800 B.C.E. Figures

Sig	Person	Dates	C/G	Description
57	David	(1040–970 B.C.)	C G	Biblical king of Israel (Jerusalem)
255	Tutankhamun	(1341–1323 B.C.)	C G	Egyptian pharaoh (King Tut)
388	Akhenaten	(1379–1349 B.C.)	C G	Pharaoh of Egypt (monotheistic reformer)
500	Hatshepsut	(?–1458 B.C.)	C G	Female pharaoh of Egypt
580	Joshua	(1550–1390 B.C.)	C G	Figure in the Torah (Book of Joshua)
888	Thutmose III	(?–1425 B.C.)	C G	Pharaoh of Egypt (stepchild of Hatshepsut)
899	Hammurabi	(1796–1750 B.C.)	C G	King of Babylon (Hammurabi's Code)
1046	Nefertiti	(1370–1330 B.C.)	C G	Wife of Egyptian Pharaoh Akhenaten
1119	Khufu	(?–2566 B.C.)	C G	Pharaoh of Egypt (pyramid)
1331	Amenhotep III	(?–1353 B.C.)	C G	Egyptian pharaoh (Father of Akhenaten)

The three pharaohs rated as most significant are a fascinating bunch. The contemporary fame of *Tutankhamun (1341–1323 B.C.)* [255] rests on the brevity and insignificance of his reign, which enabled his tomb to be forgotten and discovered in modern times. *Akhenaten (1379–1349 B.C.)* [388] was a failed religious reformer lost from history until the nineteenth century. He only became meaningful in light of the ultimate triumph of monotheism 1,600 years after his death, when the Romans accepted Christianity. *Hatshepsut (?–1458 B.C.)* [500] is renowned today as the most successful female Egyptian ruler, another distinction that only became of interest in contemporary times.

All this serves to remind us of history as a metaphor. Our connections with ancient Egyptian civilization are too distant, both temporally and culturally, for us to readily interpret events in the context of their times. Instead, we have built our own Egyptian mythology, by interpreting the material remnants of their culture in the context of our own.

7.3.1 BIRTH-DEATH PLOTS

Timelines of historic events help put into perspective when things took place, and more important, the relative ordering of different occurrences.

Who came first, *William Shakespeare* or *Isaac Newton*, or were they contemporaries?[1] To represent the chronology of a given age, we have developed a form of timeline we call a *birth-death plot*, which we will use extensively in this chapter.

Birth Death

Lycurgus of Sparta [3350]
Shalmaneser IV [83636]
Luli [165091]
Emperor Jimmu [3765]
Archilochus [5370]
Phraortes [36245]
Jeremiah [819]
Nebuchadnezzar II [633]
Anaximander [1898]
Mahavira [1129]
Gautama Buddha [52]
Confucius [134]
Aeschylus [386]
Anaxagoras [2006]
Herodotus [123]
Socrates [68]
Aristophanes [418]
Plato [25]
Laozi [512]
Aristotle [8]
Ptolemy I Soter [1100]
Alexander the Great [9]
Zeno of Citium [3062]
Ashoka [231]
Archimedes [146]
Eratosthenes [537]
Hannibal [187]
Scipio Africanus [995]
Polybius [923]
Hipparchus [818]
Tiberius Gracchus [2702]
Gaius Marius [692]
Lucius C. Sulla [383]
Julius Caesar [15]
Mark Antony [209]
Augustus [30]
Tiberius [272]
Philo [1607]
Jesus [1]
Nero [656]
Plutarch [258]
Hadrian [214]
Ptolemy [103]
Justin Martyr [3070]
Galen [324]
Septimius Severus [662]
Commodus [892]
Origen [617]
Plotinus [1047]
Porphyry [2095]
Diocletian [181]
Constantine the G. [67]
Saint George [260]
Martin of Tours [1086]
Julian the Apostate [508]
Augustine of Hippo [72]
Cyril of Alexandria [1693]
Saint Patrick [197]
Attila [193]
Odoacer [1778]
Theodoric the Great [1414]
Clovis I [666]
Justinian I [145]
Belisarius [1071]

800 B.C.E.
700 B.C.E.
600 B.C.E.
500 B.C.E.
400 B.C.E.
300 B.C.E.
200 B.C.E.
100 B.C.E.
0 B.C.E.
100 C.E.
200 C.E.
300 C.E.
400 C.E.
500 C.E.

Birth-death plots help us make sense of the flow of time by identifying the single most significant person born in a fixed interval: say every 20 years. For each person, we provide their birth and death years on the timeline. The slope of the life/line between these dates provides a visual representation of life span: the closer to vertical the line is, the longer their life was. Crossing lines means the earlier figure died prematurely, relative to his/her neighbor.

The timeline of ancient history shows the flow of figures beginning around *800 B.C.E.* and ending shortly after the fall of the Roman Empire.

Before about *500 B.C.E.*, many representative figures are near eastern or far eastern, including the founders of three eastern philosophies: *Mahavira (599–527 B.C.)* [1129] (Jainism), *Gautama Buddha (563–483 B.C.)* [52], and *Confucius (551–479 B.C.)* [134]. Several are kings of ancient Middle Eastern empires, such as *Shalmaneser IV (783–773 B.C.)* [83636], or *Phraortes (665–633 B.C.)* [36245] of the Median Empire. This may be the largest empire you've never heard of, which once stretched over 2,000 miles from modern-day Pakistan to Turkey.

[1] *William Shakespeare (1564–1616)* [4] preceded *Isaac Newton (1643–1727)* [21] by roughly 80 years.

It is the Greeks who dominate our picture after *500 B.C.E*, and they are of a decidedly different flavor. Except for *Alexander the Great (356–323 B.C.)* [9] and his general, *Ptolemy I Soter (367–283 B.C.)* [1100], the kings and general of the pre-Greek world were historically out-flanked by philosophers (*Socrates (469–399 B.C.)* [68] and *Plato (427–347 B.C.)* [25]), historians (*Herodotus (484–425 B.C.)* [123] and *Polybius (203–120 B.C.)* [923]), and playwrights (*Aristophanes (446–386 B.C.)* [418] and *Aeschylus (525–456 B.C.)* [386]). Intellectual figures dominate the historical mindshare here, more than any subsequent era.

Continuing down the timeline, we see the Greek cultural revolution eclipsed by the rising Roman Republic. The Romans defined themselves largely through conflict, with the great historical figures standing in opposition. Both sides of the Second Punic War are represented: Carthaginian *Hannibal (248–182 B.C.)* [187] versus Roman *Scipio Africanus* [995]. Same goes for the Roman civil war, between rival factions led by *Gaius Marius (157–86 B.C.)* [692] and *Lucius Cornelius Sulla (138–78 B.C.)* [383]. The final conflict, of *Augustus (63 B.C.–A.D. 14)* [30] versus *Mark Antony (83–30 B.C.)* [209], sounded the death knell of the Republic and the beginning of the Empire.

The next few centuries are dominated by Roman emperors. There is a good reason why *Edward Gibbon (1737–1794)* [573] entitled his famous history *The Decline and Fall of the Roman Empire*. It was indeed all downhill after Augustus. Figure 7.7 plots the historical significance of Roman emperors as a function of time. The trend is clear: an almost straight-line decay in greatness with a brief resurgence after Christianity took hold.

These emperors are interspersed with a few Greek-Roman cultural figures (the alliterative *Plutarch (A.D. 45–120)* [258], *Ptolemy (A.D. 90–168)* [103], *Plotinus (204–270)* [1047], and *Porphyry (234–305)* [2095]). The undercurrent of Christianity begins to overtake the Roman Empire. Martyrs and saints come to overwhelm the emperors: *Saint George (280–303)* [260], *Augustine of Hippo (354–430)* [72], and *Saint Patrick (390–460)* [197].

Finally, we see the people who pushed the Roman Empire toward its final demise, a series of distinctly different names like *Attila (406–453)* [193] (of the Huns), *Theodoric the Great (454–526)* [1414] (of the Ostrogoths), and finally *Odoacer (435–493)* [1778], the first "barbarian" king

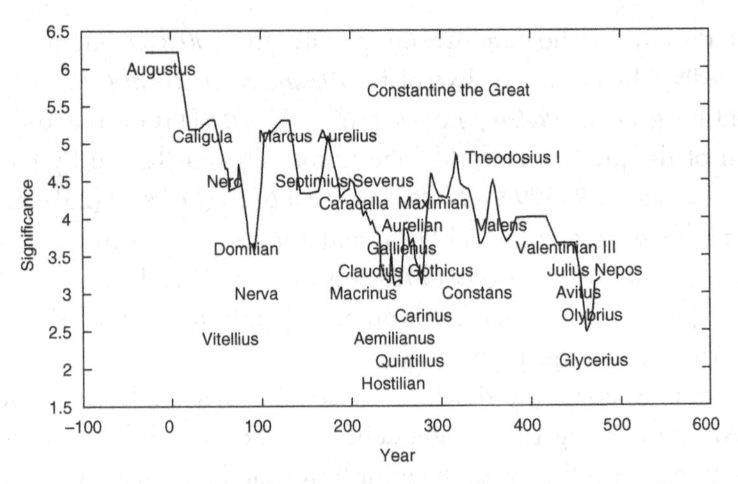

FIGURE 7.7. The significance of Roman Emperors over the course of Empire.

of Italy. As the Dark Ages descend on Western Europe, Emperor *Justinian I (483–565)* [145] and his general *Belisarius (500–565)* [1071] of the Eastern Roman (or Byzantine) Empire make failed attempts to retake Italy, signaling the end of the Roman era in Europe.

7.4 The Middle Ages

The Middle Ages timeline shows a battle between two religions. Early Christian figures *Columba (521–597)* [1052] and *Pope Gregory I (540–604)* [394] begin our timeline, but Christianity is soon drowned out by the emergence of Islam by the seventh century. *Muhammad (570–632)* [3] and his relatives (stepson *Ali (598–661)* [89] and daughter *Fatimah (614–633)* [686]) rapidly change the face of the Middle East, spreading Islam through modern Spain into Europe. The Muslim tide stops rising only in 732, at the hands of *Charles Martel (686–741)* [509]. He was the founder of the Carolingian dynasty credited with establishing the concepts of feudalism and knighthood. Later Carolingians *Pepin the Short (714–768)* [831] and *Charlemagne (742–814)* [22] unified much of Western Europe, establishing a shared Christian/European identity that stretched across France, Germany, and Italy.

Charlemagne's pan-European empire fragmented, beginning under his son *Louis the Pious (778–840)* [827]. We now begin to see the rise of

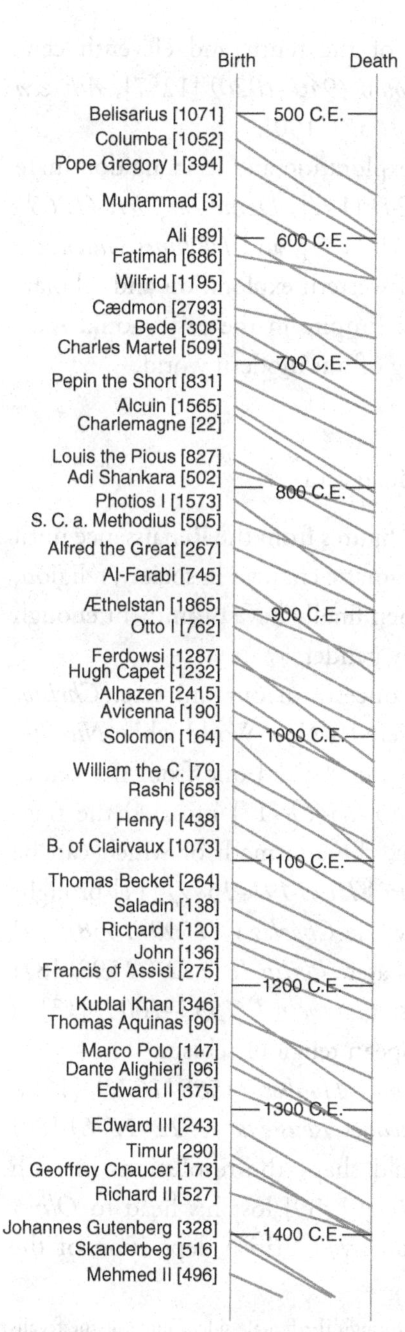

England starting at *Æthelstan (895–939)* [1865] and *William the Conqueror (1027–1087)* [70]. Five more English kings make our timeline before we exit the Middle Ages.

The Crusaders of the eleventh to thirteenth centuries go largely unrepresented in our timeline, with the notable exception of *Saladin (1138–1193)* [138], the key Muslim figure in the eventual Christian defeats. Although the Muslim world saw military success against Christian Europe, it experienced repeated failures against *Genghis Khan (1162–1227)* [38] and his descendants, two of whom make our timeline: *Kublai Khan (1215–1294)* [346] and *Timur (1336–1405)* [290]. The most famous person of every epoch tends to track the dominant civilization, but the dominant civilization has not always been Western. Following the Mongol destruction of much of the Muslim world, it again disappears from our view until the rise of the Ottoman (Turkish) Empire.

The major cultural figures also alternate between East and West. European poetry, philosophy, and music seem inextricably linked to religion, and the burgeoning monastic system. Indeed, all of our early scholars are monks: *Cædmon (657–680)* [2793], *Bede (672–735)* [308], *Alcuin (735–804)* [1565]. These cultural figures are drowned out by kings and saints until the great Muslim scholars of the tenth and eleventh centuries: *Al-Farabi (870–950)* [745], *Ferdowsi (940–1020)* [1287], *Alhazen (965–1040)* [2415], and *Avicenna (980–1037)* [190].

By the thirteenth century, figures of exploration and learning dominate Western Europe: *Marco Polo (1254–1324)* [147], *Dante Alighieri (1265–1321)* [96], *Geoffrey Chaucer (1343–1400)* [173], and *Johannes Gutenberg (1398–1468)* [328]. The reemergence of western exploration and scholarship, combined with a failing Byzantine Empire in the east, would soon trigger the Renaissance and the beginning of our modern world.

7.5 The Modern Era

Figure 7.8 shows the dominant historical figures from the Renaissance until today, a period marked by increasingly dramatic changes in society, religion, government, science, and industry. Women finally have prominent enough roles to justify separating our timelines by gender.

The Renaissance changed how we conceive of our universe: *Christopher Columbus (1451–1506)* [20] discovered a New World while *Nicolaus Copernicus (1473–1543)* [74] was removing the old one from the center of the solar system. *Martin Luther (1483–1546)* [17] initiated the fragmentation of the Roman Catholic faith, the aftermath of which can be traced through several later figures. *Henry VIII (1491–1547)* [11] brought the Protestant reformation to England, while *Charles V (1500–1558)* [84] staunchly fought it on the continent.[2] Later, *Philip II (1527–1598)* [87] sent the ill-fated Spanish Armada against *Elizabeth I (1533–1603)* [13], a conflict directly attributable to this European religious schism.

Political philosophers such as *Thomas Hobbes (1588–1679)* [167], *Voltaire (1694–1778)* [64], and *Jean-Jacques Rousseau (1712–1778)* [80] ushered in the political ideals that would shape the revolutions in both America and France. *Charles I (1600–1649)* [39] lost his head to *Oliver Cromwell (1599–1658)* [50] in the short-lived British deposition of the

[2] Henry VIII left a large personal wake in history through the female side. Our timeline recalls two of his six wives, *Anne Boleyn (1500–1536)* [154] and *Catherine Howard (1521–1542)* [1120], his two daughters *Mary I (1516–1558)* [126] and *Elizabeth I (1533–1603)* [13], his mistress *Mary Boleyn (1499–1543)* [1911], and even his nanny *Margaret Bryan (1468–?)* [25072].

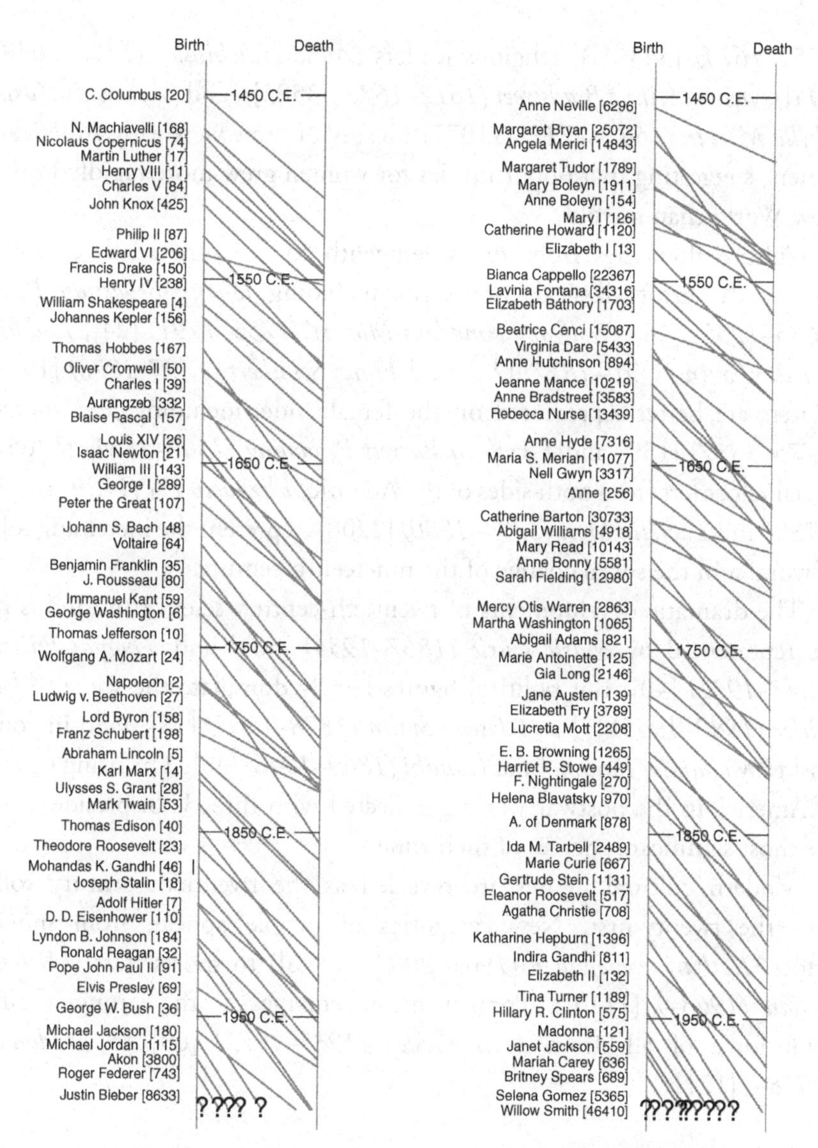

FIGURE 7.8. Timeline of recent history, separated into men (left) and women (right).

monarchy. American revolutionary figures are well represented, including *George Washington (1732–1799)* [6], *Thomas Jefferson (1743–1826)* [10], and *Benjamin Franklin (1706–1790)* [35]. A century later, political philosophy returns to the fore with *Karl Marx (1818–1883)* [14].

By the seventeenth century, women begin to make waves as something other than royalty. We encounter women as painters (*Lavinia Fontana*

(1552–1614) [34316]), religious leaders (*Anne Hutchinson (1591–1643)* [894]), writers (*Anne Bradstreet (1612–1672)* [3583]), and scientists (*Maria Sibylla Merian (1647–1717)* [11077]). Many of these were early American settlers, suggesting that opportunities for women grew more rapidly in the New World than the old.

Culture flourishes from the seventeenth to the nineteenth century. We see the great classical composers, including *Johann Sebastian Bach (1685–1750)* [48], *Wolfgang Amadeus Mozart (1756–1791)* [24], *Ludwig van Beethoven (1770–1827)* [27], and *Franz Schubert (1797–1828)* [198]. Writers are better represented on the female side, including *Jane Austen (1775–1817)* [139] and *Elizabeth Barrett Browning (1806–1861)* [1265]. Social reformers from both sides of the Atlantic, *Elizabeth Fry (1780–1845)* [3789] and *Lucretia Mott (1793–1880)* [2208], represent the increasing role of women in the social change of the nineteenth century.

The dramatic developments of twentieth-century science and industry are represented by *Marie Curie (1867–1934)* [667] and *Thomas Edison (1847–1931)* [40]. But political figures largely dominate the scene. *Adolf Hitler (1889–1945)* [7] and *Joseph Stalin (1878–1953)* [18] stand in contrast to *Mohandas Karamchand Gandhi (1869–1948)* [46]. The rising power of America in the postwar world is reflected with three U.S. presidents as the most significant memes of their time.

Modern cultural figures are revealed as the twentieth century rolls into the twenty-first. New categories of people appear, from movie actors (*Katharine Hepburn (1907–2003)* [1396]) to sports stars (*Michael Jordan (1963–)* [1115]). Popular music emerges as the strongest cultural force of all, from *Elvis Presley (1935–1977)* [69] to *Madonna (1958–)* [121].

7.5.1 THE TWENTIETH CENTURY

The past century has been so eventful that the world of 1900 is difficult to imagine today. The automobile did not exist as a consumer item; Ford Motors would not be founded until 1903. Radio communications were still an experimental medium, with broadcasting at best a distant dream. Cinema was in its very infancy, with running times of commercial films measured not in minutes but in seconds.

By contrast, all of the important elements of the modern world were in place by 1950. Certainly technology has continued to improve, and clothing styles have continued to change, but it is a time we would understand if we were magically transported there. The years between 1900 and 1950 are when our modern world appeared.

We have prepared a detailed timeline showing the single most significant birth and death for each year of the past century. Reading through it gives a distinct perspective on how the world developed. The births column illustrates which people were contemporaries, and the juxtapositions can be jarring:

- Nazi *Heinrich Himmler (1900–1945)* [671] and animator *Walt Disney (1901–1966)* [337].
- Authors *George Orwell (1903–1950)* [342], *Dr. Seuss (1904–1991)* [893], and *Ayn Rand (1905–1982)* [486].
- *Pope Benedict XVI (1927–)* [361] and Cuban revolutionary *Che Guevara (1928–1967)* [457].
- Terrorist *Osama bin Laden (1957–2011)* [765] and the entertainer *Madonna (1958–)* [121].

The clusters of related figures one sees when scanning prominent births illustrates how the times help make the man. Four U.S. presidents were born between 1908 and 1917, a direct effect on World War II. All four served with distinction, at a time when an admirable war record was a prerequisite to achieving elective office. The same men, born ten years earlier or later, would have been either too old or too young for combat.

Similarly, you pretty much had to be born during World War II in order to grow up into an early 1960s rock star. Too much younger and you didn't get the chance to be a pioneer. Too much older and you wouldn't get the sound. *John Lennon (1940–1980)* [162], *Bob Dylan (1941–)* [130], and *Jimi Hendrix (1942–1970)* [338] were all contemporaries.

So far our most prominent people born between 1965 and 1999 are all entertainers and athletes. No doubt the U.S. presidents from 2024 to 2060 will come to supplant many of them, but for now they remain the pride of their class. The cohort since 2000 is dominated by royalty. There are very few ways to achieve prominence before age ten, unless it is handed to you at birth. A special case are the *Suleman octuplets* [15710], the product of an exceptional though nonroyal birth.

Interval	Births	Sig	C/G	Deaths	Sig	C/G
1900	Heinrich Himmler	671	C G	Friedrich Nietzsche	42	C G
1901	Walt Disney	337	C G	Queen Victoria	16	C G
1902	John Steinbeck	576	C G	Cecil Rhodes	706	C G
1903	George Orwell	342	C G	Pope Leo XIII	424	C G
1904	Dr. Seuss	893	C G	Anton Chekhov	523	C G
1905	Ayn Rand	486	C G	Jules Verne	229	C G
1906	Leonid Brezhnev	1162	C G	Paul Cézanne	389	C G
1907	John Wayne	459	C G	William Thomson	450	C G
1908	Lyndon B. Johnson	184	C G	Grover Cleveland	98	C G
1909	Barry Goldwater	1535	C G	Leopold II	727	C G
1910	Mother Teresa	820	C G	Mark Twain	53	C G
1911	Ronald Reagan	32	C G	Gustav Mahler	364	C G
1912	Milton Friedman	700	C G	Bram Stoker	651	C G
1913	Richard Nixon	82	C G	J. P. Morgan	322	C G
1914	W. S. Burroughs	1561	C G	Charles S. Peirce	225	C G
1915	Frank Sinatra	357	C G	B. T. Washington	525	C G
1916	Gough Whitlam	1114	C G	Franz Joseph I	398	C G
1917	John F. Kennedy	71	C G	Edgar Degas	422	C G
1918	Nelson Mandela	356	C G	Claude Debussy	380	C G
1919	Jackie Robinson	841	C G	Theodore Roosevelt	23	C G
1920	Pope John Paul II	91	C G	Max Weber	350	C G
1921	Prince Philip	909	C G	Camille Saint-Saëns	842	C G
1922	Judy Garland	1031	C G	Alexander G. Bell	106	C G
1923	Henry Kissinger	864	C G	Warren G. Harding	242	C G
1924	George H. W. Bush	363	C G	Woodrow Wilson	47	C G
1925	Margaret Thatcher	271	C G	Sun Yat-sen	592	C G
1926	Elizabeth II	132	C G	Claude Monet	178	C G
1927	Pope Benedict XVI	361	C G	Isadora Duncan	1734	C G
1928	Che Guevara	457	C G	Thomas Hardy	378	C G
1929	Martin Luther King	221	C G	Karl Benz	840	C G
1930	Sean Connery	1077	C G	William Howard Taft	153	C G
1931	Mikhail Gorbachev	637	C G	Thomas Edison	40	C G
1932	Johnny Cash	900	C G	John Philip Sousa	922	C G
1933	Yoko Ono	1437	C G	Calvin Coolidge	370	C G
1934	Charles Manson	1307	C G	Edward Elgar	445	C G
1935	Elvis Presley	69	C G	Oliver W. Holmes	885	C G
1936	John McCain	803	C G	George V	235	C G
1937	Saddam Hussein	444	C G	John D. Rockefeller	172	C G
1938	Jerry Brown	2444	C G	Mustafa K. Atatürk	360	C G
1939	Tina Turner	1189	C G	Sigmund Freud	44	C G
1940	John Lennon	162	C G	Leon Trotsky	278	C G
1941	Bob Dylan	130	C G	Wilhelm II	200	C G
1942	Jimi Hendrix	338	C G	Franz Boas	1053	C G
1943	John Kerry	487	C G	Nikola Tesla	93	C G
1944	Jimmy Page	2167	C G	Wassily Kandinsky	618	C G
1945	Bob Marley	966	C G	Adolf Hitler	7	C G
1946	George W. Bush	36	C G	H. G. Wells	249	C G
1947	Stephen King	191	C G	Henry Ford	148	C G
1948	Al Gore	623	C G	Mohandas K. Gandhi	46	C G
1949	Meryl Streep	1490	C G	Richard Strauss	586	C G

FIGURE 7.9. Top figures born and dying (1900–1949).

Interval	Births	Sig	C/G	Deaths	Sig	C/G
1950	Stevie Wonder	1952	C G	George Bernard Shaw	213	C G
1951	Gordon Brown	1398	C G	Ludwig Wittgenstein	493	C G
1952	Vladimir Putin	1014	C G	John Dewey	417	C G
1953	Tony Blair	548	C G	Joseph Stalin	18	C G
1954	John Travolta	1692	C G	Henri Matisse	376	C G
1955	Bill Gates	904	C G	Albert Einstein	19	C G
1956	Mel Gibson	1536	C G	Bertolt Brecht	933	C G
1957	Osama bin Laden	765	C G	Joseph McCarthy	553	C G
1958	Madonna	121	C G	Pope Pius XII	294	C G
1959	Stephen Harper	1498	C G	Frank Lloyd Wright	245	C G
1960	Diego Maradona	1264	C G	Albert Camus	1623	C G
1961	Barack Obama	111	C G	Ernest Hemingway	248	C G
1962	Tom Cruise	1823	C G	Marilyn Monroe	347	C G
1963	Michael Jordan	1115	C G	John F. Kennedy	71	C G
1964	Sarah Palin	773	C G	Herbert Hoover	183	C G
1965	Dr. Dre	1895	C G	Winston Churchill	37	C G
1966	Janet Jackson	559	C G	Walt Disney	337	C G
1967	Kurt Cobain	1359	C G	Che Guevara	457	C G
1968	Celine Dion	2023	C G	Martin Luther King	221	C G
1969	Jay-Z	1367	C G	D. D. Eisenhower	110	C G
1970	Mariah Carey	636	C G	Bertrand Russell	253	C G
1971	Tupac Shakur	938	C G	Igor Stravinsky	368	C G
1972	Eminem	823	C G	Harry S. Truman	94	C G
1973	Edge	2603	C G	Pablo Picasso	171	C G
1974	Robbie Williams	2038	C G	Duke Ellington	495	C G
1975	David Beckham	1462	C G	Francisco Franco	291	C G
1976	Ronaldo	2128	C G	Mao Zedong	151	C G
1977	Kanye West	1494	C G	Elvis Presley	69	C G
1978	Kobe Bryant	1645	C G	Pope Paul VI	869	C G
1979	Pink	1922	C G	John Wayne	459	C G
1980	Christina Aguilera	1205	C G	John Lennon	162	C G
1981	Britney Spears	689	C G	Bob Marley	966	C G
1982	Kelly Clarkson	1640	C G	Ayn Rand	486	C G
1983	Carrie Underwood	2505	C G	Tennessee Williams	1213	C G
1984	LeBron James	1954	C G	Indira Gandhi	811	C G
1985	Cristiano Ronaldo	1669	C G	Orson Welles	729	C G
1986	Rafael Nadal	1946	C G	Jorge Luis Borges	989	C G
1987	Hilary Duff	1626	C G	Andy Warhol	485	C G
1988	Rihanna	1185	C G	Richard Feynman	1225	C G
1989	Chris Brown	4183	C G	Ruhollah Khomeini	676	C G
1990	Emma Watson	9272	C G	Roald Dahl	1177	C G
1991	Jamie Lynn Spears	11915	C G	Freddie Mercury	863	C G
1992	Miley Cyrus	2009	C G	Isaac Asimov	916	C G
1993	Miranda Cosgrove	10939	C G	Anthony Burgess	1430	C G
1994	Dakota Fanning	7510	C G	Richard Nixon	82	C G
1995	Luke Benward	63271	C G	Harold Wilson	1337	C G
1996	Abigail Breslin	25027	C G	Tupac Shakur	938	C G
1997	Alex Wolff	52025	C G	Mother Teresa	820	C G
1998	Jaden Smith	28426	C G	Frank Sinatra	357	C G
1999	T. D. Jakes	36901	C G	Wilt Chamberlain	1786	C G

FIGURE 7.10. Top figures born and dying (1950–1999).

Interval	Births	Sig	C/G	Deaths	Sig	C/G
2000	Noah Cyrus	42407	C G	Pierre Trudeau	870	C G
2001	Princess Elisabeth of Belgium	68561	C G	George Harrison	657	C G
2002	Prince Felix of Denmark	94431	C G	Elizabeth Bowes-Lyon	832	C G
2003	John Cassian	21483	C G	Johnny Cash	900	C G
2004	Princess Ingrid Alexandra of Norway	74255	C G	Ronald Reagan	32	C G
2005	Prince Christian of Denmark	74247	C G	Pope John Paul II	91	C G
2006	Prince Hisahito of Akishino	41250	C G	Gerald Ford	230	C G
2007	James	62562	C G	Boris Yeltsin	1913	C G
2008	Joe Haines	191742	C G	Arthur C. Clarke	1522	C G
2009	Prince Henrik of Denmark	161203	C G	Michael Jackson	180	C G
2010	Lorenzo Fertitta	99953	C G	Robert Byrd	1710	C G

FIGURE 7.11. Top figures born and dying (2000–2010).

The most prominent person to die in each given year is generally a higher gravitas figure than the representative born at that time. The juxtapositions of contemporaries can be evocative, because certain people live long past the time during which their reputation was established:

- Western gunslinger *Wyatt Earp (1848–1929)* [858] and twentieth-century president *William Howard Taft (1857–1930)* [153].
- *Mao Zedong (1893–1976)* [151] and *Elvis Presley (1935–1977)* [69].
- Experimental musician *Frank Zappa (1940–1993)* [1532] and repressed president *Richard Nixon (1913–1994)* [82].
- Actress *Marilyn Monroe (1926–1962)* [347] preceded her purported lover *John F. Kennedy (1917–1963)* [71] in their premature deaths.

The full tables of the most prominent births and deaths of the past century appear in Figures 7.9, 7.10, and 7.11.

7.6 The Historical Representation of Women

Far fewer women appear in news and historical texts than men. Our top 100 people (Figures 1.1 and 1.2) included just three women: *Elizabeth I (1533–1603)* [13], *Queen Victoria (1819–1901)* [16], and *Joan of Arc (1412–1431)* [95]. Indeed, only a hundred or so women rank among the 2,000 most significant people in history. A quarter of these make the list as royalty, because prior to the nineteenth century, the most reliable way for a woman to achieve success was to be born well and/or marry well. The modern female figures in this stratosphere comprise a more diverse group: with writers, actresses, musicians, and civil rights figures making up the bulk of the top 100 women in history.

To better understand these gender effects, we used data from the U.S. Census (2000) to identify the most common 1,219 male and 4,275 female first names. Together, these account for about 90 percent of each gender in the U.S. population.[3]

By using first name as a crude sex classifier, we can measure the relative frequency of women in the historical record. Only about 10 percent of all Wikipedia people born between 1700 and 1850 were women. This ratio increased slowly, reaching about 20 percent of those born in 1975. That proves to be an inflection point in gender balance, and about 40 percent of the youngest Wikipedia cohort are now female.

We can think of two possible explanations for the dearth of women in the historical record. The first revolves around the fact that they, as a group, were simply not in a position to contribute to the important events of their times to the same extent as men. Women's activities outside home and family have been sharply restricted for almost all of human history. They are absent from Wikipedia because they did not accumulate sufficient public accomplishments to make them encyclopedia-worthy.

The alternate hypothesis dictates that the achievements of women have been systematically ignored by the historical record. Men wrote the rough draft of history, and continue to control access today. A recent study suggests that less than 13 percent of Wikipedia contributors are women [Cohen, 2011]. Might this gender-biased contributor base impose a gender bias in the content of Wikipedia itself?

Our historical significance rankings provide a way to measure the strength of this bias hypothesis. The figures appearing in Wikipedia, by definition, represent people of outsized accomplishment, those at the very tail of human achievement. Roughly one out of every 100,000 human lives ever lived have been enshrined in Wikipedia, which means there must be vast numbers of almost equally qualified people of both genders left on the outside looking in.

[3] To minimize confusion, we eliminated gender-ambiguous names such as Ashley and Dylan, those whose dominant gender ratio in census data is less than 9:1. These gender-ambiguous names make up less than 5 percent of our dataset.

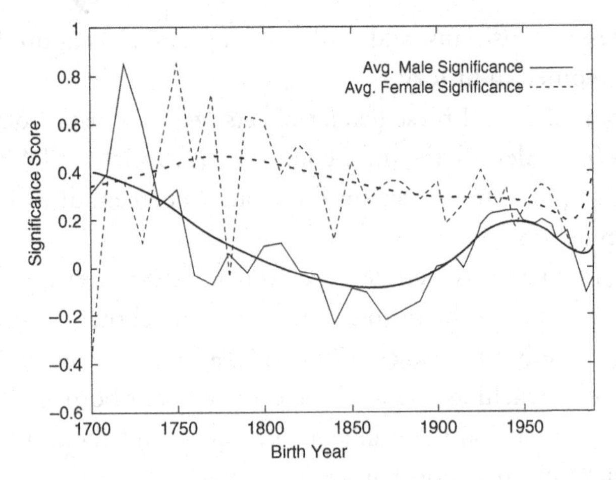

FIGURE 7.12. The average significance score of men and women grouped by birth year, from 1700 to date. Until very recently, women needed to be much more significant than their male counterparts to enter the historical record.

The key to identifying which people are missing lies in analyzing who lives in Wikipedia today. If the average female figure in Wikipedia is more significant than the average male, it would imply that there must exist a substantial reserve of absent women worthy of their own articles, or equivalently a small army of insignificant men who could be dispensed with to improve editorial standards.

This is exactly what we see in our data. Figure 7.12 plots the average significance scores of men and women born over the past 300 years. It shows a wide chasm: the women are indeed noticeably stronger than the men over most of this period. Comparing men and women of the eighteenth century, this gap works out to a quarter of a standard deviation in size. To put it another way, this difference is equivalent to requiring the average woman in Wikipedia to be roughly 4 IQ points smarter than the average man.

Fortunately, this gap between the genders has been shrinking, and has effectively disappeared for modern figures. This can be seen very clearly when we compare the distribution of significance scores for men and women from the nineteenth and twentieth centuries, shown in Figure 7.13. Examining these distributions, we can clearly see that nineteenth-century women were held to a significantly higher standard to merit inclusion in

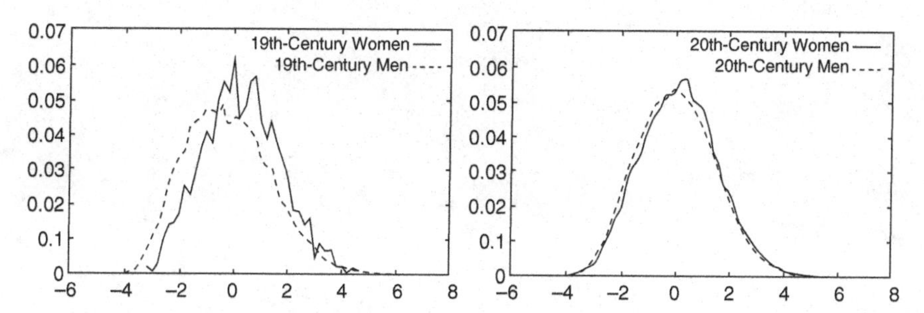

FIGURE 7.13. Significance histograms of Wikipedia-level men and women in the nineteenth (left) and twentieth (right) centuries. The nineteenth-century women achieved substantially higher significance than their male counterparts, a handicap that has dissipated over the past century.

Wikipedia. The ranks of the less significant women are severely depleted relative to their male counterparts. By comparison, the twentieth-century male and female distributions are very similar, although still slightly skewed to require stronger women than men.

PART II

Historical Rankings

8

American Political Figures

The first part of this book established the general validity of our ranking methods, and used them to illustrate grand themes and processes of history: canonization in textbooks, evaluating the precision of selection processes, measuring the flow of time, and quantifying changes in the perception of gender.

Now we will reduce our focus to the particular. We rank the significance of the world's historical figures in terms of the different niches they occupy: politicians, scientists, religious leaders, artists, actors, outlaws, and even dentists. It is instructive to see who rises to the top of each individual heap, both to refresh our memory on old historical friends and to make new ones. You have our blessing to skim through any group that you are not interested in, but sneak a peek at the ranking tables before you move on.

Some may question how we decide which figures belong in a particular group. Defining exactly who is an actor, an outlaw, or a dentist turns out to be very difficult to do in a precise way. We used the following methodology. We would start from a roster assembled in some book or Wikipedia category, and then amend the lists based on general knowledge and our sense of the nature of the category. No doubt certain omissions remain, although we believe that we have captured most of the usual suspects.

This first chapter concerns America's political leaders, from our presidents down to the mayors of our greatest cities. Besides naming names, we also explore themes like how much a president's glory reflects on those around him, and the degree to which individual states shape the historical prospects of their leaders.

8.1 Presidents of the United States

Figure 8.1 summarizes the fame and significance of all presidents of the United States. Our decay procedures[1] succeed in distinguishing between current and lasting renown: e.g., *Jimmy Carter (1924–)* [462] ranks twelfth among all presidents in current fame, but sinks into the bottom three in long-term significance.

Longevity, as measured by either the number of elected terms or the years served in office, is the most important single factor in presidential reputation. There is a difference between these two measures, as nine presidents did not complete their terms because of death or resignation, leaving nine others with partial terms to fill. Of the ten most significant presidents, eight served two complete terms. The exceptions were *Abraham Lincoln (1809–1865)* [5], martyred early in his second term, and *Theodore Roosevelt (1858–1919)* [23], who served almost two complete terms after taking office in the wake of *William McKinley's (1843–1901)* [176] assassination.

By comparison, none of the 16 least significant presidents were elected to two terms. Among this less elite group, only *Calvin Coolidge (1872–1933)* [370] won re-election, having inherited office upon the death of *Warren G. Harding (1865–1923)* [242]. Our bottom two presidents (*Millard Fillmore (1800–1874)* [446] and *Chester A. Arthur (1829–1886)* [499]) succeeded to office and were never elected president in their own right.

Presidential greatness is not exactly the same thing as historical significance, since someone who makes a complete hash of things might be significant without being great. A recent presidential ranking book, *Where They Stand* [Merry, 2012], trusts the opinion of the voters. He holds in highest esteem those elected to two full terms, who then left the office to a successor of their party. These include Washington, Jefferson, Madison, Monroe, Jackson, Grant, FDR, and Reagan. Our methods rank all of these highly except *James Monroe (1758–1831)* [220], who presided during the single-party "Era of Good Feelings."

Other observations from our rankings:

- Our significance rankings are most strongly suspect for recent presidents. Certainly we rank *George W. Bush (1946–)* [36] and *Barack Obama*

[1] Recall Section 2.4.

Sig	Fame	Poll	Person	Dates	C/G	Description
5	13	1	Abraham Lincoln	(1809–1865)		16th president (U.S. Civil War)
6	15	3	George Washington	(1732–1799)		1st president (American Revolution)
10	28	4	Thomas Jefferson	(1743–1826)		3rd president (Decl. of Independence)
23	29	5	Theodore Roosevelt	(1858–1919)		26th president (Progressive Movement)
28	64	37	Ulysses S. Grant	(1822–1885)		18th president and Civil War general
32	6	17	Ronald Reagan	(1911–2004)		40th president (Conservative Revolution)
36	1	34	George W. Bush	(1946–)		43rd president (Iraq War)
43	22	2	Franklin D. Roosevelt	(1882–1945)		32nd president (New Deal, WWII)
47	71	6	Woodrow Wilson	(1856–1924)		28th president (World War I)
51	107	13	James Madison	(1751–1836)		4th president (War of 1812)
61	124	12	John Adams	(1735–1826)		Founding Father and 2nd president
66	131	8	Andrew Jackson	(1767–1845)		7th president ("Old Hickory")
71	17	11	John F. Kennedy	(1917–1963)		35th president (Cuban Missile Crisis)
82	21	32	Richard Nixon	(1913–1994)		37th president (Watergate)
94	58	7	Harry S. Truman	(1884–1972)		33rd president (Korean War)
98	224	19	Grover Cleveland	(1837–1908)		22nd and 24th president
105	250	41	Andrew Johnson	(1808–1875)		17th president (Reconstruction)
110	60	8	Dwight D. Eisenhower	(1890–1969)		34th president and WWII general
111	3	14	Barack Obama	(1961–)		44th and current president
115	7	20	Bill Clinton	(1946–)		42nd president (Lewinsky scandal)
135	323	18	John Quincy Adams	(1767–1848)		6th president (Monroe Doctrine)
153	241	22	William Howard Taft	(1857–1930)		27th president ("Dollar Diplomacy")
176	393	20	William McKinley	(1843–1901)		25th president (Spanish-American War)
183	151	29	Herbert Hoover	(1874–1964)		31th president (Great Depression)
184	55	14	Lyndon B. Johnson	(1908–1973)		36th president ("Great Society")
220	506	14	James Monroe	(1758–1831)		5th president (Panic of 1819)
230	77	26	Gerald Ford	(1913–2006)		38th president (Cold War)
237	547	42	James Buchanan	(1791–1868)		15th president (Secession)
240	557	10	James K. Polk	(1795–1849)		11th president (Mexican-American War)
242	283	43	Warren G. Harding	(1865–1923)		29th president (Teapot Dome affair)
277	648	24	Martin Van Buren	(1782–1862)		8th president (Panic of 1837)
285	675	29	James A. Garfield	(1831–1881)		20th president (civil service reform)
288	683	38	William Henry Harrison	(1773–1841)		9th president (shortest presidency)
299	712	35	Zachary Taylor	(1784–1850)		12th president (compromise of 1850)
325	769	25	Rutherford B. Hayes	(1822–1893)		19th president (end of Reconstruction)
339	808	33	Benjamin Harrison	(1833–1901)		23rd president (McKinley Tariff)
341	818	36	John Tyler	(1790–1862)		10th president (annexation of Texas)
363	48	22	George H. W. Bush	(1924–)		41st president (Gulf War)
370	363	31	Calvin Coolidge	(1872–1933)		30th president ("Roaring Twenties")
427	1037	40	Franklin Pierce	(1804–1869)		14th president (1853–1857)
446	1079	38	Millard Fillmore	(1800–1874)		13th president (Fugitive Slave Act)
462	40	27	Jimmy Carter	(1924–)		39th president (Iranian hostage crisis)
499	1267	28	Chester A. Arthur	(1829–1886)		21st president (Chinese Exclusion Act)

FIGURE 8.1. U.S. presidents ranked by historical significance and fame, with consensus historical poll ranks.

(1961–) [111] higher than their historical record currently supports. Their presidencies will require another generation to cool before we can make more objective evaluations of them.

- That *Ronald Reagan's (1911–2004)* [32] significance exceeds his successors *Bill Clinton (1946–)* [115] and *George H. W. Bush (1924–)* [363] makes a strong case for him as the most influential recent president.
- *John F. Kennedy (1917–1963)* [71] is the most significant president who never completed a full term in office. It has been 50 years since his assassination: that he still outranks several of his successors suggests that this reputation will endure. Ngrams analysis of the previously martyred president (*William McKinley (1843–1901)* [176]) shows that interest in him decayed more quickly and substantially than has proven the case with Kennedy.
- The relatively modest significance of *Dwight D. Eisenhower (1890–1969)* [110] remains somewhat of a puzzle. Like *Ulysses S. Grant (1822–1885)* [28], he was an important general and served two complete terms. Unlike Grant, he proved a successful president whose stature seems to be rising over time.

8.1.1 HISTORICAL POLLS

Ranking the greatness of U.S. presidents has long been a cottage industry for historians. Perhaps best known are the historian polls conducted by *Arthur M. Schlesinger Sr. (1888–1965)* [27072] in 1948 and 1962. These have been repeated on a fairly regular basis – including one by his son *Arthur M. Schlesinger Jr. (1917–2007)* [6365] in 1996. These polls are generally highly correlated with each other, except for recent administrations. It takes about 20 years of distance for genuine historical consensus to emerge.

We compare our significance ranking of U.S. presidents to a consensus ranking derived from a collection of 17 historian/journalist polls taken from 1948 to 2011, which are presented as a column in Figure 8.1. There is very good agreement between how we rank the presidents and how the historians do, but the differences are telling. We see *Ulysses S. Grant (1822–1885)* [28] as more significant than his presidential poll rank indicates, because he was a great general but a lousy president. A high significance rating reflects historical renown, be it for good or evil. The Watergate scandal increased

Richard Nixon's (1913–1994) [82] significance, just as surely as it lowered his rating in historians' eyes. Similarly, *Andrew Johnson (1808–1875)* [105] and *James Buchanan (1791–1868)* [237] were weak presidents flanking the Civil War, and hence assume importance more from their times than their abilities.

8.1.2 TIME TRENDS IN THE PRESIDENCY

The times help make the man, or at least his reputation. A president becomes great when he serves effectively in times of crisis. That September 11, Hurricane Katrina, and the financial crisis all occurred on *George W. Bush's (1946–)* [36] watch suggests that his term in office will receive greater historical attention than *Bill Clinton's (1946–)* [115], independent of their individual merits as leaders.

Figure 8.2, which plots the historical significance of presidents by the time they served, shows that the times help make the man. Several trends are evident. Most notable is the steep, almost uninterrupted decline in the power of the presidency from the ratification of the Constitution until the Civil War. This is partially an artifact: the first four presidents played major roles in the founding of the republic, and much of their reputation accrued prior to their presidencies. But it also reflects the weakness of the office at that point in history, relative to other branches of government. It is telling that the only two congressional leaders whose significance approaches that

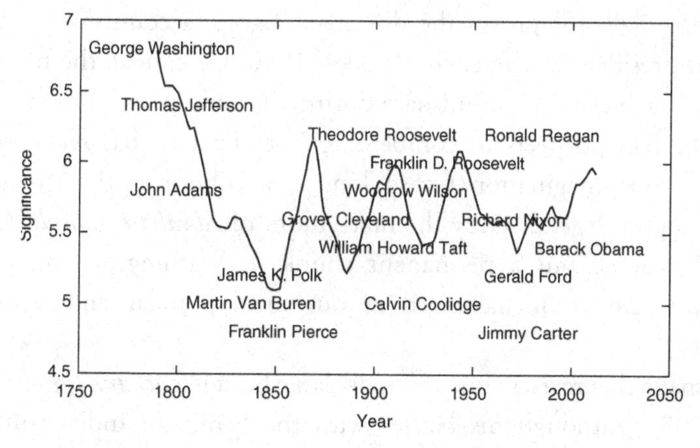

FIGURE 8.2. Significance of U.S. presidents as a function of time of office.

of presidents (*Henry Clay (1777–1852)* [252] and *Daniel Webster (1782–1852)* [453]) served during this period.

Peaks in the significance of the presidency rise during periods of national crisis: the Civil War, the Progressive Era through World War I, and the Depression through World War II. The canonical leaders here were *Abraham Lincoln (1809–1865)* [5], *Theodore Roosevelt (1858–1919)* [23], and *Franklin D. Roosevelt (1882–1945)* [43], respectively.

Finally, we see a steady rise in the significance of the presidency over the past 50 years. We posit that this may be a reflection of Wikipedia-era biases more than a fundamental surge in the clout of recent presidents.

8.1.3 PRESIDENTIAL REFERENCES

Google Ngrams enables us to monitor how much each president is discussed in the books of their day, and beyond. Presidents are generally prominent people prior to their election, establishing some preliminary presence in the literature. We would therefore expect interest to soar during their term and the immediate aftermath, before decaying according to our previously established model. Thus, a rugplot of all the presidents should lie like a nest of snakes, the heads resting around their term of office.

But at the risk of carrying this metaphor too far, some presidents are bigger snakes than others. Figure 8.3 presents presidential reference trends over the nineteenth century, from 1800 to 1900. *George Washington (1732–1799)* [6] proves the dominant figure, accounting for almost half of all presidential references through 1840. Of course, the first president *should* achieve such mindshare during the early years of the republic, because he had no peers to compete against. That he has retained such a hold on the imagination for so long is a tribute to the strength of the Washington legend. Only the martyrdom of *Abraham Lincoln (1809–1865)* [5] carved out a permanent chunk of Washington's mindshare. These two remain unchallenged as our most popular and significant presidents.

The snake that rears an unexpectedly large head is *Zachary Taylor (1784–1850)* [299]. Although his battles with the Seminole Indians in 1837 attracted attention, he became a popular hero as a military leader during the

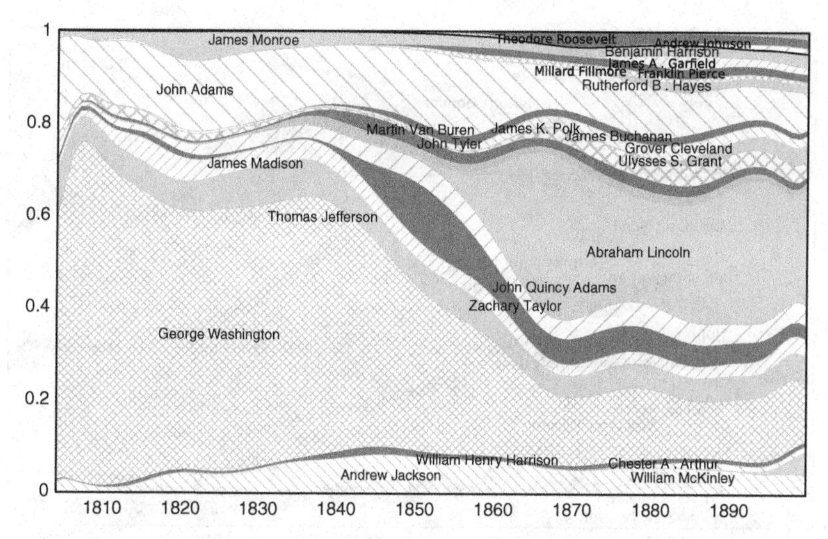

FIGURE 8.3. Reference trends for nineteenth-century U.S. presidents.

Mexican War. The Ngrams data help capture the magnitude and velocity of his rise. He was still serving as commanding officer as late as November 1847, just one year before he was elected president. But his term was brief; he died just 16 months into office.

Figure 8.4 documents the presidential reference trends of the twentieth century. We see evidence of a faster and stronger news cycle, with even minor presidents holding a visible part of the total mindshare during their term of office.

Four presidents loom disproportionately large. *Woodrow Wilson (1856–1924)* [47] was the first president whose Ngram frequency challenged Lincoln, a reflection that World War I was the dominant historical event following the Civil War. *Dwight D. Eisenhower's (1890–1969)* [110] Ngram presence has always been greater than his glamorous successor, *John F. Kennedy (1917–1963)* [71]: a tribute to his Allied leadership in World War II and the length of his administration. The Republican yin-yang of *Richard Nixon (1913–1994)* [82] and *Ronald Reagan (1911–2004)* [32] hang together like a bolus over the last half of the century.

8.1.4 FIRST LADIES AND SECOND MEN

Every president of the United States to date has been intimately bound with one man and one woman: his vice president and first lady respectively.

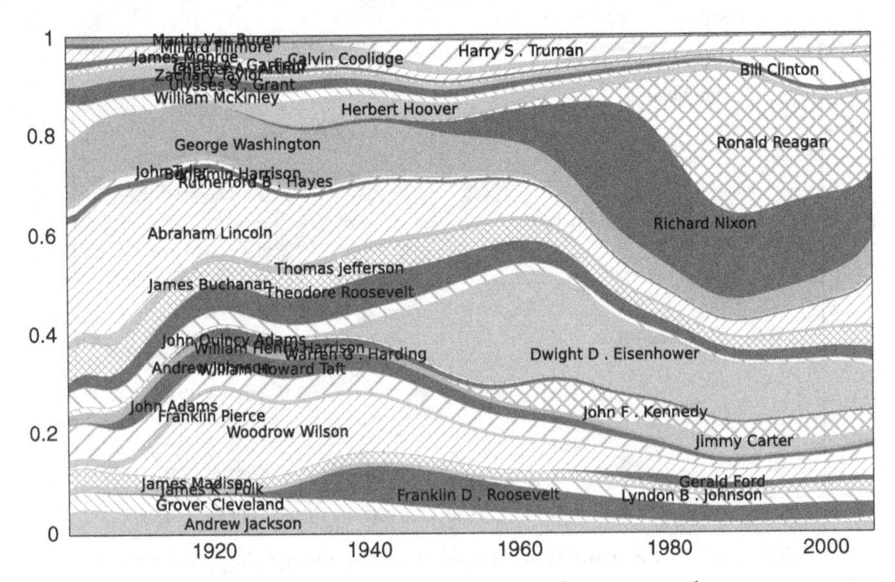

FIGURE 8.4. Reference trends for U.S. presidents, twentieth century.

Veeps: The Vice Presidents

The most prominent vice presidents were those who went on to become president. We ignore these in the following discussion. Many of the most significant remaining veeps were political heavyweights with controversial reputations. *Aaron Burr (1756–1836)* [373] is notorious for his attempt to steal the presidency in 1800, and for the duel in which he killed *Alexander Hamilton (1755–1804)* [45]. *John C. Calhoun (1782–1850)* [399] was the voice of states' rights and nullification, particularly with respect to slavery. Others achieved their greatest prominence in their failed quests for the top spot, including *Al Gore (1948–)* [623], *John C. Breckinridge (1821–1875)* [2071], and *Hubert Humphrey (1911–1978)* [2181].

The official duties of the vice president are quite minimal: breaking ties in the Senate and being ready in case the president dies. FDR's first veep, *John Nance Garner (1868–1967)* [4489], proclaimed the office as "not worth a bucket of warm spit." But the three most recent vice presidents all appear among our top 10, testimony to how the power of the office has grown in recent administrations.

Several of our least significant vice presidents (on right) have left their mark on the American landscape. Fairbanks, Alaska was named for *Charles*

Most Significant VPs

Sig	Person	Dates
373	Aaron Burr	(1756–1836)
399	John C. Calhoun	(1782–1850)
623	Al Gore	(1948–)
1059	Dick Cheney	(1941–)
1460	George Clinton	(1739–1812)
1979	Nelson Rockefeller	(1908–1979)
2071	J. C. Breckinridge	(1821–1875)
2181	Hubert Humphrey	(1911–1978)
2243	Elbridge Gerry	(1744–1814)
2303	Richard M. Johnson	(1780–1850)

Least Significant VPs

Sig	Person	Dates
13421	William A. Wheeler	(1819–1887)
11864	James S. Sherman	(1855–1912)
8860	Garret Hobart	(1844–1899)
8054	C. W. Fairbanks	(1852–1918)
6536	George M. Dallas	(1792–1864)
6511	Levi P. Morton	(1824–1920)
5963	Charles G. Dawes	(1865–1951)
5922	Thomas A. Hendricks	(1819–1885)
5690	Alben W. Barkley	(1877–1956)
5558	Henry Wilson	(1812–1875)

W. Fairbanks (1852–1918) [8054]. The source of the name for Dallas, Texas is obscure, but *George M. Dallas (1792–1864)* [6536] is the most prominent candidate.

The First Ladies

Several of our most significant first ladies have achieved substantial accomplishments independent of their spouses. *Eleanor Roosevelt (1884–1962)* [517] helped found the United Nations, while *Hillary Rodham Clinton (1947–)* [575] served her country as senator and secretary of state. *Dolley Madison (1768–1849)* [1700] rescued important government papers from a war-ravaged White House.

Most Significant First Ladies

Sig	Person	Dates
517	Eleanor Roosevelt	(1884–1962)
575	Hillary R. Clinton	(1947–)
821	Abigail Adams	(1744–1818)
1065	Martha Washington	(1731–1802)
1188	Mary Todd Lincoln	(1818–1882)
1387	J. K. Onassis	(1929–1994)
1700	Dolley Madison	(1768–1849)
2081	Nancy Reagan	(1921–)
3428	Laura Bush	(1946–)
4125	Barbara Bush	(1925–)

Least Significant First Ladies

Sig	Person	Dates
70913	Hannah Van Buren	(1783–1819)
44045	Ellen L. H. Arthur	(1837–1880)
38152	Margaret Taylor	(1788–1852)
31494	Anna Harrison	(1775–1864)
30647	Abigail Fillmore	(1798–1853)
30246	Helen Herron Taft	(1861–1943)
29812	Jane Pierce	(1806–1863)
28025	Grace Coolidge	(1879–1957)
25625	Sarah C. Polk	(1803–1891)
24539	Ida Saxton McKinley	(1847–1907)

The two least significant first ladies, *Ellen Lewis Herndon Arthur (1837–1880)* [44045] and *Hannah Van Buren (1783–1819)* [70913], died before their husbands became president – 18 years before in the case of Van Buren.

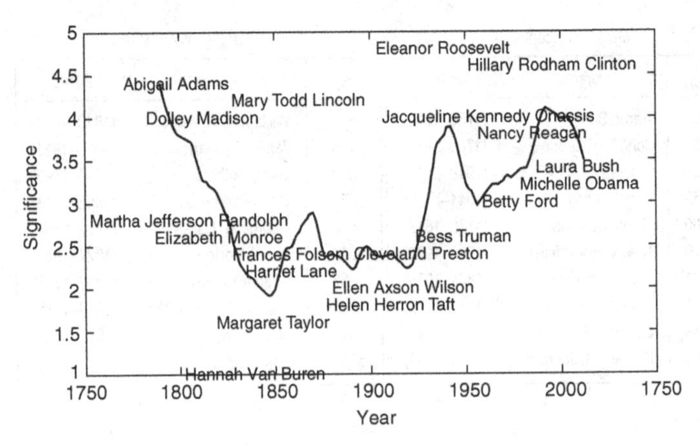

FIGURE 8.5. Time trends in significance of U.S. first ladies.

Vice presidents and first ladies both live in the shadow of their president. How much presidential glory reflects back to them? We correlated the significance of presidents with their first lady and vice president. The first ladies show a very high correlation of 0.64, because relatively few have gone on to notable achievement following their husband's term of office. The vice presidents also show a positive correlation (0.35) with the power of their mentors. Those that succeed to the presidency themselves generally become as prominent as the president they served under. Indeed, the relationship becomes adversarial: the sooner a president dies, the longer term his successor has to achieve results.

Figure 8.5 presents the time course of first lady significance over time. As previously established, this correlates strongly with presidential significance. But overlaid on that has been a strong trend toward increased significance as the position of first lady evolved with universal suffrage and the women's rights movement. First ladies began to establish independent careers and public causes to exploit their unique bully pulpit, following the model of Eleanor Roosevelt.

8.2 Congressmen

Congress is comprised of the House of Representatives and the Senate. Being senator is considered to be a better job than being in the House:

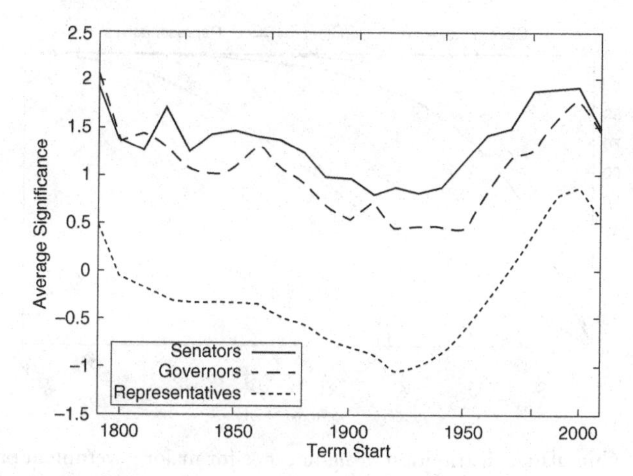

FIGURE 8.6. Significance of all major government offices since 1789. Senators are historically slightly more significant than governors. Representatives have substantially increased their rankings since World War II.

you have more job security (6 years vs. 2 years), represent more people (unless you live in one of seven states with only a single representative), and belong to the more exclusive group (100 senators vs. 435 representatives). Moreover, senators were originally appointed by state legislatures rather than popularly elected (until popular election of senators was standardized by the Seventeenth Amendment), reinforcing the notion that the Senate was a more elite club.

Figure 8.6 plots the average significance of senators, representatives, and governors as a function of the beginning of their term. Legislators generally achieve greater historical significance in the Senate than the House, though that trend has narrowed in recent years. Garden-variety senators typically rank between 20,000 and 60,000 in significance, while congressmen are scattered almost uniformly among the residents of Wikipedia; see Figure 8.7. In this section, we will identify the most significant members of the legislative branch, both by body and location.

8.2.1 SENATORS

The Senate, it is said, is full of people who think they should be president, and indeed 13 U.S. senators have gone on to hold our highest office. Even excluding these presidents, however, very few of the most significant

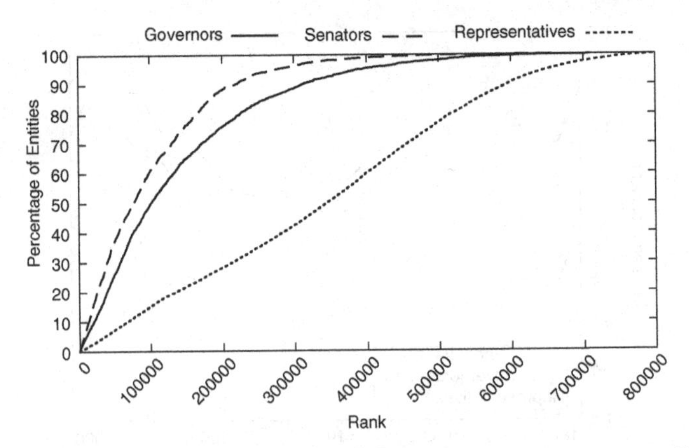

FIGURE 8.7. Cumulative distribution of significance for major government offices. Governors are a somewhat more significant cohort than senators, while representatives appear uniformly scattered among the Wikipedia population.

senators received their greatest renown as legislators. Instead they were presidential candidates, vice presidents, and even presidents of other republics (Texas and the Confederacy).

We reviewed the most significant senators in our discussion of Bonnie's textbook, back in Section 3.3.2, and limit discussion to a few others here. For almost 50 years, *Ted Kennedy (1932–2009)* [1229] was the great liberal champion of the Senate. His long tenure there permitted him to make substantial contributions to American life, enough so that he nosed out his martyred brother *Robert F. Kennedy (1925–1968)* [1355] in terms of historical significance. Anti-slavery leader *Charles Sumner (1811–1874)* [1094] was beaten nearly to death on the floor of the Senate, requiring three years convalescence before returning to his seat.

Figure 8.8 tabulates the most significant senator from each state in the union. Despite the prestige of the office, we count only twenty or so people who are most significant for their actual work in the Senate. The name of Arkansas' *J. William Fulbright (1905–1995)* [9615] lives on through the almost 300,000 people who have participated in the Fulbright scholars program. Utah senator *Reed Smoot (1862–1941)* [11013] sponsored the infamous Smoot-Hawley Tariff, widely credited with prolonging the Great Depression. Important Senate majority leaders include Nevada's *Harry Reid (1939–)* [5338] and West Virginia's *Robert Byrd (1917–2010)*

State	Person	Term	Sig	Founded	Clout	Pop. (K)
Alabama	Hugo Black	(1927–1937)	2465	1819	1.3	4802
Alaska	Ted Stevens	(1968–2009)	6740	1959	1.9	722
Arizona	John McCain	(1987–)	803	1912	2.5	6482
Arkansas	J. William Fulbright	(1945–1974)	9615	1836	1.3	2937
California	Richard Nixon	(1950–1953)	82	1850	2.1	37691
Colorado	Gary Hart	(1975–1983)	11052	1876	1.6	5116
Connecticut	Roger Sherman	(1791–1793)	1029	1788	1.7	3580
Delaware	Joe Biden	(1973–2009)	2384	1787	1.4	907
Florida	Stephen Mallory	(1851–1861)	11859	1845	1.0	19057
Georgia	William H. Crawford	(1807–1813)	3595	1788	1.8	9815
Hawaii	Daniel Inouye	(1963–2012)	11254	1959	2.1	1374
Idaho	William Borah	(1907–1940)	7729	1890	1.4	1584
Illinois	Barack Obama	(2005–2008)	111	1818	1.9	12869
Indiana	Benjamin Harrison	(1881–1887)	339	1816	1.6	6516
Iowa	Tom Harkin	(1985–)	9730	1846	1.2	3062
Kansas	Bob Dole	(1969–1996)	2842	1861	1.0	2871
Kentucky	Henry Clay	(1810–1811)	252	1792	1.4	4369
Louisiana	Huey Long	(1932–1935)	2697	1812	1.2	4574
Maine	James G. Blaine	(1876–1881)	1828	1820	1.5	1328
Maryland	Charles Carroll of Carrollton	(1789–1792)	2437	1788	1.3	5828
Michigan	Lewis Cass	(1849–1857)	2249	1837	1.7	9876
Minnesota	Hubert Humphrey	(1971–1978)	2181	1858	1.9	5344
Mississippi	Jefferson Davis	(1847–1851)	188	1817	1.5	2978
Missouri	Harry S. Truman	(1935–1945)	94	1821	1.8	6010
Montana	William A. Clark	(1901–1907)	9327	1889	1.1	998
Nebraska	George W. Norris	(1913–1937)	8906	1867	0.7	1842
Nevada	Harry Reid	(1987–)	5338	1864	1.4	2723
New Hampshire	Franklin Pierce	(1837–1842)	427	1788	1.2	1318
New Mexico	Albert B. Fall	(1912–1921)	9718	1912	1.3	2082
North Carolina	Jesse Helms	(1973–2003)	3863	1789	1.4	9656
North Dakota	Kent Conrad	(1992–)	16379	1889	1.2	683
Ohio	Warren G. Harding	(1915–1921)	242	1803	2.0	11544
Oklahoma	Robert Latham Owen	(1907–1925)	6363	1907	1.8	3791
Oregon	Mark Hatfield	(1967–1997)	12824	1859	1.4	3871
Pennsylvania	James Buchanan	(1834–1845)	237	1787	1.8	12742
Rhode Island	Ambrose Burnside	(1875–1881)	1335	1790	1.1	1051
South Carolina	John C. Calhoun	(1831–1843)	399	1788	1.6	4679
South Dakota	George McGovern	(1963–1981)	2958	1889	1.3	824
Tennessee	Andrew Jackson	(1823–1825)	66	1796	2.3	6403
Texas	Lyndon B. Johnson	(1949–1961)	184	1845	1.8	25674
Utah	Reed Smoot	(1903–1933)	11013	1896	1.7	2817
Vermont	Justin Smith Morrill	(1867–1898)	8352	1791	1.2	626
Virginia	James Monroe	(1790–1794)	220	1788	2.4	8096
Washington	Henry M. Jackson	(1953–1983)	11153	1889	0.9	6830
West Virginia	Robert Byrd	(1959–2010)	1710	1863	1.0	1855
Wisconsin	Joseph McCarthy	(1947–1957)	553	1848	1.8	5711
Wyoming	Francis E. Warren	(1890–1893)	31510	1890	0.8	568

FIGURE 8.8. Most significant senators by state.

[1710]. Many achieved greater renown when they pursued some form of higher office, such as majority leader *Lyndon B. Johnson (1908–1973)* [184] and presidential candidate *George McGovern (1922–)* [2958].

Does the State Make the Senator?

Each state in the Union sends exactly two senators to Washington. In principle, all senators have the same powers, independent of where they come from. Our significance ranks provide a way to test such hypotheses. Figure 8.9 presents our results from analyzing the significance and term length of all senators, representatives, and governors in American history,

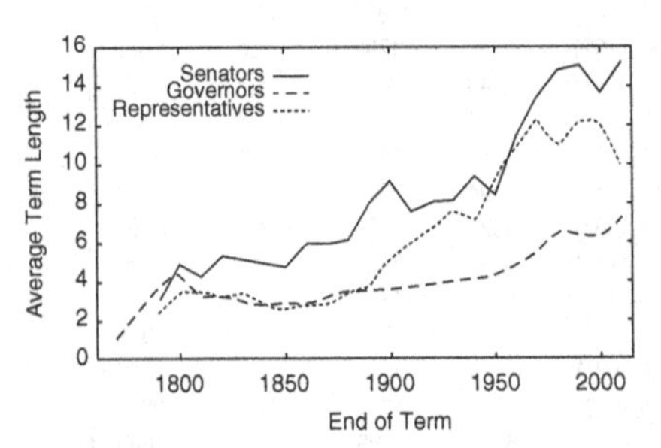

(a) Terms have been getting longer, more dramatically among congressmen than governors.

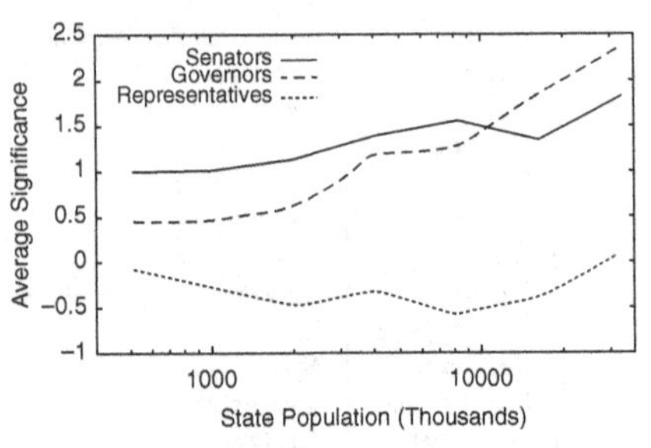

(b) Bigger states tend to produce slightly more significant senators but much more significant governors.

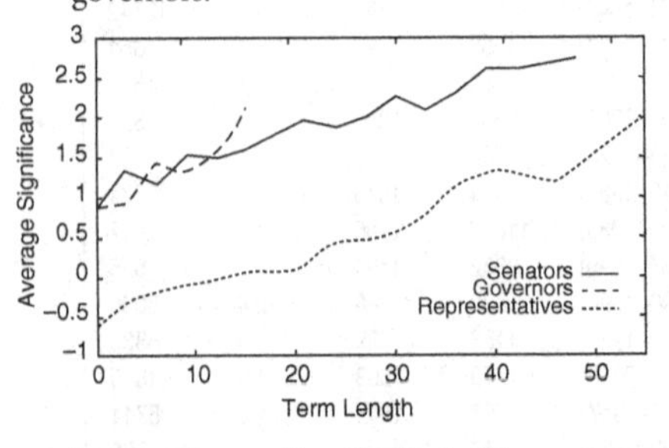

(c) Significance for all officeholders generally increases with length of tenure, although the term length of governors pales compared to that of congressmen.

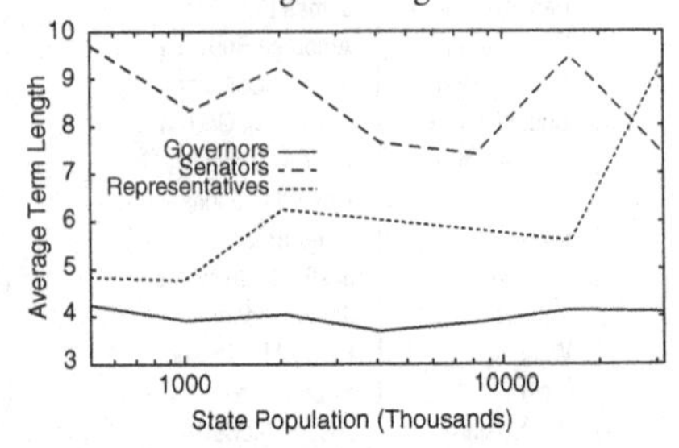

(d) The longest-serving senators tend to come from smaller states, the opposite of what is true for representatives.

FIGURE 8.9. Relationships among governmental office (senator, governor, representative), significance, length of tenure, and size of state.

as a function of state population and historical era. We uncover a variety of interesting trends in power, politics, and office.

Longer tenure in office should create more significant senators, as they accrue the advantages of seniority and have more time to accomplish things. Senate terms have indeed been getting longer with time. Figure 8.9a plots the term lengths for the cohort of senators retiring in each decade, with the average term growing progressively from only four years in 1800 to almost fifteen years today. Term length has increased sharply since 1950, with seven of the eleven longest tenured senators in U.S. history ending their career after 2001.

Increasing term length indeed results in greater historical significance. Figure 8.9b charts the almost-linear increase in historical magnitude as a function of term length. The Senate seniority system provides more powerful committee assignments to longer-served senators, another advantage that turns longer terms into greater historical significance.

Figure 8.9c plots the significance of a state's senators as a function of its population. There is an extremely weak dependence here, granting the 576,412 residents of Wyoming (2010 U.S. Census) almost as powerful senators as the 37.2 million people in California.

Term lengths favor senators from smaller states. Figure 8.9d shows that the longest-tenured senators tend to come from smaller states, where there is less competition. This trend proves counter to that of the House: representatives from populous states with larger delegations generally serve longer House terms. Here, the greater competition for higher state offices dooms more Big State congressmen to terminal House careers.

8.2.2 REPRESENTATIVES

Being a congressman was once a reputable activity. Five members of the House have gone on to become president, and many others have moved on to join the Senate. Other famous people have achieved their highest elected office in the House, including three-time presidential candidate *William Jennings Bryan (1860–1925)* [698], educator *Horace Mann (1796–1859)* [1193], and revolutionary war general *Anthony Wayne (1745–1796)* [1558].

But the most significant congressmen generally served as Speaker of the House. The most famous speaker was *Henry Clay (1777–1852)* [252], although President *James K. Polk (1795–1849)* [240] used it as a stepping stone to the White House. The most powerful speaker was *Joseph Gurney Cannon (1836–1926)* [3515]: the revolt against him in 1911 stripped the office of much of its power to appoint committee heads. Among recent speakers, we rank Democrat *Nancy Pelosi (1940–)* [2569] ahead of Republicans *Newt Gingrich (1943–)* [5753] and *John Boehner (1949–)* [8436]. Other prominent speakers include *Tip O'Neill (1912–1994)* [10964] and *Sam Rayburn (1882–1961)* [6559].

Figure 8.10 presents the most significant members of the House of Representatives by state. Many of these names should look familiar: fully 19 of them also appeared in Figure 8.8 as the most prominent senator from their state. This shows the important role the House plays as a stepping-stone to higher office.

A few representatives have interesting stories. *Shirley Chisholm (1924–2005)* [10180] became the first black woman to make a serious run for the presidency in 1972. New Jersey's *Elias Boudinot (1740–1821)* [8630] has a tenuous claim on being the first president of the United States: he served as president under the Articles of Confederation when the Treaty of Paris ended the American Revolution. As a representative, *Justin Smith Morrill (1810–1898)* [8352] sponsored the Morrill Act, the legislation that established federal funding for higher education through land grant colleges.

The relationships among term length, population, and significance developed in Figure 8.9 generally hold for representatives as they did for senators. Average term length grew sharply after the Civil War, once seniority became the decisive factor in selecting the chairmen of congressional committees [Polsby et al., 1969]. By 1950, the average House tenure exceeded that of the Senate, even though individual House terms last only two years versus six for the senior chamber.

The number of representatives has essentially been fixed at 435 since 1910, despite the increasing U.S. population. Since then the population per district has exploded from about 200,000 people to well over 700,000 today. This has increased the power and prestige of the office, also contributing to longer tenures.

State	Person	Term	Sig	Founded	Clout	Pop. (K)
Alabama	Joseph Wheeler	(1881–1900)	6004	1819	-0.5	4802
Alaska	Don Young	(1973–)	35123	1959	0.4	722
Arizona	John McCain	(1987–)	803	1912	0.6	6482
Arkansas	J. William Fulbright	(1945–1974)	9615	1836	-0.4	2937
California	Richard Nixon	(1950–1953)	82	1850	0.1	37691
Colorado	Ben Nighthorse Campbell	(1993–2005)	22686	1876	-0.4	5116
Connecticut	Roger Sherman	(1791–1793)	1029	1788	-0.5	3580
Delaware	Louis McLane	(1827–1829)	10042	1787	0.3	907
Florida	Stephen Mallory	(1851–1861)	11859	1845	0.0	19057
Georgia	Anthony Wayne	(1791–1792)	1558	1788	-0.0	9815
Hawaii	Daniel Inouye	(1963–2012)	11254	1959	1.2	1374
Idaho	Larry Craig	(1991–2009)	22491	1890	0.0	1584
Illinois	Abraham Lincoln	(1847–1848)	5	1818	-0.6	12869
Indiana	James Wilson	(1857–1860)	1330	1816	-0.6	6516
Iowa	Tom Harkin	(1985–)	9730	1846	-0.3	3062
Kansas	Bob Dole	(1969–1996)	2842	1861	-0.8	2871
Kentucky	Henry Clay	(1810–1811)	252	1792	-0.5	4369
Louisiana	Bobby Jindal	(2005–2008)	7953	1812	-0.0	4574
Maine	Stephen Foster	(1857–1860)	1576	1820	-0.2	1328
Maryland	Daniel Carroll	(1789–1790)	8640	1788	-0.0	5828
Massachusetts	John F. Kennedy	(1947–1952)	71	1788	0.2	6587
Michigan	Gerald Ford	(1949–1974)	230	1837	-0.4	9876
Minnesota	Eugene McCarthy	(1959–1971)	6342	1858	-0.0	5344
Mississippi	Jefferson Davis	(1847–1851)	188	1817	-0.5	2978
Missouri	Sterling Price	(1845–1846)	3426	1821	-0.8	6010
Montana	Jeannette Rankin	(1917–1942)	7826	1889	-0.1	998
Nebraska	William Jennings Bryan	(1891–1894)	698	1867	-0.2	1842
Nevada	Harry Reid	(1987–)	5338	1864	-0.0	2723
New Hampshire	Franklin Pierce	(1837–1842)	427	1788	-0.4	1318
New Jersey	Elias Boudinot	(1789–1794)	8630	1787	-0.4	8821
New Mexico	Bill Richardson	(1983–1998)	7463	1912	0.0	2082
New York	Millard Fillmore	(1833–1842)	446	1788	-0.4	19465
North Carolina	William R. King	(1848–1852)	4295	1789	-0.5	9656
North Dakota	Byron Dorgan	(1992–2001)	25947	1889	0.0	683
Ohio	James A. Garfield	(1863–1880)	285	1803	-0.7	11544
Oklahoma	Tom Coburn	(2005–)	16020	1907	-0.5	3791
Oregon	Ron Wyden	(1996–)	20580	1859	-0.0	3871
Pennsylvania	James Buchanan	(1834–1845)	237	1787	-0.6	8821
Rhode Island	Nelson W. Aldrich	(1881–1911)	4662	1790	-0.5	1051
South Carolina	John C. Calhoun	(1831–1843)	399	1788	-0.4	4679
South Dakota	George McGovern	(1963–1981)	2958	1889	-0.0	824
Tennessee	Andrew Jackson	(1823–1825)	66	1796	-0.3	6403
Texas	Lyndon B. Johnson	(1949–1961)	184	1845	-0.2	25674
Utah	George Sutherland	(1905–1917)	15644	1896	-0.5	2817
Vermont	Justin Smith Morrill	(1867–1898)	8352	1791	-0.3	626
Virginia	James Madison	(1789–1796)	51	1788	-0.2	8096
Washington	Henry M. Jackson	(1953–1983)	11153	1889	-0.2	6830
West Virginia	Robert Byrd	(1959–2010)	1710	1863	-0.2	1855
Wisconsin	Robert M. La Follette	(1925–1947)	1431	1848	-0.3	5711
Wyoming	Dick Cheney	(1979–1990)	1059	1890	-0.0	568

FIGURE 8.10. Most significant representatives by state.

8.3 Governors

Governors are the executive leaders of individual states. Being governor is like serving as the president of a small country, making it excellent preparation for higher office. Indeed, a typical modern governor ranks between 30,000 and 100,000 in historical significance, putting them in the same league as the leader of small European nations.

Figure 8.11 presents the most significant governor by state. In general, the most significant governors earned their greatest fame elsewhere: as a revolutionary patriot (*Patrick Henry (1736–1799)* [472]), military general (*George B. McClellan (1826–1885)* [554]), Supreme Court justice (*John Jay (1745–1829)* [411]), or movie star (*Arnold Schwarzenegger (1947–)* [959]). My (Charles) home state of Alabama is ignominiously represented by *George Wallace* [2620], most famous for his "Stand in the schoolhouse door" against desegregation.

It is surprising that few states had their greatest governors double as senators. The distinction here is between the executive branch of government and the legislative branch; ambitious figures would generally seek national office instead of a lateral shift between state offices. The only name to appear in both tables is Gen. *Ambrose Burnside (1824–1881)* [1335] of Rhode Island, and he is more famous for his facial hair than either his government or military service. *Strom Thurmond (1902–2003)* [1856] served as both governor and senator of South Carolina, but was outranked as senator by *John C. Calhoun (1782–1850)* [399].

A particular curiosity here is *Joshua Chamberlain (1828–1914)* [2234], a Bowdoin College professor and four-time governor of Maine who was the last Civil War soldier to die of his wounds, almost 50 years after suffering them.

Governors by Power

The longest gubernatorial term in U.S. history is only 16 consecutive years, less than half that of the longest senatorial term. Why is there such a difference between these offices? To some extent, it reflects the difference between the legislative and executive branches of government. The buck stops at the governor, making him responsible for controversial decisions affecting local

State	Person	Term	Sig	Founded	Clout	Pop. (K)
Alabama	George Wallace	(1983–1987)	2620	1819	1.1	4802
Alaska	Sarah Palin	(2006–2009)	773	1959	1.2	722
Arizona	Janet Napolitano	(2003–2009)	13463	1912	0.6	6482
Arkansas	Bill Clinton	(1983–1992)	115	1836	1.1	2937
California	Ronald Reagan	(1967–1975)	32	1850	2.3	37691
Colorado	Bill Ritter	(2007–2011)	42205	1876	0.5	5116
Connecticut	Samuel Huntington	(1786–1796)	7111	1788	1.0	3580
Delaware	George Read	(1777–1778)	3804	1787	1.5	907
Florida	Jeb Bush	(1999–2007)	6823	1845	1.2	19057
Georgia	Jimmy Carter	(1971–1975)	462	1788	1.1	9815
Hawaii	Linda Lingle	(2002–2010)	26529	1959	1.4	1374
Idaho	Frank Steunenberg	(1897–1901)	23649	1890	0.8	1584
Illinois	Adlai Stevenson	(1875–1880)	2641	1818	1.4	12869
Indiana	Thomas R. Marshall	(1909–1913)	3206	1816	1.4	6516
Iowa	Samuel J. Kirkwood	(1866–1867)	15959	1846	0.4	3062
Kansas	Alf Landon	(1933–1937)	8012	1861	0.5	2871
Kentucky	John J. Crittenden	(1842–1848)	4407	1792	1.6	4369
Louisiana	Huey Long	(1932–1935)	2697	1812	1.2	4574
Maine	Joshua Chamberlain	(1867–1871)	2234	1820	0.5	1328
Maryland	Spiro Agnew	(1967–1969)	3271	1788	1.3	5828
Massachusetts	Calvin Coolidge	(1919–1921)	370	1788	2.2	6587
Michigan	Frank Murphy	(1937–1939)	4551	1837	0.8	9876
Minnesota	Jesse Ventura	(1999–2003)	6951	1858	1.5	5344
Mississippi	Adelbert Ames	(1870–1874)	6187	1817	0.7	2978
Missouri	Sterling Price	(1845–1846)	3426	1821	0.9	6010
Montana	Brian Schweitzer	(2005–)	42108	1889	-0.1	998
Nebraska	Ben Nelson	(2001–)	27828	1867	0.4	1842
Nevada	Paul Laxalt	(1974–1987)	39430	1864	0.6	2723
New Hampshire	John Sullivan	(1789–1790)	3871	1788	1.0	1318
New Jersey	Woodrow Wilson	(1911–1913)	47	1787	1.2	8821
New Mexico	John Burroughs	(1959–1961)	7422	1912	0.6	2082
New York	Theodore Roosevelt	(1899–1900)	23	1788	2.7	19465
North Carolina	Zebulon Baird Vance	(1879–1894)	5970	1789	1.3	9656
North Dakota	William Langer	(1941–1959)	32504	1889	0.7	683
Ohio	William McKinley	(1892–1896)	176	1803	1.3	11544
Oklahoma	William H. Murray	(1913–1916)	30555	1907	1.1	3791
Oregon	Mark Hatfield	(1967–1997)	12824	1859	1.2	3871
Pennsylvania	Gifford Pinchot	(1931–1935)	2994	1787	1.7	12742
Rhode Island	Ambrose Burnside	(1875–1881)	1335	1790	0.8	1051
South Carolina	Strom Thurmond	(1954–1956)	1856	1788	0.6	4679
South Dakota	Joe Foss	(1955–1959)	37406	1889	0.2	824
Tennessee	Andrew Johnson	(1875–1875)	105	1796	1.4	6403
Texas	George W. Bush	(1995–2000)	36	1845	1.9	25674
Utah	Mike Leavitt	(1993–2003)	38906	1896	0.7	2817
Vermont	Howard Dean	(1991–2003)	4510	1791	0.3	626
Virginia	Thomas Jefferson	(1779–1781)	10	1788	2.0	8096
Washington	Christine Gregoire	(2005–)	19586	1889	0.9	6830
West Virginia	Jay Rockefeller	(1985–)	22521	1863	0.9	1855
Wisconsin	Robert M. La Follette	(1925–1947)	1431	1848	1.0	5711
Wyoming	Nellie Tayloe Ross	(1925–1927)	21636	1890	0.3	568

FIGURE 8.11. Most significant governor for each state. In cases of nonconsecutive terms, the most recent is given.

voters. The act of balancing constituencies implies they make more enemies over time, and have usually worn out their welcome by the end of their second term.

But there is also a constitutional reason. The laws regulating gubernatorial elections are fixed by each individual state, generally through state constitutions. Fully 36 of the 50 states subject their governors to term limits, generally to at most two consecutive terms. On the other hand, the Supreme Court, in *U.S. Term Limits, Inc. v. Thornton (1995)*, ruled that states cannot impose stricter qualifications on congressional elections than the U.S. Constitution. Hence, no state can impose limits on the length of congressional terms.

Governors generally rival U.S. senators in terms of historical significance. However, their clout rises sharply as a function of population. Indeed, Figure 8.9b shows that governors of states with about 10 million people[2] become more significant than their state's senators.

8.4 Mayors

Many national politicians got their start in local politics as mayors. *Grover Cleveland (1837–1908)* [98] served as both mayor and sheriff, which earned him the nickname "The Buffalo Hangman." More recently, *Sarah Palin's (1964–)* [773] stint as mayor of Wasilla, Alaska led to her election as governor and on to national prominence. Several famous people have served stints as mayor during interludes in their real careers, including circus promoter *P. T. Barnum (1810–1891)* [767] (Bridgeport, CT) and actor *Clint Eastwood (1930–)* [1254] (Carmel-by-the-Sea, CA).

Five of the top seven mayors ruled New York, which is understandable because the Big Apple is vastly larger than any other American city. The U.S. Census (2011) puts the population of New York at 8,244,910, meaning that New York is bigger than the *sum* of the second- (Los Angeles) third- (Chicago), and fifth-largest (Philadelphia) U.S. cities, by almost the population of Salt Lake City. The city would rank as the twelfth largest state in the country, just ahead of Virginia. And population matters: our experiments show that mayoral significance correlates with city population.

[2] According to the 2010 Census: California, Texas, New York, Florida, Illinois, Pennsylvania, and Ohio.

Mayors of Major U.S. Cities

Sig	Person	Dates	C/G	City
2463	Fiorello La Guardia	(1882–1947)	C G	New York City
2657	Rudy Giuliani	(1944–)	C G	New York City
3021	Michael Bloomberg	(1942–)	C G	New York City
4859	Richard J. Daley	(1902–1976)	C G	Chicago
6581	Rahm Emanuel	(1959–)	C G	Chicago
6963	Ed Koch	(1924–2013)	C G	New York City
7644	John Lindsay	(1921–2000)	C G	New York City
7958	Anton Cermak	(1873–1933)	C G	Chicago
8481	John F. Fitzgerald	(1863–1950)	C G	Boston
9010	Asa Griggs Candler	(1851–1929)	C G	Atlanta
10093	Gavin Newsom	(1967–)	C G	San Francisco
11015	Seth Low	(1850–1916)	C G	New York City
11856	Harold Washington	(1922–1987)	C G	Chicago
11888	Jerry Springer	(1944–)	C G	Cincinnati
12427	Newton D. Baker	(1871–1937)	C G	Cleveland

Thus, the mayor of New York is a national figure as much as a local one, sometimes considered a potential presidential candidate. Two recent mayors, *John Lindsay (1921–2000)* [7644] and *Rudy Giuliani (1944–)* [2657], unsuccessfully sought their party's presidential nomination. Current mayor *Michael Bloomberg (1942–)* [3021] is often mentioned as a possible independent candidate. Yet no mayor of New York has ever been elected president.

Chicago mayors also appear prominently in our rankings. Some have served very lengthy terms: *Richard J. Daley's (1902–1976)* [4859] 21 years on the job were eclipsed only by his son *Richard M. Daley's (1942–)* [17921] 22 years in power. Other Chicago mayors died prematurely in office: the first black mayor *Harold Washington (1922–1987)* [11856] of a heart attack, and *Anton Cermak (1873–1933)* [7958] from a bullet intended for *Franklin D. Roosevelt (1882–1945)* [43]. By comparison, Los Angeles mayors seem to be slighted, with longtime mayors *Tom Bradley (1917–1998)* [15342] and *Sam Yorty (1909–1998)* [46737] ranking as their most significant political leaders. Los Angeles water baron *William Mulholland (1855–1935)* [8962] and newspaper publisher *Harrison Gray Otis (1837–1917)* [10689] score higher in our rankings than any of their city's mayors.

Of those from other cities, Atlanta mayor *Asa Griggs Candler (1851–1929)* [9010] achieved his greatest renown as the founder of the Coca-Cola Company. *Newton D. Baker (1871–1937)* [12427] served as Secretary of

War during World War I. Finally, *Jerry Springer (1944–)* [11888] parlayed his term as mayor of Cincinnati into a notorious career as TV talk show host.

8.5 Judges and Justices

The Supreme Court is the primary symbol of federal jurisprudence. The chief justice is considered the leader of the court, but our rankings place associate justice *Antonin Scalia (1936–)* [2144] as more significant than the current chief *John Roberts (1955–)* [2677].

The primary revelation among the top justices is how many great justices had previous lives in other branches of government before they went on the court. They were governors, senators, and even a former president. This is no longer the case, for the court has become exclusively the province of legal scholars with extensive judicial experience.

Sandra Day O'Connor (1930–) [5221] was the last Supreme Court justice to have ever held elective office. The institution seems considerably poorer without this practical experience in government. It is telling that only one of the nine active justices ranks among the twenty most significant in history.

Supreme Court Justices

Sig	Person	Dates	C/G	Description
153	William Howard Taft	(1857–1930)	C G	27th president ("Dollar Diplomacy")
401	John Marshall	(1755–1835)	C G	U.S. chief justice (Judicial review)
411	John Jay	(1745–1829)	C G	First chief justice of U.S. Supreme Court
885	Oliver Wendell Holmes	(1841–1935)	C G	American jurist (Clear and present danger)
1134	Louis Brandeis	(1856–1941)	C G	Supreme Court justice ("People's Lawyer")
1164	Salmon P. Chase	(1808–1873)	C G	Supreme Court chief justice, governor of Ohio
1253	Roger B. Taney	(1777–1864)	C G	5th chief justice of the U.S. Supreme Court
1330	James Wilson	(1742–1798)	C G	Declaration signer, Supreme Court justice
1707	Thurgood Marshall	(1908–1993)	C G	First African American Supreme Court justice
1760	Charles Evans Hughes	(1862–1948)	C G	American statesman, Supreme Court justice
1908	John Rutledge	(1739–1800)	C G	Supreme Court chief justice, governor of S.C.
2034	Earl Warren	(1891–1974)	C G	Chief justice (*Brown v. Board of Education*)
2144	Antonin Scalia	(1936–)	C G	Conservative Supreme Court justice
2190	Oliver Ellsworth	(1745–1807)	C G	Senator from Connecticut, Supreme Court justice
2465	Hugo Black	(1886–1971)	C G	Senator from Alabama, Supreme Court justice

8.6 Founding Mothers and Fathers

We previously discussed colonial and revolutionary figures in the context of Bonnie's history textbook. Here we will focus on two particular groups associated with the American Revolution: women and the signers of our founding documents.

8.6.1 DAUGHTERS OF THE REVOLUTION

Men dominated the records of historical events of colonial and revolutionary times. Almost exclusively it was men who founded the colonies, ran the plantations, served in Congress, and fought in the Revolution. Society was simply not organized in a way for women to take a visible role in these events.

More modern times, however, ushered in the need for revolutionary heros of the opposite gender. The Google Ngrams data presented in Figure 8.12 chart an exciting tale of shifting reputations.

One significant trend has been the growth of *Abigail Adams (1744–1818)* [821] at the expense of *Martha Washington (1731–1802)* [1065] and *Betsy Ross (1752–1836)* [2430]. Our second first lady proved her intellect and ability in the course of her famous letters to John Adams, and seems a more appropriate role model for modern times than the first first lady. The story of Betsy Ross as seamstress of the American flag dates back only to 1870, on the strength of a claim made by her grandson. She satisfied a need

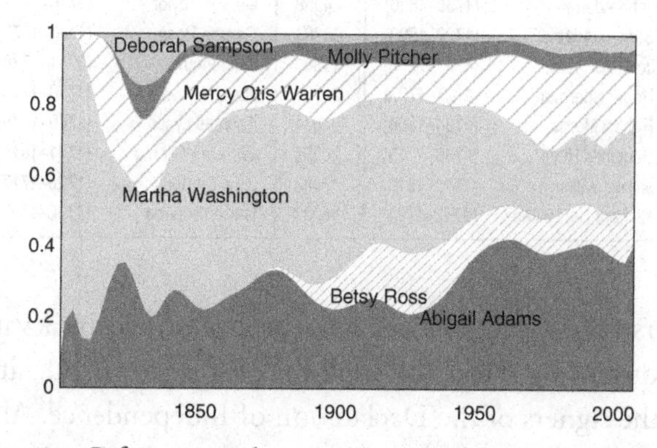

FIGURE 8.12. Reference trends among women of the American Revolution.

for more female Revolutionary War heros, but has somewhat fallen from grace since World War II.

Mercy Otis Warren (1728–1814) [2863] is a generally under-appreciated figure, a political writer who authored one of the first histories of the American Revolution. Her contributions to independence were substantially more important than those of *Deborah Sampson (1760–1827)* [3797] and the semi-mythical *Molly Pitcher (1754–1832)* [2926]: an Amazonian pair of women who fought in Revolutionary War battles. Warren's historical status was recognized in her lifetime and has only grown with the years, to the point that she is now beginning to approach *Martha Washington (1731–1802)* [1065] in scale.

8.6.2 SIGNERS OF THE DECLARATION

The 56 signers of the Declaration of Independence are members of the most exclusive club in American history. They are the real Founding Fathers, present at the creation of our country. Examining the most and least significant among them, we see that the top four rank among the 500 most significant figures in history. However, the fifth, *John Hancock* [616], derives most of his contemporary notoriety from the flourish with which he signed the document.

Biggest Declaration Signers

Sig	Person	Dates
10	Thomas Jefferson	(1743–1826)
35	Benjamin Franklin	(1706–1790)
61	John Adams	(1735–1826)
467	Samuel Adams	(1722–1803)
616	John Hancock	(1737–1793)
1029	Roger Sherman	(1721–1793)
1273	Robert Morris	(1734–1806)
1283	Benjamin Rush	(1746–1813)
1330	James Wilson	(1742–1798)
1362	Richard Henry Lee	(1732–1794)

Smallest Declaration Signers

Sig	Person	Dates
27303	Thomas Heyward	(1747–1809)
26953	James Smith	(1719–1806)
24958	George Taylor	(1716–1781)
23301	George Ross	(1730–1779)
21840	John Penn	(1741–1788)
21521	Joseph Hewes	(1730–1779)
20634	Thomas Lynch	(1749–1779)
20255	William Williams	(1731–1811)
18883	John Morton	(1725–1777)
18237	Carter Braxton	(1736–1797)

The signers of the Declaration have been glorified in a variety of art forms and monuments. Near the Mall in Washington, DC, sits an island memorial to the signers of the Declaration of Independence. Although the location is called Constitution Gardens, only the Declaration signers are individually honored there. *John Trumbull's (1756–1843)* [1697] painting "The Declaration of Independence" sits proudly in the Capitol Rotunda of the United States and is broadly recognized by the public, having appeared on the reverse of the $2 bill. Trumbull's painting is much better (and better known) than the Capitol's painting of the "Signing of the Constitution" by *Howard Chandler Christy (1873–1952)* [50589]. There was even a wonderful Broadway musical, *1776*, devoted to the signers and shenanigans of the Continental Congress. Yet our analysis shows the Constitution signers to be a more historically significant group than the better-known signers of the Declaration.

Biggest Constitution Signers			Smallest Constitution Signers		
Sig	Person	Dates	Sig	Person	Dates
6	George Washington	(1732–1799)	22912	David Brearley	(1745–1790)
35	Benjamin Franklin	(1706–1790)	21489	D. of S. T. Jenifer	(1723–1790)
45	Alexander Hamilton	(1755–1804)	20747	Jacob Broom	(1752–1810)
51	James Madison	(1751–1836)	19144	Richard D. Spaight	(1758–1802)
1029	Roger Sherman	(1721–1793)	16515	Gunning Bedford	(1747–1812)
1273	Robert Morris	(1734–1806)	14504	James McHenry	(1753–1816)
1330	James Wilson	(1742–1798)	13995	Thomas Fitzsimons	(1741–1811)
1564	John Dickinson	(1732–1808)	12863	William Few	(1748–1828)
1901	Rufus King	(1755–1827)	10112	Hugh Williamson	(1735–1819)
1908	John Rutledge	(1739–1800)	9862	Nicholas Gilman	(1755–1814)

The main reason for this is generational. The Constitution was signed on September 17, 1787, more than 11 years after the Declaration. Only six men signed both documents: *George Read* [3804], *Roger Sherman* [1029], *Benjamin Franklin* [35], *Robert Morris* [1273], *George Clymer* [6798], and *James Wilson* [1330]. The signers of the Declaration were the cream of the Colonies, but a new group of younger, battle-hardened leaders emerged after the Revolution. It is these men who would become the first governors and senators of the new republic, bringing greater historical significance to the group.

9

Modern World Leaders

The previous chapter presented the popular history of American political leadership. Here we apply our analytical tools to study the reputations of the rest of the world's leaders. We are particularly interested in understanding the factors that shape our perceptions of other places, such as size, economics, and culture.

Knowing the who's who of a country is helpful to understand what's what. In the same way that speaking a trivial amount of the local language ("excuse me") can pay big dividends, we have found that knowing the name of the current or iconic political leader opens doors to greater understanding. Asking immigrants or locals "What do you think of . . . ?" is an excellent way to start a revealing conversation. We provide tables identifying the dominant political figure from essentially every country on earth. These can be seen as crib notes in cultural literacy; a chance to refresh our memories about the people who built or led nations.

9.1 Analyzing Political Leadership

In the course of our study, we have created a dataset of more than 3,000 modern national leaders from more than 150 different countries. But any meaningful attempt to identify the "most significant" leader of a nation faces several issues of definition and bias. We discuss some of these considerations here.

9.1.1 CULTURAL AND REGIONAL BIAS

Our analysis relies on datasets that inherently display linguistic and cultural biases. Both our Wikipedia analysis and book Ngrams data are based exclusively on English-language sources, and hence do not equally reflect the contributions of other languages and cultures.

This limitation becomes particularly problematic in comparing international leaders. English-speaking countries like Great Britain, Australia, and Canada will be artificially favored over those of the rest of the world. Similarly, our methods slant toward foreign leaders who have interacted with the English-speaking world. Much of the historical perception of *Ho Chi Minh (1890–1969)* [779] is a result of American involvement in the Vietnam War. By comparison, *Sheikh Mujibur Rahman (1920–1975)* [4731] was also the founding father of a nation that achieved its independence in the 1970s. Bangladesh is a country twice as populous as Vietnam, yet its *George Washington* [6] measures as substantially less significant in our data.

9.1.2 HEADS OF GOVERNMENT VS. HEADS OF STATE

In many countries, the burdens of leadership are divided between administrative and ceremonial figures. The *head of government* is the executive leader charged with running the mechanics of government. The *head of state* represents the authority of the nation and its people. The prime minister of England serves as head of government, while the king or queen serves as head of state. In the United States, both roles are invested in the president. However, certain ceremonial duties have traditionally been assigned to the vice president, such as representing the nation at the funerals of foreign dignitaries.

Identifying the canonical leader of a country requires choosing between the heads of government and state. The balance varies between different countries. Although the head of government typically holds the real power, they sometimes serve at the pleasure of the head of state. In Thailand, King *Bhumibol Adulyadej*'s *(1927–)* [3926] prestige was high enough to force the resignation of two military governments and facilitate his nation's transition to democracy.

Hereditary monarchs serve as head of state for many modern nations. They generally serve for life: at this writing the Thai king has reigned for 66 years and Queen *Elizabeth II (1926–)* [132] of England, for 60 years. Such long reigns enable heads of state to accumulate historical significance at odds with the limits of their actual power. Another wild card is that certain heads of state represent more than one nation. In particular, Elizabeth II serves as head of state for each nation in the British Commonwealth.

We have generally selected the most significant member of either leadership role to represent its nation, with the explicit removal of only a few problematic cases, such as the British queen.

9.1.3 PERIOD OF NATIONHOOD

A second concern marks the date when the modern nationhood of a country begins. For example, Egypt's cultural history dates back to the pharaohs, but its modern political history begins only with the fall of the Ottoman Empire. So who should we consider to be the most significant Egyptian ruler, *Tutankhamun (1341–1323 B.C.)* [255] or *Gamal Abdel Nasser (1918–1970)* [1404]?

For us, the decision is clear. We are interested in the modern nation-state, and so have collected lines of succession back from the present day. The length of these lines varies substantially between nations. Our oldest line is Liechtenstein, dating back to 1608, followed by the United Kingdom in 1762. We include our national start date in the figures in this chapter, to help you make better sense of possible earlier omissions.

These start dates reveal the historical periods of most active nation-building. Founded almost 250 years ago, the United States is old relative to most of its peers. The political consolidation of Europe into modern nation-states began in the second half of the nineteenth century.

World War I marked the end of the Austrian-Hungarian, Ottoman, and Russian empires, and created an unstable mix of new nations in its wake. World War II redrew the map again. The end of colonialism in the 1950s and 1960s created dozens of independent countries around the world. Most recently, the fall of the Soviet Union created 15 post-Soviet states in 1991. It also sparked the merger of certain artificially partitioned nations (the two Germanies) and the dissolution of others (Czechoslovakia and Yugoslavia).

9.1.4 LENGTH OF REIGN

Certain leaders rule only for days, some for years, and others for decades. Historical significance is very much a function of term of service: longer lasting rulers are usually more significant because more history happens on their watch.

We hypothesize that the average reign in a particular country is at least partially a function of population. Smaller places should result in longer terms of office. There are several reasons why this should be the case.

First, a smaller populace means that there will be fewer potential challengers to the current leader. Not everyone wants to be king, but the probability that an able, ambitious opponent emerges grows rapidly with size. The power (and hence attractiveness) of a leadership position increases with the number of people you get to serve. Smaller places also tend to be more homogeneous, making it easier to keep everybody happy. Finally, more personal communication makes it easier to keep control of a smaller place.

These are testable conjectures. Figure 9.1 presents our analysis of the relationships between leadership, term length, population size, and historical significance for the nations of the world. This mirrors the analysis we performed in the previous chapter for the United States.

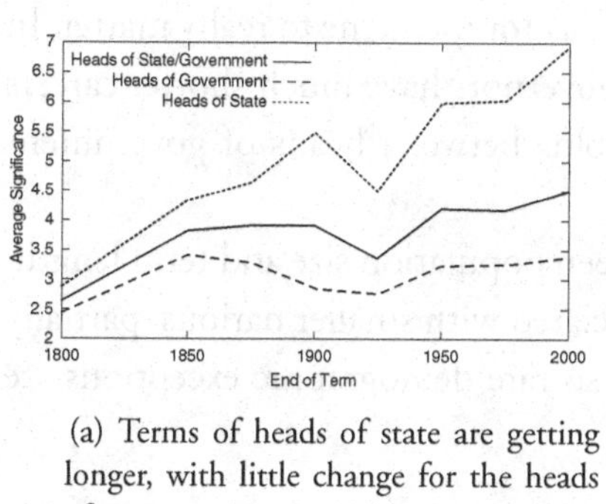

(a) Terms of heads of state are getting longer, with little change for the heads of government.

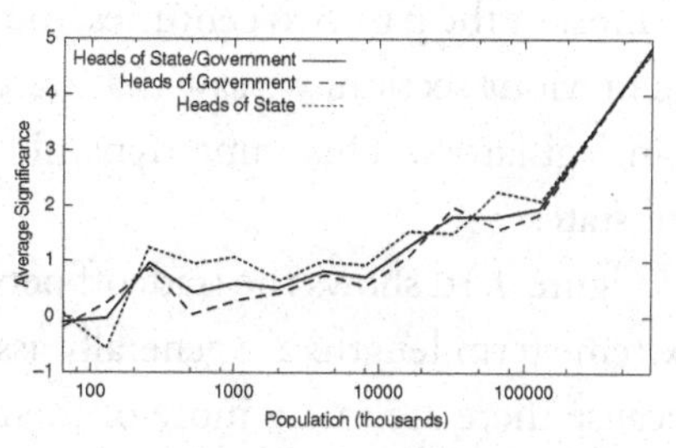

(b) Bigger nations produce much more significant leaders.

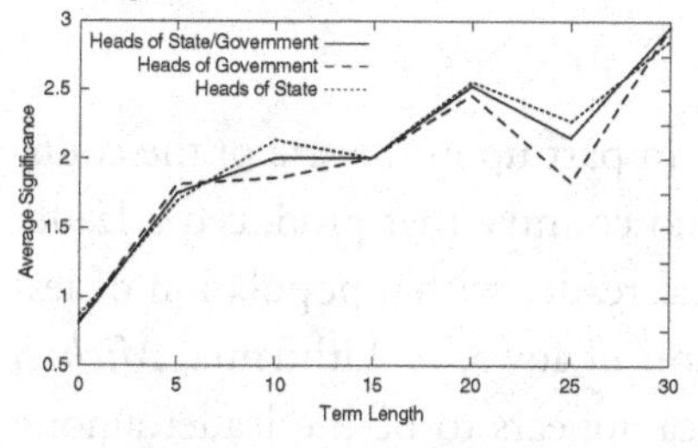

(c) Significance for all leaders increases with length of tenure.

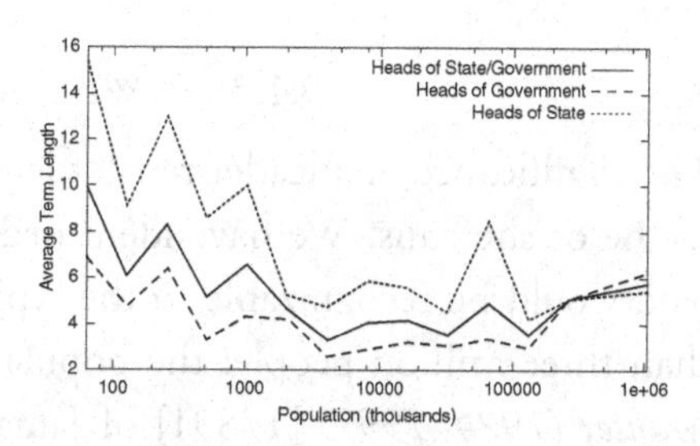

(d) The longest-serving leaders tend to come from smaller nations.

FIGURE 9.1. Relationships between leadership position (head of government vs. head of state), significance, length of tenure, and size of nation.

Three of these plots directly involve term length. The most direct association is between term length and significance, shown in Figure 9.1c. The longer you stay in office, the more significant you will be. The biggest jump seems to occur after the first five years, meaning you must stay in power at least that long for anything to happen.

In Figure 9.1a, we look at changes in term length for heads of state/government over the past two centuries. The longevity of heads of state has increased dramatically during this period, for two reasons: decreasing power and better health. In the shift from kings to prime ministers, the role of the head of state has become increasingly ceremonial. The heads of state now cruise above the fray, so they evoke greater affection. Also, there is little benefit to replacing them. This makes for a more stable position. And of course, these honored leaders have benefited from improvements in health care and technology to at least the same extent as their subjects.

Today's heads of government also have good medical plans, but are much more accountable to a restive public. Our chart shows that their average political life span has held remarkably steady at between three and four years for most of the past two centuries, too short for medicine to really matter. In the previous section, we saw that state governors have much shorter careers than legislators. This same dynamic holds between heads of government and state.

Figure 9.1d shows the tradeoff between population size and term length. Extreme term lengths are generally associated with smaller nations, partially because there are many more of them, so rare demographic exceptions are more likely to occur there.

9.1.5 POWER AND PLACES

The significance of a leader rests at least in part upon the size of the country he or she runs. We have identified no country that produced a leader who would be recognizable to the typical reader with a population of less than three million people: the population of Iowa, or Lithuania. *Michael Manley (1924–1997)* [17891] of Jamaica appears to be the leader among the countries of this size.

Figure 9.2 presents a cartogram of the world, with each country sized according to the relative significance of its leaders. This provides a way to

FIGURE 9.2. Cartogram of countries based on the significance of their leaders in our dataset. Cuba, Israel, and the nations of Western Europe loom large relative to their actual size, at the expense of the republics of the former Soviet Union.

visualize which leaders create their own significance, as opposed to having it predetermined by the geography of the nation they serve. An undistorted map would show the significance of leaders as a simple linear function of each nation's landmass, while a warped map highlights regions generating leaders of unexpected significance.

The cartogram of Figure 9.2 is wildly distorted, sometimes in surprising ways. The basic shapes of North and South America are largely unchanged, but the image struggles to contain a wildly outsized version of Cuba. Cuba's modern political leadership has effectively been monopolized by *Fidel Castro (1926–)* [506], the revolutionary leader who liberated Cuba and then ruled it for 48 years. He parlayed Cuba, a nation of only 11 million people, into worldwide fame and recognition. His longevity, coupled with his unique stature on the world scene, makes Cuba a leading power in leader power. Canada holds up fairly well under this measure, despite having just a California-sized population in the world's largest landmass.

Greater distortions are present overseas. Western Europe swells to the point that it crushes north Africa, while the entirety of Eastern Europe shrivels to a Spain-sided appendage. The low profile of the former Soviet republics so warps Asia that China is reduced to a narrow stick. The relatively low profile of China's leaders is owing to the nation's oligarchic structure: the party is bigger than any particular leader.

The Middle East achieves an outsized significance, due to the lengthy tenure of most Arab leaders as well as the region's tendency to drive world events.

9.2 The Supreme Leaders

We have used our leader dataset to identify the most significant heads of state/government in modern world history. The presidents of the United States have been excluded from consideration here. The result is a diverse set of figures.

The three most significant names in the table led their respective nations (Germany, Russia, Great Britain) during World War II, a testament to how large the war still looms in the collective consciousness. Curiously, all rank ahead of then U.S. president *Franklin D. Roosevelt (1882–1945)* [43]. Other highly ranked European leaders during this war include *Charles de Gaulle (1890–1970)* [396] (France) and *Francisco Franco (1892–1975)* [291] (Spain).

Several figures here are revered (or reviled) as the founders of important nation-states, including *Mao Zedong (1893–1976)* [151] (China), *John A. Macdonald (1815–1891)* [159] (Canada), *Mustafa Kemal Atatürk (1881–1938)* [360] (Turkey), *Chiang Kai-shek (1887–1975)* [440] (Taiwan), and *Fidel Castro (1926–)* [506] (Cuba).

Heads of State/Government

Sig	Person	Dates	C/G	Description
7	Adolf Hitler	(1889–1945)	C G	Fuehrer of Nazi Germany (WWII)
18	Joseph Stalin	(1878–1953)	C G	Premier of USSR (WWII)
37	Winston Churchill	(1874–1965)	C G	Prime minister of Britain (WWII)
75	Vladimir Lenin	(1870–1924)	C G	Soviet revolutionary and premier of USSR
132	Elizabeth II	(1926–)	C G	Present queen of the United Kingdom
151	Mao Zedong	(1893–1976)	C G	Chinese communist revolutionary and dictator
159	John A. Macdonald	(1815–1891)	C G	First prime minister of Canada
174	Arthur Wellesley	(1769–1852)	C G	Duke of Wellington (Battle of Waterloo)
271	Margaret Thatcher	(1925–2013)	C G	U.K. prime minister ("The Iron Lady")
279	William Ewart Gladstone	(1809–1898)	C G	Four-time British prime minister
291	Francisco Franco	(1892–1975)	C G	Spanish Civil War general and dictator
303	Benjamin Disraeli	(1804–1881)	C G	Two-time British prime minister
360	Mustafa Kemal Atatürk	(1881–1938)	C G	Founder and president of modern Turkey
369	David Lloyd George	(1863–1945)	C G	British prime minister (WWI)
396	Charles de Gaulle	(1890–1970)	C G	French general (WWII) and prime minister

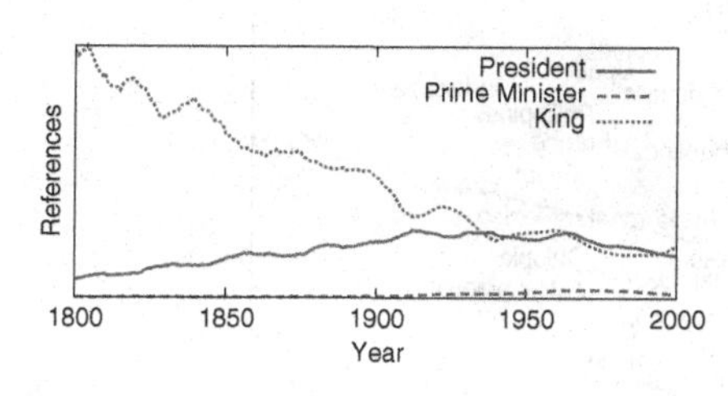

One observation from this table is the modern decline of royalty. There are no kings, and only one of the twenty most significant modern leaders serves as queen. The book Ngrams dataset illustrates how steadily elected/appointed officials have replaced royal sovereigns. Indeed, these data overrepresent the mindshare of royalty, because *king* is both a common last name and used colloquially like the "King of Rock and Roll."

Let's now analyze each region's governance in greater detail, identifying the most significant leader of essentially every country on earth. For each nation, we include the significance and term of office of its most significant figure.

We include four properties of each nation in these tables. We present our notion of what year the nation was founded, meaning that we have analyzed all leaders from beyond this point until the present day. Interesting leaders often served before this date, so check before complaining that we missed somebody. Unfortunately, our lines of succession do not always go as far back as they should, for technical reasons.

To assess the strength of a nation's leadership, we provide a *clout* score for each country. This represents the average significance of the heads of government/state in our database, weighted proportionally to their term of service.

We also provide each country's population, measured in millions of people. Greater clout from a smaller population implies that a nation's leadership is punching above its weight. This becomes clear in the dotplot of Figure 9.3, where countries above the line have leaders with more significance than expected from a simple fit. Generally, they are better off economically than the countries below the line, although it is difficult to know whether this is cause or effect.

Finally, we provide a coarse review of the political situation in each country. Is it currently rated as free, partially free, or not free, based on the freedom indices maintained by several organizations? These freedom indices will be discussed in greater detail in Section 9.3.

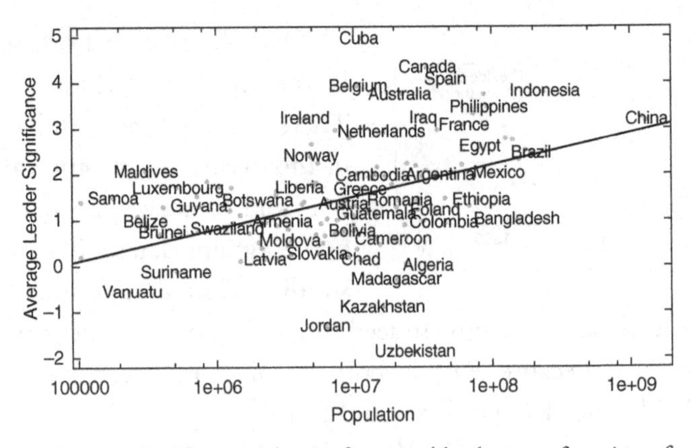

FIGURE 9.3. Average significance (clout) of national leaders as a function of populations. High-clout countries are generally economically advantaged over lower-clout nations.

9.2.1 THE NEW WORLD

The western hemisphere consists of North, South, and Central America, plus the Caribbean nations.

North America

The previous chapter dealt extensively with the history of the American presidency. Here we briefly review the leadership of our neighbors, Canada and Mexico.

North American Leaders

Country	Person	Dates/Term	Sig	Founded	Clout	Pop. (M)	Freedom
Canada	John A. Macdonald	(1878–1891)	159	1867	4.3	34.0	Free
Mexico	Vicente Fox	(2000–2006)	4325	1920	2.0	113.4	Partly
United States	Abraham Lincoln	(1861–1865)	5	1789	5.7	310.3	Free

John A. Macdonald (1815–1891) [159] was the leading figure behind the Canadian Confederation in 1867, when several British colonies united to form the country of Canada. Macdonald went on to serve 19 years as Canada's first prime minister: a cohort that historically serves long terms of office. Two others are particularly prominent. *William Lyon Mackenzie King (1874–1950)* [668] spent a total of 22 years in office, most notably as Canada's leader during World War II. *Pierre Trudeau (1919–2000)* [870] served 16 years (from 1968 to 1984) as the first French-Canadian prime minister, and helped preserve the Confederation in the face of Quebec separatism.

The history of Mexico has been much more turbulent. The First Federal Republic of Mexico was founded in 1824. Its best known president was *Antonio López de Santa Anna (1794–1876)* [682], who served 11 noncontiguous terms of office (ranging from 11 days to slightly more than two years) between 1833 and 1855. He styled himself "The Napoleon of the West" for his self-proclaimed military prowess. Santa Anna was forced into exile three times, an achievement Napoleon only managed twice. Santa Anna is best known as the victorious villain at the Battle of the Alamo, but lost California, Nevada, and Utah to the United States in the Mexican War.

A European attempt at establishing an empire under *Maximilian I (1832–1867)* [798] (who "reigned" 1864–67) ended in disaster. He was defeated and executed by forces led by Republican president *Benito Juárez (1806–1872)* [1035], who served five terms as president of Mexico. Stability but repression marked the 35-year tenure of *Porfirio Díaz (1830–1915)* [1527].

The end of the Mexican Revolution in 1920 constrained the president to a single six-year term, and marks the beginning of our analysis. We find the most significant modern Mexican president to be *Vicente Fox (1942–)* [4325], whose 2006 victory made him the first opposition party member elected since 1910.

South America

Instability has been the most salient fact about the politics of South America, and the most significant leaders have generally been the most durable. Representative democracy has been established over the past 20 years or so throughout most of the continent. This is why 5 of 12 South American countries have had their most significant leader during the past 15 years, despite national histories going back to the nineteenth century.

Bolivia survived 80 different administrations following its establishment in 1825 by *Simón Bolívar (1783–1830)* [448]. Most left little trace. After Bolívar, the most significant presidents of Bolivia have been recent: *Hugo Banzer (1926–2002)* [18817] and the current *Evo Morales (1959–)* [6614].

South American Leaders

Country	Person	Dates/Term	Sig	Founded	Clout	Pop. (M)	Freedom
Argentina	Domingo Faustino Sarmiento	(1868–1874)	2075	1854	1.9	40.4	Free
Bolivia	Simón Bolívar	(1825–1825)	448	1825	0.7	9.9	Partly
Brazil	Luiz Inácio Lula da Silva	(2003–2010)	6905	1985	2.4	194.9	Free
Chile	Augusto Pinochet	(1974–1990)	639	1817	1.9	17.1	Free
Colombia	Gustavo Rojas Pinilla	(1953–1957)	20032	1886	0.9	46.2	Partly
Ecuador	Eloy Alfaro	(1906–1911)	7766	1883	0.5	14.4	Partly
Guyana	Cheddi Jagan	(1992–1997)	26923	1970	1.3	0.7	Free
Paraguay	Francisco Solano López	(1862–1869)	8477	1844	1.0	6.4	Free
Peru	Alberto Fujimori	(1990–2000)	5509	1931	1.2	29.0	Free
Suriname	Johan Ferrier	(1975–1980)	62720	1975	-0.1	0.5	Free
Uruguay	Tabaré Vázquez	(2005–2010)	45386	1972	0.8	3.3	Free
Venezuela	Hugo Chávez	(1999–2013)	1841	1811	1.4	28.9	Partly

Bolívar himself served just six months as the first president of the country named after him, but other South American nations have equal or greater claims on him. At various times, he was president of Venezuela, Peru, and Gran Colombia (comprising present-day Colombia, Venezuela, Ecuador, and Panama): all this before dying at the age of 47.

Our significance measure turns up a few surprises. In Argentina, *Domingo Faustino Sarmiento (1811–1888)* [2075] edges out *Juan Perón (1895–1974)* [3726], the three-time president played by the male lead of the Broadway musical *Evita*. Sarmiento was an intellectual leader as well as Argentina's seventh president. In Paraguay, *Alfredo Stroessner (1912–2006)* [9317] ruled with an iron fist from 1954 to 1989. Yet he is outranked by *Francisco Solano López (1826–1870)* [8477], who led his nation through the War of the Triple Alliance against Argentina, Brazil, and Uruguay. This war is considered the most proportionally destructive conflict of modern times [Pinker, 2011]. It killed more than 60 percent of the population of Paraguay, one of the last victims being López himself in 1870.

Central America and the Caribbean

The nations of Central America and the Caribbean are generally small, and few leaders from the region have played major roles on the world stage.

Certain names are surprising at first glance but make sense upon reflection. President *Óscar Arias (1940–)* [53166] of Costa Rica won the Nobel Peace Prize in 1987 for his efforts to mediate several conflicts in the region. Yet by our measure, three-time president *José Figueres Ferrer (1906–1990)*

Central American Leaders

Country	Person	Dates/Term	Sig	Founded	Clout	Pop. (M)	Freedom
Belize	George Cadle Price	(1989–1993)	52222	1981	0.9	0.3	Free
Costa Rica	José Figueres Ferrer	(1970–1974)	27331	1948	0.3	4.6	Free
El Salvador	Maximiliano Hernández Martínez	(1935–1944)	55673	1841	0.3	6.1	Free
Guatemala	Rafael Carrera	(1844–1865)	16912	1844	1.1	14.3	Partly
Honduras	Manuel Bonilla	(1912–1913)	40341	1839	1.0	7.6	Partly
Mexico	Vicente Fox	(2000–2006)	4325	1920	2.0	113.4	Partly
Nicaragua	Anastasio Somoza Debayle	(1974–1979)	12657	1854	0.9	5.7	Partly
Panama	Guillermo Endara	(1989–1994)	31792	1904	0.2	3.5	Free

Caribbean Leaders

Country	Person	Dates/Term	Sig	Founded	Clout	Pop. (M)	Freedom
Antigua and Barbuda	Vere Bird	(1981–1994)	73505	1981	0.7	13.0	N/A
Barbados	Errol Barrow	(1986–1987)	48185	1966	1.0	0.2	Free
Cuba	Fidel Castro	(1976–2008)	506	1976	4.9	11.2	Not
Dominica	Roosevelt Skerrit	(2004–)	79776	1978	-0.8	0.0	Free
Dominican Republic	Rafael Trujillo	(1942–1952)	2497	1924	1.7	9.9	Free
Grenada	Keith Mitchell	(1995–2008)	101396	1984	0.2	0.1	Free
Haiti	Jean Pierre Boyer	(1818–1843)	9346	1806	1.1	9.9	Partly
Jamaica	Michael Manley	(1989–1992)	17891	1962	1.6	2.7	Free
Saint Kitts and Nevis	Denzil Douglas	(1995–)	129176	1983	-0.6	17.0	N/A
Saint Lucia	Kenny Anthony	(2011–)	70836	1979	0.0	0.1	Free
Saint Vincent and the G.	Ralph Gonsalves	(2001–)	103595	1979	-0.4	0.1	N/A
Trinidad and Tobago	Eric Williams	(1962–1981)	16371	1962	1.3	1.3	Free

[27331] comes in ahead of him as the nation's most significant leader. Ferrer abolished the Costa Rican army after a painful civil war, creating a demilitarized state that persists today. This provided Arias the credibility he needed to succeed in his mediation efforts.

Sandinista leader *Daniel Ortega (1945–)* [16410] of Nicaragua was a controversial cause celebre in the 1980s, who has been serving a second term as president since 2007. Yet our rankings opt for *Anastasio Somoza Debayle (1925–1980)* [12657], the last representative of the Somoza family dynasty, which ruled Nicaragua from 1936 until his fall in 1979.

Most Caribbean nations are small islands whose leaders have never played a significant geopolitical role. The primary exception is *Fidel Castro (1926–)* [506], the revolutionary who liberated Cuba and then ruled it for 48 years. Other prominent leaders emerge on Haiti and the Dominican Republic, the two nations sharing the island of Hispanola. Both were ruled by longtime political bosses: "Papa Doc" *François Duvalier (1907–1971)* [14577] and *Rafael Trujillo (1891–1961)* [2497]), respectively. "Papa Doc," however, is edged out in our rankings by *Jean Pierre Boyer (1776–1850)* [9346], who briefly united the whole of Hispanola.

9.2.2 EUROPE

Figure 9.4 summarizes the most significant leader of each European country. World War II produced many iconic leaders. On the Allied side, England and France are represented by *Winston Churchill (1874–1965)* [37] and *Charles de Gaulle (1890–1970)* [396], respectively. The most significant leader of several other European countries are World War II–era Fascists, including *Adolf Hitler (1889–1945)* [7] in Germany. Italy would have been represented by *Benito Mussolini (1883–1945)* [101] had we deemed their modern state to be founded before 1946. Spain (*Francisco Franco (1892–1975)* [291]) and Portugal (*António de Oliveira Salazar (1889–1970)* [3277]) both suffered under nearly 40 years of dictatorial rule, and did not become modern democracies until the 1970s.

Certain countries are still represented by hated leaders from the Soviet era. *Enver Hoxha (1908–1985)* [2921], the Communist Party leader of Albania,[1] closed off his country from the rest of the world like a North Korea in Europe. *Nicolae Ceauşescu (1918–1989)* [2420] ruled Romania from 1965 until his much celebrated execution. Others made their reputation fighting the Communist establishment. Trade union organizer *Lech Walesa (1943–)* [8108] led Poland's Solidarity movement, becoming president following the fall of the Soviet Union. *Viktor Yushchenko (1954–)* [9198] survived a widely reported assassination attempt via dioxin poisoning to lead Ukraine's Orange Revolution.

The iconic leaders of two European countries are essentially failed revolutionaries. *Michael Collins (1890–1922)* [1240] ruled Ireland's provisional government for only eight months prior to his assassination. *Imre Nagy (1896–1958)* [6774] led Hungary for 11 days during the failed Revolution

[1] Hoxha was nominally prime minister from 1944 until 1954, though he retained effective control of the country until his death in 1985.

Country	Person	Dates/Term	Sig	Founded	Clout	Pop. (M)	Freedom
Albania	Enver Hoxha	(1944–1954)	2921	1944	1.2	3.2	Partly
Andorra	Ramon Iglesias i Navarri	(1943–1969)	108027	1907	-0.3	26.0	Free
Armenia	Serzh Sargsyan	(2008–)	36520	1991	0.9	3.0	Partly
Austria	Engelbert Dollfuss	(1932–1934)	6983	1918	1.3	8.3	Free
Belarus	Alexander Lukashenko	(1994–)	10960	1994	2.7	9.5	Not
Belgium	Leopold II	(1865–1909)	727	1831	3.9	10.7	Free
Bosnia and Herzegovina	Alija Izetbegović	(2000–2000)	35846	1996	-0.6	3.7	Partly
Bulgaria	Todor Zhivkov	(1962–1971)	8655	1946	0.7	7.4	Free
Croatia	Franjo Tuman	(1990–1999)	9891	1990	1.4	4.4	Free
Cyprus	Makarios III	(1974–1977)	9135	1960	1.7	1.0	Free
Czech Republic	Václav Klaus	(2003–)	17786	1993	1.1	10.4	Free
Denmark	Christian IX	(1863–1906)	1671	1863	2.2	5.5	Free
Estonia	Lennart Meri	(1992–2001)	34819	1938	0.9	1.3	Free
Finland	Carl Gustaf Emil Mannerheim	(1944–1946)	2225	1917	1.8	5.3	Free
France	Charles de Gaulle	(1959–1969)	396	1959	3.0	62.7	Free
Germany	Adolf Hitler	(1933–1945)	7	1919	3.0	82.3	Free
Greece	Konstantinos Karamanlis	(1974–1980)	6629	1974	1.6	11.3	Free
Hungary	Imre Nagy	(1956–1956)	6774	1920	0.7	9.9	Free
Iceland	Ólafur Ragnar Grímsson	(1996–)	47737	1917	0.8	0.3	Free
Ireland	Michael Collins	(1922–1922)	1240	1919	3.2	4.4	Free
Italy	Silvio Berlusconi	(2008–2011)	2073	1946	2.1	60.5	Free
Latvia	Vaira Vīķe-Freiberga	(1999–2007)	77473	1990	0.1	2.2	Free
Liechtenstein	Johann I Joseph	(1805–1836)	17765	1608	0.9	0.1	Free
Lithuania	Vytautas Landsbergis	(1990–1992)	29501	1990	0.4	3.3	Free
Luxembourg	Adolphe	(1890–1905)	6974	1890	1.7	0.5	Free
Macedonia	Nikola Gruevski	(2006–)	63325	1991	0.5	2.0	Free
Malta	Gerald Strickland	(1927–1932)	19847	1921	1.2	0.4	Free
Moldova	Vladimir Voronin	(2001–2009)	29533	1990	0.5	3.5	Free
Monaco	Albert I	(1889–1922)	6809	1814	1.9	0.1	Free
Netherlands	William I	(1839–1840)	1648	1839	2.9	16.6	Free
Norway	Haakon VII	(1905–1957)	2538	1905	2.3	4.8	Free
Poland	Lech Wałęsa	(1990–1995)	8108	1944	1.1	38.2	Free
Portugal	António de Oliveira Salazar	(1932–1968)	3277	1932	2.0	10.6	Free
Republic of Macedonia	Kiro Gligorov	(1991–1999)	23524	1991	1.5	8.0	Free
Romania	Ion Iliescu	(2000–2004)	9053	1989	1.4	21.4	Free
Russia	Vladimir Putin	(2012–)	1014	1991	2.7	142.9	Not
Serbia	Boris Tadić	(2004–2012)	25188	1991	0.6	7.3	Free
Serbia and Montenegro	Slobodan Milošević	(1997–2000)	4760	1992	1.3	3.0	Free
Slovakia	Robert Fico	(2012–)	53442	1993	0.2	5.4	Free
Slovenia	Janez Drnovšek	(2002–2007)	34485	1990	1.3	2.0	Free
Soviet Union	Joseph Stalin	(1941–1953)	18	1922	3.4	293.0	Not
Spain	Francisco Franco	(1936–1975)	291	1931	4.0	46.0	Free
Sweden	Charles XIV John of Sweden	(1818–1844)	1289	1818	2.7	9.3	Free
Ukraine	Viktor Yushchenko	(2010–)	9198	1990	1.4	45.8	Partly
United Kingdom	Winston Churchill	(1951–1955)	37	1762	4.4	62.0	Free
Vatican City	Pope John Paul II	(1978–2005)	91	1922	3.3	0.0	N/A

FIGURE 9.4. Most significant European leaders, by country.

of 1956, which was crushed by the Soviet Union. Nagy was later executed by the Soviets.

Several European nations are represented by relatively recent kings. *Leopold II (1835–1909)* [727] of Belgium is notorious for the African colony he founded as the Congo Free State. Free in name only, millions of Africans died to satisfy his lust for profits from ivory and rubber [Hochschild, 1999]. More heroic was King *Haakon VII (1872–1957)* [2538], who led Norwegian resistance to the Germans during World War II and reigned for 52 years.

9.2.3 AFRICA AND THE MIDDLE EAST

Only a handful of African countries have national histories that significantly pre-date the senior author of this book. Ethiopia always maintained independent rule, except for a period as an Italian colony under Mussolini during World War II. Liberia was founded by former American slaves in 1847. Its time as a relative oasis of stability ended in 1980, with the coup of *Samuel Doe (1951–1990)* [20763], and it descended into the horror of a civil war from which it is only beginning to recover. Swaziland did not fully gain independence from Britain until 1968, although its line of chieftains predates independence.

Only two modern African leaders rank among the thousand most significant historical figures. From his prison cell, *Nelson Mandela (1918–)* [356] became the symbol of South Africa's fight against apartheid. His personal dignity, and conciliatory treatment toward the whites who imprisoned him for 27 years, unified the country and stirred the world. The last Ethiopian emperor, *Haile Selassie I (1892–1975)* [738], defended his nation against invasion from fascist Italy and served as a messianic figure to followers of Jamaica's Rastafari movement.

The African leaders in Figure 9.5 generally fall into three batches. The first group led their nations from colonialism to independence, such as *Kwame Nkrumah (1909–1972)* [3774] of Ghana, *Jomo Kenyatta (1893–1978)* [3551] of Kenya, and *Julius Nyerere (1922–1999)* [8093] of Tanzania. The second wave consisted of later military strongmen, whose long rule by force pillaged their countries. These include *Mobutu Sese Seko (1930–1997)* [4736] (Zaire), *Idi Amin (1928–2003)* [1926] (Uganda), and *Robert Mugabe (1924–)* [2012] (Zimbabwe). Finally, a clique of current leaders appear to be bringing a somewhat better class of government to their nations, such as *Paul Kagame (1957–)* [18286] of Rwanda.

Country	Person	Dates/Term	Sig	Founded	Clout	Pop. (M)	Freedom
Angola	José Eduardo dos Santos	(1979–)	20489	1975	1.7	19.0	Not
Benin	Mathieu Kérékou	(1996–2006)	40001	1990	1.1	8.8	Free
Botswana	Seretse Khama	(1966–1980)	29815	1966	1.4	2.0	Free
Burkina Faso	Thomas Sankara	(1984–1987)	18216	1971	–0.4	16.4	Partly
Burundi	Pierre Buyoya	(1996–2003)	87854	1966	0.7	8.3	Partly
Cameroon	Paul Biya	(1982–)	26597	1960	0.5	19.5	Not
Cape Verde	Pedro Pires	(2001–2011)	68638	1975	0.5	0.4	Free
Central African Republic	David Dacko	(1979–1981)	59551	1979	-0.3	4.4	Not
Chad	Idriss Déby	(1990–)	53141	1991	0.1	11.2	Not
Comoros	Azali Assoumani	(2002–2006)	76467	2001	0.7	0.7	Partly
Dem. Rep. of the Congo	Mobutu Sese Seko	(1972–1997)	4736	1964	2.4	65.9	N/A
Djibouti	Hassan Gouled Aptidon	(1977–1999)	53134	1977	0.7	0.8	Not
Equatorial Guinea	Teodoro O. N. Mbasogo	(1979–)	16431	1968	–0.8	0.7	Not
Eritrea	Isaias Afewerki	(1993–)	29931	1993	1.9	5.2	Not
Ethiopia	Haile Selassie I	(1927–1936)	738	1909	1.4	82.9	Not
Gabon	Omar Bongo	(1967–2009)	18306	1960	1.1	1.5	Not
Ghana	Kwame Nkrumah	(1960–1966)	3774	1960	2.2	24.3	Free
Guinea	Ahmed Sékou Touré	(1958–1984)	29042	1958	0.9	9.9	Not
Guinea-Bissau	João Bernardo Vieira	(2005–2009)	123766	1973	0.1	1.5	Not
Kenya	Jomo Kenyatta	(1964–1978)	3551	1963	2.9	40.5	Partly
Lesotho	Letsie III	(1996–)	63698	1966	0.3	2.1	Partly
Liberia	Joseph Jenkins Roberts	(1872–1876)	8455	1848	1.7	3.9	Partly
Madagascar	Marc Ravalomanana	(2002–2009)	36494	1959	–0.3	20.7	Partly
Malawi	Hastings Banda	(1966–1994)	9441	1966	2.1	14.9	Partly
Mali	Alpha Oumar Konaré	(1992–2002)	63350	1960	0.3	15.8	Free
Mauritania	Moktar Ould Daddah	(1960–1978)	59157	1960	0.9	3.4	Partly
Mauritius	Anerood Jugnauth	(2000–2003)	30287	1968	1.7	1.2	Free
Mozambique	Samora Machel	(1975–1986)	18584	1974	0.8	23.3	Partly
Namibia	Sam Nujoma	(1990–2005)	39704	1990	0.9	2.2	Free
Niger	Mamadou Tandja	(1999–2010)	40873	1960	0.4	15.5	Partly
Nigeria	Olusegun Obasanjo	(1999–2007)	9291	1963	2.3	158.4	Not
Republic of the Congo	Denis Sassou Nguesso	(1997–)	35836	1992	0.9	4.0	Not
Rwanda	Paul Kagame	(2000–)	18286	1961	0.3	10.6	Not
Senegal	Léopold Sédar Senghor	(1960–1980)	10594	1960	1.2	12.4	Partly
Seychelles	France-Albert René	(1977–2004)	66378	1976	1.0	27.0	Partly
Sierra Leone	Siaka Stevens	(1971–1985)	31329	1971	1.4	5.8	Partly
South Africa	Nelson Mandela	(1994–1999)	356	1961	0.9	50.1	Free
Swaziland	Mswati III	(1968–)	16235	1899	0.8	1.1	Not
Tanzania	Julius Nyerere	(1964–1985)	8093	1964	0.8	44.8	Partly
Togo	Sylvanus Olympio	(1960–1963)	42025	1960	0.5	6.0	Not
Uganda	Idi Amin	(1971–1979)	1926	1963	2.0	33.4	Partly
Zambia	Kenneth Kaunda	(1991–2002)	10157	1991	1.9	13.0	Free
Zimbabwe	Robert Mugabe	(1987–)	2012	1980	1.1	12.5	Not

FIGURE 9.5. Most significant African leaders, by country.

9.2.4 MIDDLE EAST AND GREATER ARABIA

Many of the Middle Eastern nations, quite broadly defined, were created in the dissolution of the Ottoman Empire, others at the end of colonialism after World War II.

The most significant leader here is *Mustafa Kemal Atatürk (1881–1938)* [360], the founder of modern Turkey. He instituted enormous cultural reforms that created a modern secular state. Ayatollah *Ruhollah Khomeini (1902–1989)* [676] turned the opposite trick, creating an Islamic Republic from the secular state of Iran.

Israel is the only non-Islamic country in this group. *David Ben-Gurion (1886–1973)* [1269] is the founder of modern Israel and served as his nation's first prime minister. Several other prime ministers are of almost equal stature, including pioneering woman *Golda Meir (1898–1978)* [2429] (who served 1969–74) and the Nobel Peace Prize recipients *Yitzhak Rabin (1922–1995)* [2611] and *Shimon Peres (1923–)* [4418]. This is an amazing accomplishment for a nation with fewer people than New York City.

Middle Eastern and Greater Arabian Leaders

Country	Person	Dates/Term	Sig	Founded	Clout	Pop. (M)	Freedom
Afghanistan	Hamid Karzai	(2001–)	5258	1973	1.9	31.4	Not
Algeria	Ahmed Ben Bella	(1963–1965)	16589	1979	0.0	35.4	Not
Azerbaijan	Heydar Aliyev	(1993–2003)	17917	1991	0.1	9.1	Not
Bahrain	Hamad bin Isa Al Khalifa	(1999–)	17320	1971	1.2	1.2	Not
Egypt	Gamal Abdel Nasser	(1956–1970)	1404	1953	2.5	81.1	Partly
Iran	Ruhollah Khomeini	(1979–1989)	676	1979	3.3	73.9	Not
Iraq	Saddam Hussein	(1979–2003)	444	2006	3.2	31.6	Not
Israel	David Ben-Gurion	(1955–1963)	1269	1948	2.9	7.4	Free
Jordan	Abdullah I	(1951–1951)	2609	1946	−1.3	6.1	Partly
Kuwait	Jaber A. A. Al-Sabah	(1977–2006)	38822	1961	1.5	2.7	Partly
Lebanon	Rafic Hariri	(2000–2004)	10267	1943	1.3	4.2	Partly
Libya	Muammar Gaddafi	(1969–2011)	4961	1951	1.4	6.3	Not
Morocco	Hassan II	(1961–1999)	12497	1927	0.9	31.9	Not
Oman	Thuwaini bin Said	(1856–1866)	68465	1806	0.8	2.7	Partly
Pakistan	Zulfikar Ali Bhutto	(1971–1973)	2801	1947	2.5	173.5	Partly
Qatar	Hamad bin K. A. Thani	(1995–)	25646	1850	0.7	1.7	Partly
Saudi Arabia	Faisal of Saudi Arabia	(1964–1975)	6223	1932	2.7	27.4	Not
Somalia	Siad Barre	(1969–1991)	7055	1956	1.2	9.3	Not
Syria	Hafez al-Assad	(1971–2000)	7025	1961	1.2	20.4	Not
Tunisia	Habib Bourguiba	(1957–1987)	11118	1957	1.3	10.4	Partly
Turkey	Mustafa Kemal Atatürk	(1923–1938)	360	1923	1.9	72.7	Partly
United Arab Emirates	Zayed bin S. A. Nahyan	(1971–2004)	9794	1971	1.8	7.5	Partly
Yemen	Ali Abdullah Saleh	(1994–2012)	37914	1990	0.0	24.0	Not

9.2.5 ASIA AND OCEANIA

Several of the dominant leaders in Asia were the country's revolutionary founders, including *Mao Zedong (1893–1976)* [151] (China), *Ho Chi Minh (1890–1969)* [779] (Vietnam), and *Sheikh Mujibur Rahman (1920–1975)* [4731] (Bangladesh).

Mohandas Karamchand Gandhi (1869–1948) [46] was the spiritual leader of India's quest for independence, but he never held a political leadership position. *Jawaharlal Nehru (1889–1964)* [530] served as the first prime minister of India from 1947 to 1964, and proved far more instrumental in establishing the modern nation.

Emperor Meiji (1852–1912) [1043] ruled Japan for 45 years, during which it emerged from a feudal shogunate and turned into a world power. He edges out *Hirohito (1901–1989)* [1083], whose 63-year reign included Japan's defeat in World War II and postwar economic resurrection.

Asian Leaders

Country	Person	Dates/Term	Sig	Founded	Clout	Pop. (M)	Freedom
Bangladesh	Sheikh Mujibur Rahman	(1975–1975)	4731	1971	0.9	148.6	Partly
Bhutan	Jigme Singye Wangchuck	(1972–2006)	21710	1907	1.5	0.7	Partly
Brunei	Hassanal Bolkiah	(1967–)	28333	1906	0.7	0.3	Not
Burma	Ne Win	(1974–1981)	8058	1948	0.9	48.3	Not
Cambodia	Pol Pot	(1976–1979)	2475	1860	1.9	14.1	Partly
China	Mao Zedong	(1949–1959)	151	1949	3.1	1317.9	Not
India	Jawaharlal Nehru	(1947–1964)	530	1947	3.2	1224.6	Free
Indonesia	Suharto	(1967–1998)	1836	1945	3.7	239.8	Free
Japan	Emperor Meiji	(1867–1912)	1043	1867	2.7	126.5	Free
Kazakhstan	Nursultan Nazarbayev	(1990–)	17066	1990	–0.9	16.0	Not
Kyrgyzstan	Askar Akayev	(1990–2005)	36888	1990	0.7	5.3	Partly
Laos	Kaysone Phomvihane	(1991–1992)	57207	1975	0.6	6.2	Not
Malaysia	Mahathir Mohamad	(1981–2003)	4688	1957	2.1	28.4	Partly
Maldives	Maumoon Abdul Gayoom	(1978–2008)	19787	1953	2.0	0.3	Partly
Mongolia	Tsakhiagiin Elbegdorj	(2009–)	34109	1992	0.1	2.7	Free
Nepal	Gyanendra of Nepal	(2005–2006)	11423	1990	1.4	29.9	Partly
North Korea	Kim Il-sung	(1972–1994)	1666	1948	2.2	24.3	Not
Philippines	Ferdinand Marcos	(1965–1986)	1298	1899	3.4	93.2	Partly
Republic of China	Chiang Kai-shek	(1950–1975)	440	1948	2.8	25.0	Free
Singapore	Lee Kuan Yew	(1959–1990)	2315	1959	1.7	5.0	Partly
Sri Lanka	Junius Richard Jayewardene	(1978–1989)	11479	1972	1.9	20.8	Partly
Tajikistan	Emomalii Rahmon	(1992–)	37853	1990	–1.3	6.5	Not
Thailand	Thaksin Shinawatra	(2006–2006)	5965	1925	1.3	69.1	Partly
Turkmenistan	Saparmurat Niyazov	(1990–2006)	15944	1990	2.6	5.0	Not
Uzbekistan	Islam Karimov	(1990–)	14382	1990	–1.8	27.4	Not
Vietnam	Ho Chi Minh	(1945–1969)	779	1945	3.7	87.8	Not

Gyanendra of Nepal (1947–) [11423] was the last king of Nepal. The monarchy was abolished following a massacre of the royal family by the Crown Prince in 2001, and Gyanendra's own misrule.

Five of the former Soviet states that became independent states in 1990 appear in the accompanying table. In four cases the founding national ruler remains in power more than 20 years following independence. In the fifth, Kyrgyzstan, the founding ruler died young, after only 15 years at the helm.

The island nations of the Pacific are often economically linked to Asia. The two countries of greatest clout are Australia and New Zealand. Australia is represented by Prime Minister *Gough Whitlam (1916–)* [1114]. New Zealand rests on one of its founding figures, *George Grey (1812–1898)* [2282], present at creation when the British colony split off from Australia.

Other islands, like Fiji and Papua New Guinea, are represented by independence-era figures who long retained power. Tonga's *Taufa'ahau Tupou IV (1918–2006)* [27521] was a larger-than-life figure (weighing in at 400 pounds) who exploited all the unique economic opportunities that come from running a sovereign state, from selling stamps to collectors to claiming domain over scarce locations for geosynchronous satellites.

Leaders of Pacific Oceania

Country	Person	Dates/Term	Sig	Founded	Clout	Pop. (M)	Freedom
Australia	Gough Whitlam	(1972–1975)	1114	1901	3.7	22.2	Free
East Timor	José Ramos-Horta	(2007–2012)	23389	2002	1.5	1.1	Partly
Federated S. of Micronesia	Manny Mori	(2007–)	310072	1979	−0.7	0.1	N/A
Fiji	Kamisese Mara	(1993–2000)	17842	1967	1.8	0.8	Not
Kiribati	Anote Tong	(2003–)	128006	1979	−0.5	0.1	Free
Marshall Islands	Litokwa Tomeing	(2008–2009)	208853	1979	−0.4	0.1	N/A
Nauru	Marcus Stephen	(2007–2011)	48665	2007	1.5	0.0	Free
New Zealand	George Grey	(1877–1879)	2282	1856	2.5	4.3	Free
Palau	Johnson Toribiong	(2009–)	152565	1981	−0.4	0.0	N/A
Papua New Guinea	Michael Somare	(2002–2010)	32422	1975	1.2	5.8	Free
Samoa	Malietoa Tanumafili II	(1962–2007)	25818	1959	1.4	0.1	Free
Solomon Islands	Derek Sikua	(2007–2010)	118135	1978	0.2	0.5	Partly
Tonga	Tåufa'åhau Tupou IV	(1965–2006)	27521	1845	1.3	0.1	Partly
Tuvalu	Maatia Toafa	(2010–2010)	159138	1978	−0.2	0.0	N/A
Vanuatu	Edward Natapei	(2001–2004)	107604	1980	−0.5	0.2	Free

9.3 Dictators and Despots

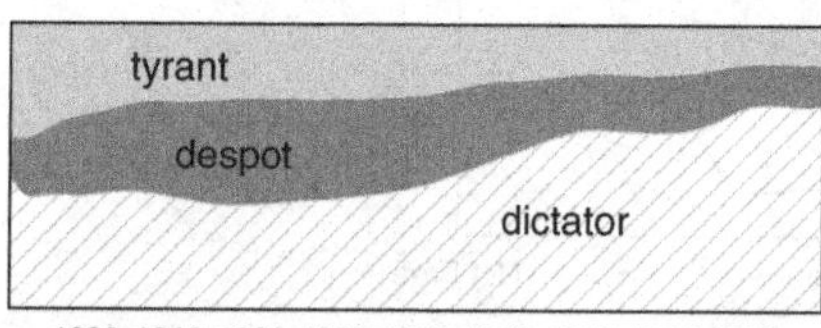

There have always been authoritative rulers, but the language used to describe them has changed with time. The book Ngram data shows the transition from "tyrant" to "despot" to "dictator" over the past 200 years, reflecting the shift from inherited monarchies to military governments.

It is tempting to compare the relative significance of dictatorial figures against elected leaders. Which is the surer path to glory? The case can be made that maybe dictators *deserve* greater historical significance than their democratically elected peers from comparable nations. In some sense, they represent the executive, legislative, and judicial branches rolled up in a single figure.

But the challenge is identifying exactly who is a dictator. Nobody admits to being a repressive leader. Many hold elections where they win by acclamation. The first countries to emerge in response to a Wikipedia query for "Democratic Republic of" prove to be (in order) the Congo, East Germany, Madagascar, North Korea, Ethiopia, North Vietnam, Serbia, Azerbaijan, Georgia, Armenia, Afghanistan, Belarus, and Somalia. Clearly, self-reported descriptions don't work here.

To study this question, we employed national freedom ratings produced by relevant nongovernmental organizations. The Democracy Index, compiled by the Economist Intelligence unit, uses a zero to ten scale to measure the state of national democracy. In 2011, Norway was rated the most democratic country, while North Korea lagged the rest of the world. The Press Freedom Index, by Reporters without Borders, measures the extent of government obstruction on the ability of journalists to do their work. Here Norway ranks tied for first and North Korea next to last. The Heritage Foundation/Wall Street Journal's index of economic freedom measures the degree to which businesses are free to operate. The latest (2012) ranking puts Hong Kong on top. Norway sinks to 40th place, while North Korea

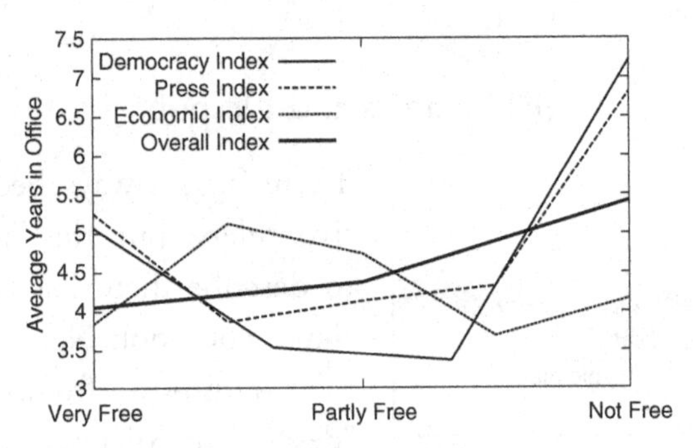

FIGURE 9.6. Longevity of leaders as a function of the freedom of their countries. Leaders in dictatorships survive longer than in countries with free elections.

again brings up the rear. Our final index is the overall evaluation by the U.S.-based Freedom House.

These indices generally but imperfectly correlate with each other. They can all be mapped onto a similar freedom scale, with nations grouped into four or five buckets that range from very free to not free. Figure 9.6 charts the average longevity of leaders against these indices of political, press, and economic freedom [Wikipedia, 2012b]. And indeed, leaders of nations rated "not free" enjoy more robust terms than their democratically elected brethren. Interestingly, it is the leaders of nations that are partially free who have the shortest careers of all.

9.4 Military Leaders

It also proves hard to distinguish military leaders from political leaders. The separation of military and civilian leadership is a relatively recent innovation. Kings were expected to lead their countries' troops into battle, even well into the eighteenth century. In addition, great generals have always tended to seek the reins of power. *Julius Caesar (100–44 B.C.)* [15] is the ancient, quintessential example of a leader who took power in a military coup. In modern democracies, generals have often found success in battle a useful prerequisite to seeking elected office.

In Figure 9.7 we summarize the most significant military leaders of history. We restrict our scope to those who actually led troops in the field, as

Sig	Person	Dates	C/G	Description
2	Napoleon	(1769–1821)	C G	Emperor of France (Battle of Waterloo)
9	Alexander the Great	(356–323 B.C.)	C G	Greek king and conqueror of the known world
15	Julius Caesar	(100–44 B.C.)	C G	Roman general and statesman ("Et tu, Brute?")
38	Genghis Khan	(1162–1227)	C G	Founder of the Mongol Empire
50	Oliver Cromwell	(1599–1658)	C G	Lord Protector of England (English Civil War)
76	Robert E. Lee	(1807–1870)	C G	Confederate general (U.S. Civil War)
95	Joan of Arc	(1412–1431)	C G	French military leader and saint
120	Richard I	(1157–1199)	C G	King of England (Third Crusade)
138	Saladin	(1138–1193)	C G	Sultan of Egypt and Syria (Crusade)
163	Frederick the Great	(1712–1786)	C G	Prussian king and military leader
174	Arthur Wellesley	(1769–1852)	C G	Duke of Wellington (Battle of Waterloo)
187	Hannibal	(248–182 B.C.)	C G	Carthaginian military commander
193	Attila	(406–453)	C G	Ruler of the Hunnic Empire
217	Gilbert du Motier	(1757–1834)	C G	French military officer (Revolutionary War)
219	Horatio Nelson	(1758–1805)	C G	Royal Navy officer (Napoleonic Wars)
222	Benedict Arnold	(1741–1801)	C G	General and traitor (Revolutionary War)
233	William Tecumseh Sherman	(1820–1891)	C G	Union general (U.S. Civil War)
257	Cyrus the Great	(576–529 B.C.)	C G	Founder of the First Persian Empire
283	Stonewall Jackson	(1824–1863)	C G	Confederate general (U.S. Civil War)
290	Timur	(1336–1405)	C G	Mongol founder of Timurids (a.k.a. Tamerlane)

FIGURE 9.7. The most significant military leaders (excluding U.S. presidents).

opposed to figures such as *Adolf Hitler (1889–1945)* [7]. Even so restricted, about half of our top 25 went on to lead their country.

The aura of great military leaders is often defined in terms of great opposition. But it is not always the victor whose reputation survives most gloriously. *Napoleon (1769–1821)* [2] was eventually vanquished by *Arthur Wellesley (1769–1852)* [174], otherwise known to history as the Duke of Wellington. The fame of *Hannibal (248–182 B.C.)* [187] endures far beyond his defeat at the hands of *Scipio Africanus* [995].

Other figures are remembered for winning the battle, but losing the war. *Richard I (1157–1199)* [120] ("The Lionheart") edges out his opponent *Saladin (1138–1193)* [138] in our remembrance, on the strength of his tactical victories, although he failed to take Jerusalem in the Third Crusade. *Robert E. Lee (1807–1870)* [76] is remembered as a brilliant general: but he, too, lost his war.

A few military figures are remembered for even less positive reasons. *Benedict Arnold (1741–1801)* [222] was an excellent military leader, but is remembered primarily as a traitor. And, of course, *George Armstrong Custer (1839–1876)* [379] is remembered for his spectacular defeat at the Little Bighorn.

10

Science and Technology

The human ability to understand and manipulate our world is what separates us from the rest of the animals. In this chapter we will revisit those responsible for breakthroughs that changed our lives and shaped our notions of what is possible: the scientists, inventors, businessmen, and explorers.

10.1 Scientists

Science is both an ancient and modern undertaking. Prehistoric man pondered the night sky and wondered why. Ancient Greek thinkers such as *Euclid* [149] and *Archimedes (287–212 B.C.)* [146] laid the foundations for science and mathematics, although philosophy trumped observation. The modern scientific method, marked by hypothesis and experimentation, did not emerge until *Francis Bacon (1561–1626)* [81] and the Renaissance. Investigations into natural philosophy began to reveal the workings of the universe. Indeed, the word *scientist* was not invented until about 1800, and does not really pick up in the Ngrams corpus until the 1860s.

10.1.1 GROWTH OF THE SCIENCES

Figure 10.1 enables us to trace the relative importance of different branches of the sciences through the past 200 years. *Natural philosophy* began to fade well before 1850, and had vanished by 1900. Astronomy, a strong claimant as the oldest science, remained the preeminent scientific discipline until the

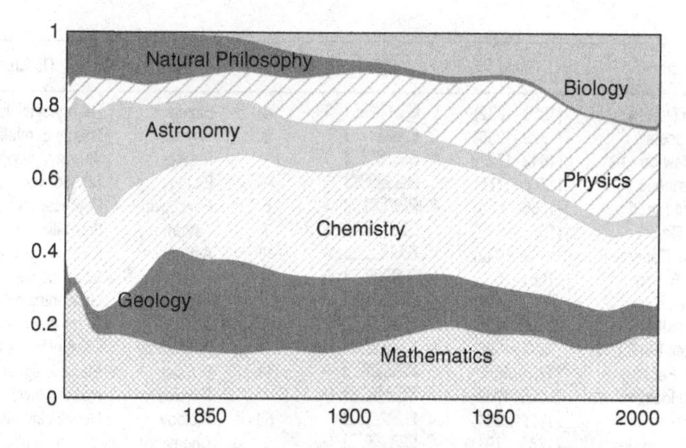

FIGURE 10.1. Ngrams data of scientific disciplines.

beginning of the nineteenth century, but was then crowded out by newer sciences.

Modern geology began with *William Smith's (1769–1839)* [3166] map of British rock layers, or *strata*. The observation that strata could be identified by fossils embedded within pointed to a world much older than the 6,000 years since biblical creation. *Charles Lyell's (1797–1875)* [1107] *Principles of Geology* revealed that the earth was shaped by slow-moving forces, such as erosion and uplift. Geology quickly became the most important scientific field, but this popularity slowly eroded with the uplift of physics and then biology.

With the revolution in modern physics of the twentieth century, physics became the hot scientific field through the atomic era. It made famous such names as *Albert Einstein (1879–1955)* [19], *Erwin Schrödinger (1887–1961)* [1987], and *Max Planck (1858–1947)* [539]. The life sciences have taken off since the 1950s, particularly molecular biology in the wake of the discovery of the structure of DNA by *Francis Crick (1916–2004)* [1429] and *James D. Watson (1928–)* [3619].

10.1.2 THE GREAT SCIENTISTS

We highlight the 15 most significant scientists across an assortment of disciplines in Figure 10.2. All rank among the top 300 people in history, a stature similar to the presidents of the United States. But the fundamental

Sig	Person	Dates	C/G	Sci100	Field	Description
12	Charles Darwin	(1809–1882)	C G	4	Biology	Theory of evolution
19	Albert Einstein	(1879–1955)	C G	2	Physics	Theory of relativity
21	Isaac Newton	(1643–1727)	C G	1	Physics	Universal gravitation
31	Carl Linnaeus	(1707–1778)	C G	76	Biology	Taxonomical classification
44	Sigmund Freud	(1856–1939)	C G	6	Psychology	Psychoanalysis
49	Galileo Galilei	(1564–1642)	C G	7	Astronomy	Scientific revolution
74	Nicolaus Copernicus	(1473–1543)	C G	10	Astronomy	Heliocentric cosmology
81	Francis Bacon	(1561–1626)	C G		Philosophy	Scientific method
103	Ptolemy	(A.D. 90–168)	C G		Astronomy	Geocentric model
112	Louis Pasteur	(1822–1895)	C G	5	Biology	Germ theory of disease
156	Johannes Kepler	(1571–1630)	C G	9	Astronomy	Laws of planetary motion
175	Michael Faraday	(1791–1867)	C G	11	Physics	Electromagnetism
216	Robert Hooke	(1635–1703)	C G		Physics	Hooke's law
250	Gregor Mendel	(1822–1884)	C G	60	Biology	Genetic inheritance
276	Antoine Lavoisier	(1743–1794)	C G	8	Chemistry	Discovery of elements

FIGURE 10.2. The 15 most significant scientists: with field, discovery, and Science 100 rank [Simmons, 2000].

laws governing the natural world can only be discovered once, granting each of these figures a unique and forever untarnishable achievement.

The top scientists of the ancient world made fundamental errors as they explored the darkness. That *Ptolemy's (A.D. 90–168)* [103] universe, where the sun traveled around the earth, endured for more than 1,500 years until the Renaissance shows just how far ahead of his time Ptolemy was.

As shown in Figure 10.2, the three most significant scientists are the physicists *Isaac Newton (1643–1727)* [21] and *Albert Einstein (1879–1955)* [19] and the evolutionary biologist *Charles Darwin (1809–1882)* [12]. These men, along with the discoverer of the heliocentric universe (*Nicolaus Copernicus (1473–1543)* [74]) and the germ theory of disease (*Louis Pasteur (1822–1895)* [112]), revolutionized the way we understand our world.

There are more scientists working today than at any point in human history, which makes the relative antiquity of our most significant figures particularly striking. Einstein is the only one who lived into the twentieth century. This is because the lowest-hanging scientific fruit gets picked first: macroscopic questions about the physical world were identified relatively early in human intellectual history, and proved addressable using techniques in place by the nineteenth century. Several early scientists made important contributions to multiple disciplines, because they were the first ones to be rummaging around the neighborhood. Today's scientists generally focus on deeper but less accessible problems, and work in increasingly large research teams.

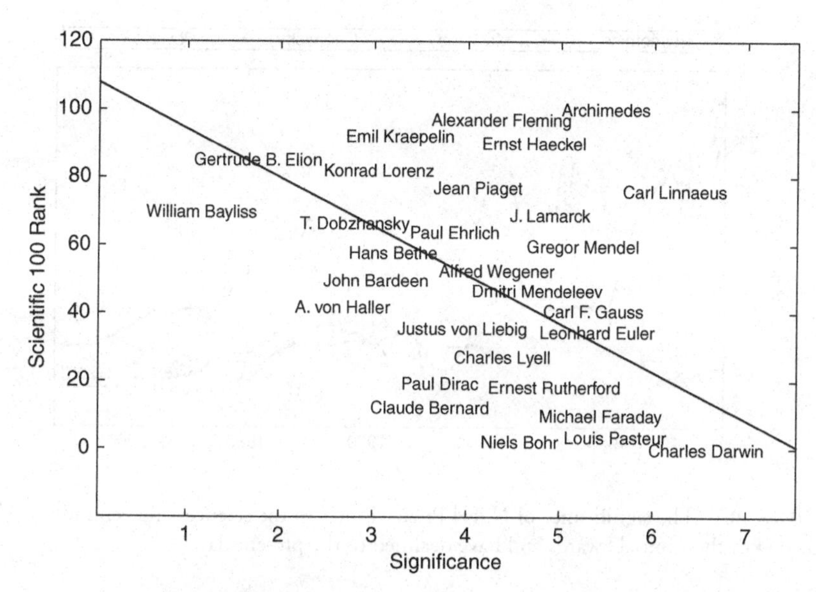

FIGURE 10.3. Significance of scientists by their Scientific 100 rank.

To provide an alternate historical perspective, we include rankings of influential scientists from "The Scientific 100" [Simmons, 2000], who generally do a good job across many disciplines. Figure 10.3 shows the strong correlation between the Science 100 rankings and our significance scores. We tend to rate *Archimedes (287–212 B.C.)* [146] and *Carl Linnaeus (1707–1778)* [31] somewhat more highly than he does, whereas the reverse is true for theoretical physicist *Sheldon Lee Glashow (1932–)* [22885] and physiologist *Claude Bernard (1813–1878)* [3278]. Our top scientists not appearing in Simmons' rankings include the closely associated figures of *Robert Hooke (1635–1703)* [216], whose early work on mechanics influenced Newton, and *Robert Boyle (1627–1691)* [310], the transitional figure between alchemy and chemistry.

10.1.3 THE NOBEL PRIZES

The Nobel Prize winners in physics, chemistry, and medicine provide an interesting window into the popular renown of contemporary scientists. The prize announcements are always among the biggest scientific news stories of the year. There is widespread public perception that the winners of the Nobel Prize rank among the greatest scientists of the day.

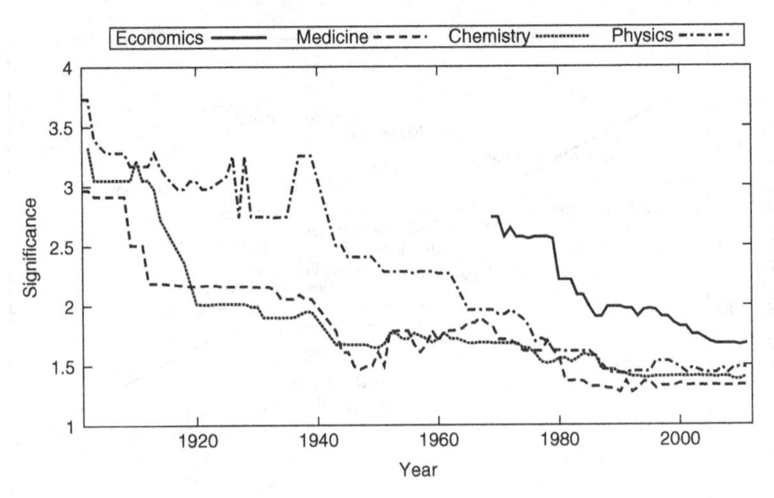

FIGURE 10.4. The significance of Nobel Prize winners in the sciences and economics. All peaked with their initial awards and have declined to the present day.

Figure 10.4 plots the average significance of Nobel Prize winners in each of the science disciplines, as a function of award year. In all these fields, the earliest awardees proved to be the most highly significant group, reflecting the backlog of worthy, older recipients awaiting recognition. Only after this had been cleared could a steady state be reached, where each new year brings in a fresh crop of young scientists, the very best of whom mature into Nobel-caliber figures.

The trends by discipline are interesting and revealing. Physicists have dominated the significance of the winners in the hard sciences since the first prizes were awarded in 1901. They started from a higher peak and

Physics			Chemistry		
Sig	Person	Dates	Sig	Person	Dates
19	Albert Einstein	(1879–1955)	431	Ernest Rutherford	(1871–1937)
514	J. J. Thomson	(1856–1940)	667	Marie Curie	(1867–1934)
539	Max Planck	(1858–1947)	1286	Linus Pauling	(1901–1994)
667	Marie Curie	(1867–1934)	2703	Svante Arrhenius	(1859–1927)
759	Niels Bohr	(1885–1962)	3588	Wilhelm Ostwald	(1853–1932)
777	Guglielmo Marconi	(1874–1937)	3628	Fritz Haber	(1868–1934)
1351	Enrico Fermi	(1901–1954)	3681	Otto Hahn	(1879–1968)
1659	Werner Heisenberg	(1901–1976)	4061	Hermann E. Fischer	(1852–1919)
1987	Erwin Schrödinger	(1887–1961)	4810	William Ramsay	(1852–1916)
2033	Max Born	(1882–1970)	5655	Walther Nernst	(1864–1941)

maintained a deeper bench. *Marie Curie (1867–1934)* [667] sits near the top of the rankings in both physics and chemistry, a tribute to the two prizes she received. Only the first three chemists would be recognizable to the general public, as opposed to most of the top ten physicists.

But the significance of the winners in the hard sciences has decayed at a steady rate, leveling off since about 1980. Despite the continuous breakthroughs in these disciplines, the winners remain a fairly anonymous bunch. There are three primary reasons. First, science has become so highly technical and specialized that it is difficult to inform the public about the achievements being commemorated. Although some awards go to advances that directly touch people's lives, such as the invention of MRI medical imaging technology (awarded to *Paul Lauterbur (1929–2007)* [50006] and *Peter Mansfield (1933–)* [76670]) or integrated circuits (awarded to *Robert Noyce (1927–1990)* [14824] and *Jack Kilby (1923–2005)* [14595]), most scientific research is too obscure to resonate with the public.

Second, contemporary science is a highly collaborative enterprise, making the "Great Man" model of science less tenable. The paper first describing the sequencing of the human genome had more than a hundred authors, from over a dozen different institutions. Today's famous scientists tend to be controversial, at least within the academic community. Indeed, the most publicly visible scientists of recent years, such as physicist *Stephen Hawking (1942–)* [1656] or biologist *Craig Venter (1946–)* [27548], have not received Nobel Prizes as of this writing.

Finally, the scientific community does not venture into public discourse as it did in years past, failing to exploit its Nobel Prizes as a bully pulpit

Medicine

Sig	Person	Dates
962	Robert Koch	(1843–1910)
1124	Alexander Fleming	(1881–1955)
1429	Francis Crick	(1916–2004)
2259	Paul Ehrlich	(1854–1915)
2310	Thomas Hunt Morgan	(1866–1945)
3619	James D. Watson	(1928–)
3938	Camillo Golgi	(1843–1926)
4869	S. R. Y. Cajal	(1852–1934)
4999	Frederick Banting	(1891–1941)
5418	Barbara McClintock	(1902–1992)

Economics

Sig	Person	Dates
700	Milton Friedman	(1912–2006)
1156	Friedrich Hayek	(1899–1992)
3452	Paul Krugman	(1953–)
3654	Paul Samuelson	(1915–2009)
5360	Amartya Sen	(1933–)
5880	Herbert Simon	(1916–2001)
7800	Kenneth Arrow	(1921–)
8607	Gunnar Myrdal	(1898–1987)
8953	Ronald Coase	(1910–)
10212	Robert Solow	(1924–)

for political or social issues. By contrast, chemistry winner *Linus Pauling (1901–1994)* [1286] later won a Peace Prize for his work on nuclear disarmament, and became a controversial advocate for the power of vitamin C. With the possible exception of *James Hansen (1941–)* [19567], no scientist has emerged to become the face of global climate change. Instead, former vice president *Al Gore (1948–)* [623] remains the public figure most closely associated with that issue.

We will see in the next section that the Nobel Prize winners in economics have higher significance today than those of any of the hard sciences, but even they have been trumped by the winners of the prizes for literature and peace. The public profiles of these awardees have remained constantly higher than the science winner, and have not decayed over the past century of Nobel prizes.

10.1.4 ECONOMISTS

Our Nobel Prize significance rankings identify economics as a discipline of greater public stature than today's hard sciences. Modern economics rests on a foundation of advanced mathematics, yet is substantially influenced by political and social forces, making economists more visible public figures than other scientists.

The great classical economists dominate the landscape of our top ranked figures: *Adam Smith (1723–1790)* [56], *John Maynard Keynes (1883–1946)*

Most Significant Economists

Sig	Person	Dates	C/G	Description
56	Adam Smith	(1723–1790)	C G	Economist (*The Wealth of Nations*)
201	John Stuart Mill	(1806–1873)	C G	British philosopher (*On Liberty*)
265	Thomas Robert Malthus	(1766–1834)	C G	English political economist (Malthusianism)
469	John Maynard Keynes	(1883–1946)	C G	British economist (Keynesian economics)
561	David Ricardo	(1772–1823)	C G	English economist (comparative advantage)
700	Milton Friedman	(1912–2006)	C G	Influential American economist (monetarism)
1156	Friedrich Hayek	(1899–1992)	C G	Austrian economist and political thinker
1235	John von Neumann	(1903–1957)	C G	American polymath (Von Neumann architecture)
1538	Henry George	(1839–1897)	C G	American political economist (land value tax)
1717	Ludwig von Mises	(1881–1973)	C G	Philosopher, economist (Austrian School)
2123	Thorstein Veblen	(1857–1929)	C G	Economist ("The Theory of the Leisure Class")
2983	William Stanley Jevons	(1835–1882)	C G	British economist, logician (Jevons paradox)
3264	John Kenneth Galbraith	(1908–2006)	C G	American economist (Keynesian)
3452	Paul Krugman	(1953–)	C G	Economist / *NYT* liberal columnist
3654	Paul Samuelson	(1915–2009)	C G	American economist / Nobelist (econ textbook)

[469], and *David Ricardo (1772–1823)* [561] developed theories governing the behavior of markets, money, and trade. Another thread revolves around political economists, philosophers seeking to design governing systems that maximize output and utility. These include philosophers such as *John Stuart Mill (1806–1873)* [201] and *Thomas Robert Malthus (1766–1834)* [265].

Acolytes of modern conservative economists such as *Milton Friedman (1912–2006)* [700] and *Friedrich Hayek (1899–1992)* [1156] have succeeded in lifting them above their liberal counterparts *Paul Samuelson (1915–2009)* [3654] and *Paul Krugman (1953–)* [3452].

10.1.5 MATHEMATICIANS

In *A Mathematician's Apology*, number theorist *G. H. Hardy (1877–1947)* [3270] writes that "mathematical fame, if you have the cash to pay for it, is one of the soundest and steadiest investments." There are philosophical questions of whether mathematical research is more discovery or invention, but publication standards and definitional clarity ensure that new mathematics gets credited to the appropriate person.

We have identified 20 of the most historically significant mathematicians. We have omitted mathematical physicists such as *Isaac Newton (1643–1727)* [21] and philosophers of logic like *Aristotle (384–322 B.C.)* [8]. What remains are several figures from the ancient world (*Pythagoras (570–495 B.C.)* [133] and *Euclid* [149]) plus others on the boundary of mathematics and philosophy (*René Descartes (1596–1650)* [92] and *Gottfried Leibniz (1646–1716)* [119]).

Most Significant Mathematicians

Sig	Person	Dates	C/G	Description
92	René Descartes	(1596–1650)	C G	French philosopher ("I think, therefore I am")
119	Gottfried Leibniz	(1646–1716)	C G	German philosopher, mathematician (Calculus)
133	Pythagoras	(570–495 B.C.)	C G	Greek mathematician (Pythagorean Theorem)
146	Archimedes	(287–212 B.C.)	C G	Greek mathematician ("Eureka!")
149	Euclid	(?–?)	C G	Greek mathematician (Euclidean geometry)
157	Blaise Pascal	(1623–1662)	C G	Mathematician, philosopher (Pascal's Wager)
185	Leonhard Euler	(1707–1783)	C G	Swiss mathematician (mathematical constant *e*)

(continued)

207	Carl Friedrich Gauss	(1777–1855)	C	G	German mathematician (Gaussian distribution)
273	Charles Babbage	(1791–1871)	C	G	Mathematician, inventor (difference engine)
556	Thales	(635–543 B.C.)	C	G	Greek philosopher ("the first mathematician")
638	Georg Cantor	(1845–1918)	C	G	German mathematician (set theory)
659	Pierre-Simon Laplace	(1749–1827)	C	G	Mathematician, astronomer (Laplace transform)
683	Joseph Louis Lagrange	(1736–1813)	C	G	Mathematician, astronomer (Lagrange points)
785	Henri Poincaré	(1854–1912)	C	G	French mathematician (Poincaré conjecture)
886	M. i. M. al-Khwârizm'	(780–850)	C	G	Persian mathematician (algebra)
924	Pierre de Fermat	(1601–1665)	C	G	French lawyer (Fermat's last theorem)
994	Ada Lovelace	(1815–1852)	C	G	English mathematician (Analytical Engine)
1091	Bernhard Riemann	(1826–1866)	C	G	German mathematician (Riemann Hypothesis)
1127	Alan Turing	(1912–1954)	C	G	British "Father of Computer Science"
1170	George Boole	(1815–1864)	C	G	English mathematician (Boolean algebra)

The most significant modern mathematicians are associated with computers. *Charles Babbage (1791–1871)* [273] made a valiant but unsuccessful attempt to build a nineteenth-century computing engine out of shafts and gears. *Blaise Pascal (1623–1662)* [157] also invented a mechanical calculator, but his greater accomplishment was laying the foundations of probability theory. The most prominent twentieth-century mathematicians, *Alan Turing (1912–1954)* [1127] and *John von Neumann (1903–1957)* [1235], made important early contributions to computer architecture as a consequence of their theoretical work.

The Fields Medal, awarded since 1936, serves as the equivalent of the Nobel Prize in Mathematics. But the medal's rules ultimately limit its public visibility. Instead of an annual prize, it is awarded only once every four years. It is restricted to mathematicians under the age of 40, meaning that recipients have not had enough time to establish a public presence prior to selection. Many important mathematicians have been overlooked during this narrow window (such as the peripatetic combinatorialist *Paul Erdos (1913–1996)* [5640]) or achieved their central result later (like *Andrew Wiles (1953–)* [12376], conqueror of Fermat's last theorem). And finally, the cash prize of $15,000 is only 1 percent of the value of a Nobel Prize, and so does not signal its importance outside the mathematical community. Norway's *Niels Henrik Abel (1802–1829)* [2155] Prize, inaugurated in 2003, addresses all these deficiencies, and we anticipate that it will soon be recognized as the greatest honor in mathematics.

Among the most famous living mathematicians is game theorist *John Forbes Nash (1928–)* [8513], but his popular renown is partly a consequence of his battles with schizophrenia, dramatized in the movie *A Beautiful Mind*. Discrete mathematician and computer scientist *Donald Knuth (1938–)* [5917] appears to be the most historically significant living mathematician.

10.1.6 MEDICAL SCIENTISTS

Modern medicine is a not a science, though it interacts strongly with the biological sciences. This distinction was much less clear during the early history of scientific medicine, when the exploration of human anatomy depended upon access to former patients for dissection. Our rankings try to recognize healing physicians instead of the scientists who unraveled the mysteries of disease, such as *Louis Pasteur (1822–1895)* [112].

The great physicians of ancient times, *Hippocrates (460–370 B.C.)* [239] and *Galen (129–201)* [324], influenced medical treatment through the Renaissance. Prohibitions against human dissection froze knowledge of human anatomy until the Renaissance, when *Andreas Vesalius (1514–1564)* [754] published a new anatomy based on direct observation. Only then could *William Harvey (1578–1657)* [468] and *Marcello Malpighi (1628–1694)* [3415] unravel the mysteries of blood circulation.

Most Significant Physicians

Sig	Person	Dates	C/G	Description
239	Hippocrates	(460–370 B.C.)	C G	Ancient Greek physician (Hippocratic Oath)
324	Galen	(129–201)	C G	Influential Roman physician, philosopher
468	William Harvey	(1578–1657)	C G	English physician (systemic circulation)
746	Edward Jenner	(1749–1823)	C G	English physician (smallpox vaccine)
754	Andreas Vesalius	(1514–1564)	C G	Anatomist ("De humani corporis fabrica")
805	Rudolf Virchow	(1821–1902)	C G	German doctor, biologist (cellular pathology)
901	Joseph Lister	(1827–1912)	C G	British surgeon (antiseptic surgery)
962	Robert Koch	(1843–1910)	C G	German physician (Koch's postulates)
998	Paracelsus	(1493–1541)	C G	Physician, alchemist, botanist (zinc)
1471	Ignaz Semmelweis	(1818–1865)	C G	Hungarian physician (antiseptic obstetrics)
1942	John Snow	(1813–1858)	C G	English physician (anesthesia pioneer)
2441	William Osler	(1849–1919)	C G	Physician (one of Johns Hopkins "Big Four")
2883	Jean-Martin Charcot	(1825–1893)	C G	Neurologist ("Founder of modern neurology")
3415	Marcello Malpighi	(1628–1694)	C G	Italian doctor ("Malpighian tubule system")
3775	Jonas Salk	(1914–1995)	C G	American medical researcher (polio vaccine)

The staggering toll of maternal deaths from infection during childbirth did not abate until antiseptic surgical procedures were widely adopted.

Ignaz Semmelweis (1818–1865) [1471] and *Joseph Lister (1827–1912)* [901] pioneered antiseptic surgery. *John Snow (1813–1858)* [1942] established the importance of epidemiology, famously tracing the source of a cholera outbreak to one particular water pump. Standards for medical training in the nineteenth century were remarkably low, reflecting the limited state of real knowledge. *William Osler (1849–1919)* [2441] is the seminal figure in modern medical education.

With complete understanding of human anatomy and the germ theory of disease, medical research has now largely shifted from the clinic to the laboratory. *Robert Koch (1843–1910)* [962] and *Jonas Salk (1914–1995)* [3775] are transitional figures, trained as physicians, but famous for fundamental research leading to effective therapies. The days of the heroic, world-famous surgeon appear to have ended after pioneering cardiac surgeons *Christiaan Barnard (1922–2001)* [10749] and *Michael E. DeBakey (1908–2008)* [9258] made heart transplants a reality. Today, the Nobel Prize in Medicine is seldom awarded to a clinical physician.

Dentists

By comparison, few of the most prominent dentists earned their reputation working on teeth. Foremost were *William T. G. Morton (1819–1868)* [7233] and *Horace Wells (1815–1848)* [25753], whose demonstrations of surgical anesthesia made possible today's (relatively) painless dentistry. *Pierre Fauchard (1678–1761)* [16496] is the father of modern dentistry, responsible for introducing dental fillings and the notion that sugar caused cavities.

Most Significant Dentists

Sig	Person	Dates	C/G	Description
2147	Doc Holliday	(1851–1887)	C G	American gambler, gunfighter (O.K. Corral)
4066	Zane Grey	(1872–1939)	C G	Western author (*Riders of the Purple Sage*)
7233	William T. G. Morton	(1819–1868)	C G	Dentist and pioneer of anesthesia
16496	Pierre Fauchard	(1678–1761)	C G	French physician ("father of modern dentistry")
16968	Fatima Jinnah	(1893–1967)	C G	Dental surgeon and Founding Mother of Pakistan

The remaining dentists are a diverse group who generally used their stable, prosperous, and respectable dental position as a springboard for their real career: in writing (*Zane Grey (1872–1939)* [4066]), politics

(*Gurbanguly Berdimuhamedow (1957–)* [36382] of Turkmenistan), or gun slinging (*Doc Holliday (1851–1887)* [2147]).

Psychologists

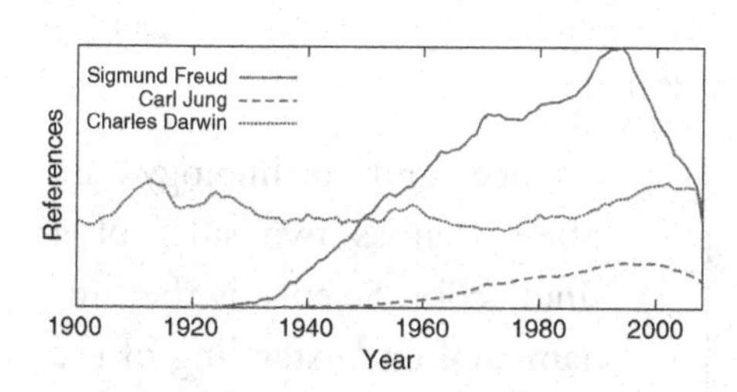

Philosophers and medical doctors began to study the workings of the human mind during the second half of the nineteenth century. *William James's (1842–1910)* [442] book *Principles of Psychology* assembled the foundations of an emerging discipline. On the medical side, *Wilhelm Wundt (1832–1920)* [1464] can be considered as the father of experimental psychology.

The pioneers in medical psychoanalysis were *Sigmund Freud (1856–1939)* [44] and *Carl Jung (1875–1961)* [333]. Freud was one of the dominant figures of his century. However, his reputation rapidly declined once effective psychotropic drugs replaced psychoanalysis as the primary treatment for mental illness. Ngram time series shows that *Charles Darwin (1809–1882)* [12] was more prominent than *Sigmund Freud* [44] until 1950. Freud's mindshare continued to grow until his plunge to Darwinian levels began in 1990. Our analysis suggests that by 2015, Darwin will reemerge as the more relevant scientist. Jung's reputation has grown at Freud's expense, and appears to be holding on more tenaciously.

Most Significant Psychologists

Sig	Person	Dates	C/G	Description
44	Sigmund Freud	(1856–1939)	C G	Neurologist and creator of psychoanalysis
333	Carl Jung	(1875–1961)	C G	Swiss psychologist (Jungian psychotherapy)
417	John Dewey	(1859–1952)	C G	American philosopher (pragmatism)
442	William James	(1842–1910)	C G	Psychologist, philosopher (pragmatism)
1464	Wilhelm Wundt	(1832–1920)	C G	Father of experimental psychology
1686	B. F. Skinner	(1904–1990)	C G	American psychologist (behaviorism)
1834	Jacques Lacan	(1901–1981)	C G	French psychoanalyst (post-structuralist)
2824	Wilhelm Reich	(1897–1957)	C G	Austrian psychoanalyst (Freudo-Marxism)
2831	Timothy Leary	(1920–1996)	C G	Psychologist ("Turn on, tune in, drop out")

Several famous psychologists were controversial. The popularizing interests of *Wilhelm Reich (1897–1957)* [2824] and *Timothy Leary (1920–1996)*

[2831] in sexual energy (organe) and psychedelic drugs led to a general discrediting of their scientific work. *Jacques Lacan*'s *(1901–1981)* [1834] philosophical ideas on post-structuralist theory may have had a greater influence on literary analysis than psychoanalysis.

10.2 Inventors

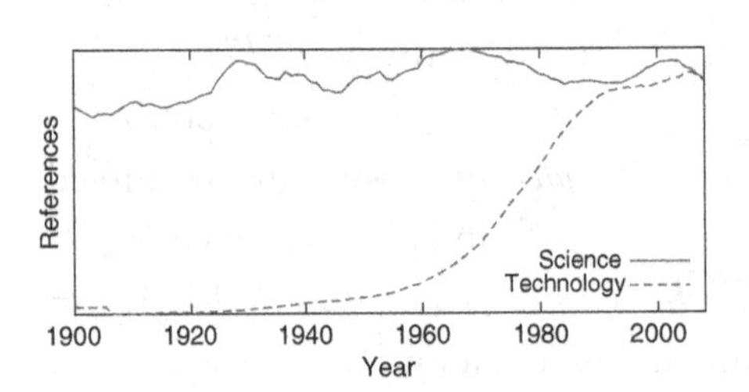

Science and technology are often seen as two sides of a single coin. Science is the fundamental understanding of the natural world. Technology is the extent to which we can manipulate the natural world for our betterment. Although technology is an old word, the Ngram dataset shows that it was rarely employed until World War II. Now its use has exploded to where it rivals that of science itself.

10.2.1 THE GOLDEN AGE OF INVENTION

The iconic image of the inventor is of a tinkerer: the mad genius puttering around in a laboratory or garage. However, many of the best known "inventors" made their reputation protecting and selling the better mousetrap, independent of niceties about whether they really invented it.

Most Significant Inventors

Sig	Person	Dates	C/G	Description
40	Thomas Edison	(1847–1931)	C G	Inventor (lightbulb, phonograph)
93	Nikola Tesla	(1856–1943)	C G	Inventor (alternating current)
106	Alexander Graham Bell	(1847–1922)	C G	Inventor (telephone)
112	Louis Pasteur	(1822–1895)	C G	French chemist and founder of microbiology
328	Johannes Gutenberg	(1398–1468)	C G	German printer (movable type)
337	Walt Disney	(1901–1966)	C G	American film animation ("Mickey Mouse")
532	Samuel Morse	(1791–1872)	C G	Inventor (telegraph system)
547	Alfred Nobel	(1833–1896)	C G	Founded Nobel Prizes, invented dynamite
758	Eli Whitney	(1765–1825)	C G	American inventor (cotton gin)
777	Guglielmo Marconi	(1874–1937)	C G	Italian inventor (Marconi's law)
887	George Washington Carver	(1864–1943)	C G	Agricultural scientist (uses for peanuts)
954	Robert Fulton	(1765–1815)	C G	Engineer and inventor (steamboat)

(continued)

Sig	Person	Dates	C/G	Description
1015	Eadweard Muybridge	(1830–1904)	C G	English photographer (Yosemite Valley)
1351	Enrico Fermi	(1901–1954)	C G	American physicist (first nuclear reactor)
1461	Gottlieb Daimler	(1834–1900)	C G	Automotive pioneer (Daimler Motors)
1584	George Eastman	(1854–1932)	C G	Inventor, entrepreneur (Eastman Kodak)
1812	George Westinghouse	(1846–1914)	C G	American entrepreneur (Westinghouse Electric)
1929	Rudolf Diesel	(1858–1913)	C G	German inventor (diesel engine)
1984	Charles Goodyear	(1800–1860)	C G	American inventor (vulcanized rubber)
2027	Louis Comfort Tiffany	(1848–1933)	C G	American artist and designer (Favrile glass)

The most famous Golden Age inventor figures were usually successful industrialists, who built long-lasting companies. These include *Thomas Edison (1847–1931)* [40] (General Electric), *Henry Ford (1863–1947)* [148] (Ford Motor Company), and *Alexander Graham Bell (1847–1922)* [106] (Bell Telephone). On the other hand, *Walter Hunt (1796–1859)* [16684] was a tinkerer: he patented the safety pin, the repeating rifle, and the fountain pen without making a killing on any of them.

Elias Howe (1819–1867) [6927] may be most responsible for inventing the modern sewing machine, but it was *Isaac Singer (1811–1875)* [4114] who got them into almost every home. By contrast, *George Westinghouse (1846–1914)* [1812] achieved success by inventing the railway air brake, but his fame was derived from building the Westinghouse Electric Corporation. Other inventors died poor, but later had companies named after them, such as *Charles Goodyear (1800–1860)* [1984].

We consider the most significant members of the National Inventors Hall of Fame, located on the grounds of the U.S. Patent and Trademark Office in Alexandria, Virginia. Since their establishment in 1973, they have diligently elected about ten new members every year. They have amassed an eclectic bunch ranging from *Thomas Edison (1847–1931)* [40] to *Robert Adler (1913–2007)* [67242], the couch-potato-creating inventor of the TV remote control.

The scientists prove more distinguished, with 20 ranking among the 500 most significant people in history compared to seven for the inventors. Several members of the Inventors Hall of Fame are better known as scientists (*Enrico Fermi (1901–1954)* [1351] and *Louis Pasteur (1822–1895)* [112]), artists (*Louis Comfort Tiffany (1848–1933)* [2027]), explorers (*Jacques Cousteau (1910–1997)* [2799]), or even entertainers (guitarist *Les*

Paul (1915–2009) [2214] and animator *Walt Disney (1901–1966)* [337]) – but all are unquestionably innovative and creative thinkers.

10.2.2 MODERN INVENTION

The most significant members of the Inventors Hall of Fame have one thing in common: all are long dead. Despite living in a technological age, we seem to lack the great inventors of yesteryear.

We can get a perspective on contemporary times through the winners of the Lemelson-MIT Prize for invention. Endowed by *Jerome H. Lemelson (1923–1997)* [58245], himself holder of 605 patents, the prize has been awarded annually since 1995 and now provides $500,000 to the recipient. So who are the lucky winners?

Most Significant Lemelson-MIT Prize Winners

Sig	Person	Dates	C/G	Description
5796	Douglas Engelbart	(1925–2013)	C G	American inventor (computer mouse, hypertext)
14636	David Packard	(1912–1996)	C G	American businessman (Hewlett-Packard)
15345	Ray Kurzweil	(1948–)	C G	Inventor (OCR), futurist (transhumanism)
19057	William R. Hewlett	(1913–2001)	C G	American engineer (co-founded Hewlett-Packard)
35915	Dean Kamen	(1951–)	C G	American entrepreneur, inventor (Segway)

The most significant winner has been *Douglas Engelbart (1925–2013)* [5796], recognized as the inventor of the soon-to-vanish but still-ubiquitous computer mouse. More important, he deserves much of the credit for today's networked, interactive computing environments. The two founders of Hewlett-Packard, *David Packard (1912–1996)* [14636] and *William Reddington Hewlett (1913–2001)* [19057], created a culture of invention at HP and in the Silicon Valley, but are more renowned as businessmen than inventors.

Two other names are perhaps recognizable to the broader public. Futurist *Ray Kurzweil (1948–)* [15345] rose to prominence by inventing music synthesizers and book-reading machines for the blind. *Dean Kamen (1951–)* [35915] invented the Segway human transporter, to enable security personnel to zip through shopping malls and airport terminals. Both fit the public conception of an "inventor," but neither seem destined to rival *Alexander Graham Bell (1847–1922)* [106] or even *Lee De Forest*

(1873–1961) [2525] (inventor of a vacuum tube amplifier important for early radios) in the history books.

One consequence of the increasing complexity of technology is that contemporary inventions no longer have recognizable inventors. We live in an age of large industrial research teams, led by management figures less responsible for the inspiration than for coordinating the perspiration.

Who invented television? Was it *Philo Farnsworth (1906–1971)* [4812], who first demonstrated an electronic television system to the public? *Vladimir K. Zworykin (1888–1982)* [8005], who developed the cathode ray tube (CRT) display, only recently displaced by flat screen televisions? Or *David Sarnoff (1891–1971)* [5939], who created the commercial television industry at the Radio Corporation of America (RCA) and the National Broadcast Corporation (NBC)? The answer is yes to all of them.

The computer has an equally murky origin. Information processing technology dates back to *Herman Hollerith's (1860–1929)* [2544] punched card machines for the U.S. Census. *Alan Turing (1912–1954)* [1127] and *John von Neumann (1903–1957)* [1235] were the intellectual godfathers of contemporary stored-program machines. *John Mauchly (1907–1980)* [7494] and *J. Presper Eckert (1919–1995)* [18998] built ENIAC, the first electronic computer, but perhaps stole key ideas from *John Vincent Atanasoff (1903–1995)* [16594]. *Steve Jobs (1955–2011)* [2051] and *Bill Gates (1955–)* [904] arguably invented the personal computer and the microcomputer software industry, respectively.

The passing of Apple founder *Steve Jobs (1955–2011)* [2051] brought forth eulogies comparing Jobs to Edison. Both men changed the way people encounter technology, but Edison's contributions were vastly deeper. The incandescent bulb changed night into day; the phonograph turned music from a personal to a public experience; motion pictures transformed us from a written to a visual culture. The personal computing technologies Jobs championed will quickly fade from view as new ones arise to take their place. We predict Jobs will eventually be remembered more like the inventor of the telegraph *Samuel Morse (1791–1872)* [532]: pioneering a historically important but transitional technology that will soon be remote from contemporary experience.

So who should be credited as the inventor of the computer in Bonnie's history book?

10.2.3 PATENTS AND INVENTION

In *Bambi vs. Godzilla* [Mamet, 2007], screenwriter David Mamet boils all the sins of Hollywood down into a single factoid. In 1958, 230 Hollywood producers created 2,000 films. By 2003, the ratio had shifted to 1,200 producers for only 240 films. What has been lost was a sense of personal auteurship, resulting in less creativity and innovation despite seemingly greater available resources.

A study of invention through the statistical lens of patents would reveal a similar story. Although there appear to be few truly seminal new inventions, the number of patents awarded has skyrocketed, along with the number of patented inventors. What is the reason for this disconnect?

The modern era of invention could not get started before the creation of a patent system. The experience of *Johannes Gutenberg (1398–1468)* [328] with his printing press was probably typical. His printing technology rapidly spread through Europe, without anyone paying him royalties. He promptly went bankrupt, although late in life the royal court honored him with a stipend.

The British patent system dates back to the reign of *James I (1566–1625)* [41], and slowly evolved into its modern form. The industrial revolution sparked a greater demand for innovation and the need to protect it. The United States set up its patent system early in the republic, and issued the first patent to *Samuel Hopkins (1743–1818)* [30255] on July 31, 1790: it was signed by both President *George Washington (1732–1799)* [6] and Secretary of State *Thomas Jefferson (1743–1826)* [10].

Identifying the most significant patent holder proves to be relatively easy. *Abraham Lincoln (1809–1865)* [5] received U.S. Patent number 6,469 for "A Device for Buoying Vessels Over Shoals" on May 22, 1849. What should stand out is how small his patent number was. During its first 60 years of operation, the patent office awarded an average of only 110 patents per year.

This slow rate of growth is apparent from the patent numbers of iconic inventions. *Eli Whitney's (1765–1825)* [758] cotton gin received patent 72X on March 14, 1794. *Samuel Morse's (1791–1872)* [532] telegraph received patent number 1,647 on June 20, 1840, meaning that the total number of patents issued tripled during the 1840s before Lincoln got his.

10.3 Business Leaders

Business propagates technological improvements to society at large. Many prominent corporations have been built on the strength of a particular idea or invention, which is often named for its creator: *Philip Danforth Armour (1832–1901)* [15273], meatpacking; *Richard Warren Sears (1863–1914)* [26813], department stores; *William Wrigley (1861–1932)* [12068], chewing gum. The U.S. patent office now awards patents for business processes – more evidence of the tight ties between business and innovation.

Figure 10.5 presents our rankings of the most historically significant business leaders. Almost half made their fortunes during the Gilded Age, running roughly from 1870 to 1900. This is when the railroads were built and the financial structure of modern corporations established. It was a period of rapid industrial growth after the destruction of the Civil War. The Progressive Era soon followed, bringing with it antitrust actions to break up the giant monopolies that built these great fortunes. Today, we are living in a second Gilded Age. Several of the historically most significant businessmen in our table are still alive and active.

Sig	W100	Person	Dates	Business	Company
104	5	Andrew Carnegie	(1835–1919)	Steel	Carnegie Steel
148	11	Henry Ford	(1863–1947)	Cars	Ford Motor Company
172	1	John D. Rockefeller	(1839–1937)	Oil	Standard Oil
322	23	J. P. Morgan	(1837–1913)	Finance	JP Morgan Chase
337		Walt Disney	(1901–1966)	Media	The Walt Disney Company
565		William Randolph Hearst	(1863–1951)	Media	Hearst Corporation
706		Cecil Rhodes	(1853–1902)	Mining	De Beers
767		P. T. Barnum	(1810–1891)	Media	Barnum and Bailey Circus
898	81	Howard Hughes	(1905–1976)	Oil	Hughes Aircraft
904	31	Bill Gates	(1955–)	Software	Microsoft
925	2	Cornelius Vanderbilt	(1794–1877)	Railroad	Accessory Transit Company
1281	69	Johns Hopkins	(1795–1873)	Railroad	Johns Hopkins Hospital
1413	8	Jay Gould	(1836–1892)	Finance	Union Pacific
1455	3	John Jacob Astor	(1763–1848)	Trading	American Fur Company
1549	79	Joseph Pulitzer	(1847–1911)	Media	New York World
1584	44	George Eastman	(1854–1932)	Photography	Eastman Kodak
1766		Rupert Murdoch	(1931–)	Media	News Corporation
1858	39	Warren Buffett	(1930–)	Finance	Brekshire Hathaway
1920	34	Leland Stanford	(1824–1893)	Railroad	Southern Pacific Company
2051		Steve Jobs	(1955–2011)	Computer	Apple Computer

FIGURE 10.5. The most significant business leaders, with Wealthy 100 ranking and associated companies.

Many are known for their philanthropy, giving back to society much of what they arguably stole during the making of their fortunes. Steel baron *Andrew Carnegie (1835–1919)* [104] was among the richest men of all time, yet gave away all of his fortune to build 3,000 libraries and countless other institutions. Oil magnate *John D. Rockefeller (1839–1937)* [172] provided major funding to schools from Spelman College to the University of Chicago, plus countless New York institutions like Rockefeller University. *Bill Gates (1955–)* [904] is very much a modern parallel, using the first half of his life to ruthlessly accumulate great wealth, and the second half seeing that it is used effectively for noble causes such as curing disease.

10.3.1 WEALTH AND STATURE

Business leaders tend to be a wealthy bunch. We present the relative significance of the wealthiest people in U.S. history in Figure 10.6, as identified in [Klepper and Gunther, 1996]. Comparing economic clout across different timescales is challenging. One reason is the need to correct for inflation, which makes it meaningless to compare absolute dollar amounts from different periods. In principle, this can be done by converting wealth into purchasing power, the amount of goods that could be bought using one dollar in (say) 1970. A second factor concerns the relative size of the national

Person	Dates	W100	GDP-R	Sig	Description
John D. Rockefeller	(1839–1937)	1	65	172	American oil magnate (Standard Oil)
Cornelius Vanderbilt	(1794–1877)	2	87	925	American industrialist, philanthropist
John Jacob Astor	(1763–1848)	3	107	1455	American business magnate (Astor family)
Stephen Girard	(1750–1831)	4	150	15469	American banker, philanthropist (1812 bailout)
Andrew Carnegie	(1835–1919)	5	166	104	Steel magnate and philanthropist
Alexander Turney Stewart	(1803–1876)	6	178	8732	Irish entrepreneur (dry goods)
Friedrich Weyerhäuser	(1834–1914)	7	182	106684	German American timber mogul
Jay Gould	(1836–1892)	8	185	1413	American railroad robber baron
Stephen Van Rensselaer	(1764–1839)	9	194	9463	Lieutenant gov. of NY (founded RPI)
Marshall Field	(1834–1906)	10	205	5578	American businessman (department stores)
Henry Ford	(1863–1947)	11	231	148	American industrialist (Ford Motors)
Andrew W. Mellon	(1855–1937)	12	258	2341	Banker, philanthropist (Carnegie Mellon Univ.)
Richard B. Mellon	(1858–1933)	13	258	51321	Brother, business partner of Andrew Mellon
Sam Walton	(1918–1992)	14	275	4923	American businessman (Walmart)
James Graham Fair	(1831–1894)	15	280	34520	Politician (part owner of Comstock Lode)

FIGURE 10.6. Historical significance of the wealthiest U.S. figures from the Wealthy 100 [Klepper and Gunther, 1996].

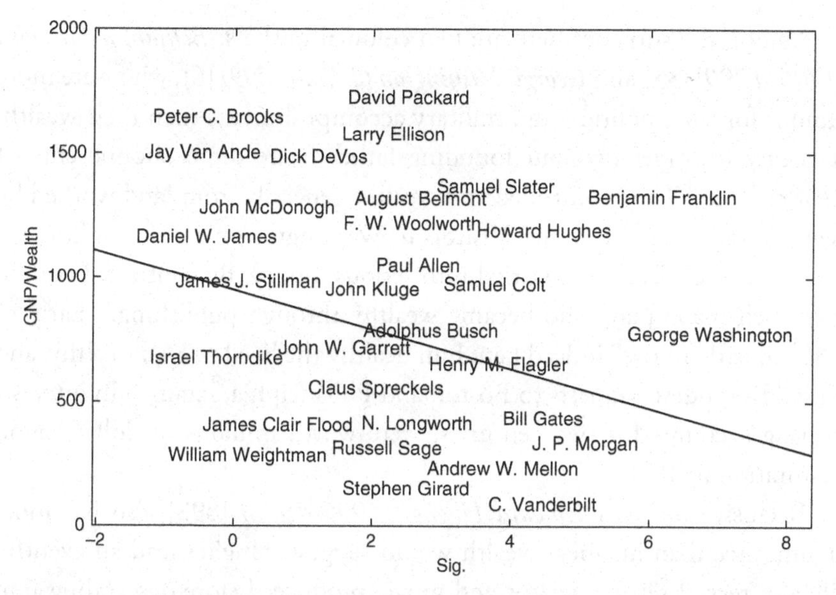

FIGURE 10.7. Historical significance versus GDP/Wealth for members of the Wealthy 100, with regression line. Wealthier people generally prove more historically significant.

economy. My economic clout depends upon how my wealth stacks up to the people around me. Those who become famous for their wealth must stand out among their peers in what they own and control.

To identify the Wealthy 100, they normalized accumulated wealth by taking the ratio with respect to the Gross Domestic Product (GDP) of that time. When *Bill Gates (1955–)* [904] led the Forbes 400 with an estimated wealth of $59 billion, the U.S. GDP was $14.5 *trillion*, or roughly equal to 246.2 times Bill Gates.[1] In comparison, they credit *John D. Rockefeller (1839–1937)* [172] with a ratio of 65, or almost four times the clout of Bill Gates.

Figure 10.7 plots historical significance versus normalized wealth for members of the Wealthy 100. The regression line confirms that historical significance generally increases with riches. It also identifies outliers. Those below the regression line were more wealthy than their significance indicates, while those above the line achieved greater renown with less lavish living.

[1] *The Wealthy 100* was published in 1996, at which point Gates was credited with a ratio of 425. For consistency, we will use the published numbers throughout this section.

Among the surprises here are two colonial outliers, *Benjamin Franklin (1706–1790)* [35] and *George Washington (1732–1799)* [6], who were more famous for their political and military accomplishments than their wealth. It is easy to forget that our founding fathers were the economic elite of their colonies. They controlled vast wealth, generally from land worked by slaves or indentured servants. Visitors to Washington's magnificent home at Mount Vernon learn he married into serious money. By contrast, Franklin was a self-made man who became wealthy through publishing. "Early to bed and early to rise" indeed kept him healthy (he lived to 84), wealthy, and wise. His modest bequests to Boston and Philadelphia, retained in interest-bearing accounts for two centuries, were worth millions of dollars upon maturation in 1990.

Reclusive oil man *Howard Hughes (1905–1976)* [898] also has more significance than his mere wealth would suggest. Hughes lead an eventful life as a record-setting aviator and movie producer before descending into mental illness. His massive estate was left to the Howard Hughes Medical Institute, where it funds research in the life sciences.

Those below the regression line are either overvalued by wealth or undervalued by significance. Extreme points include *Stephen Girard (1750–1831)* [15469], a banker who purchased the remnants of the First Bank of United States in 1811 and personally financed much of the War of 1812. He is posited to be the fourth-wealthiest American of all time. *Friedrich Weyerhäuser (1834–1914)* [106684] was a timber mogul whose eponymous company (Weyerhauser) remains the world's largest seller of wood and paper products. Yet he is surprisingly little known, despite having been the eighth-wealthiest American in history.

10.3.2 PHILANTHROPIC IMPACT OF UNIVERSITIES

Weyerhäuser's relative anonymity today is a reflection of his limited philanthropic activities. Money can't buy you love, but it can buy you fame.

Figure 10.8 provides a ranking of the most significant figures who have North American institutions of higher learning named after them.[2]

[2] Note that *William Legge's (1731–1801)* [12631] title was the second Earl of Dartmouth.

Person	Dates	Sig	School	Students
Andrew Carnegie	(1835–1919)	104	Carnegie Mellon University	10970
Cornelius Vanderbilt	(1794–1877)	925	Vanderbilt University	12714
Johns Hopkins	(1795–1873)	1281	Johns Hopkins University	19020
Andrew W. Mellon	(1855–1937)	2341	Carnegie Mellon University	10970
Peter Cooper	(1791–1883)	4644	Cooper Union	918
John Harvard	(1607–1638)	5045	Harvard University	21225
Bernard Baruch	(1870–1965)	5962	Baruch College	12870
Charles F. Kettering	(1876–1958)	8308	Kettering University	1922
Elihu Yale	(1649–1721)	8948	Yale University	11993
Stephen Van Rensselaer	(1764–1839)	9463	Rensselaer Polytechnic Institute	7523
David Lipscomb	(1831–1917)	10422	Lipscomb University	3742
Ezra Cornell	(1807–1874)	11059	Cornell University	20939
James Bowdoin	(1726–1790)	11167	Bowdoin College	1777
Oral Roberts	(1918–2009)	12399	Oral Roberts University	3790
William Legge	(1731–1801)	12631	Dartmouth College	4196

FIGURE 10.8. Historical significance of funders of namesake American colleges and universities.

With few exceptions, they received this honor to acknowledge financial contributions to the institution.

Johns Hopkins (1795–1873) [1281] bequeathed $7 million in 1876 to create the first real research institution in the United States. But the donations founding several Ivy League institutions seem fairly modest by comparison. *Ezra Cornell (1807–1874)* [11059], founder of Western Union, attained his university for $400,000 in 1865. *John Harvard (1607–1638)* [5045] bequeathed £780 to the university, plus 320 books, in 1638. *Elihu Yale (1649–1721)* [8948] received an even better deal, donating goods worth £800 in 1718.

A small positive correlation exists between the magnitude of the educational institution (as measured by its total enrollment) and the historical significance of its namesake. This is true even though many donations are made to honor family members. For example, railroad magnate *Leland Stanford (1824–1893)* [1920] founded Stanford University as a memorial for his young departed son *Leland Stanford Jr. (1868–1884)* [46110], yet the senior Stanford receives much of the recognition himself.

Here is a note we address to any spectacularly wealthy reader of this book. Perhaps you have reached a stage in life where you are thinking about your legacy. How will you be remembered? What can you do to ensure that you will be honored by future generations? Your strongest asset would be establishing a durable institution devoted to perpetuating your memory. Founding a successful self-named company like Westinghouse ensures name propagation, but only until the company fails or is acquired. It would be far better to get your name attached to a major university.

But building a new university is expensive and problematic, typified by the difficulty Domino's Pizza founder *Tom Monaghan (1937–)* [22015] has experienced in founding his new Ave Maria University. Instead, we believe that you are better off buying the naming rights to an existing state university. The precedent here is the former Glassboro State College in New Jersey, which became Rowan University in the wake of a $100 million donation by *Henry Rowan (1923–)* [166156], in 1997.

The budgetary pressures of state universities would make many places amenable to such a deal. We believe that you could start a serious discussion on renaming our university (Stony Brook) for about $1 billion. No promises, but we are willing to serve as liaison between management and any interested investors.

The notorious robber baron *Jay Gould (1836–1892)* [1413] was one of the richest men in the world, yet today has an Ngram reference frequency roughly half that of "Stony Brook." Let this be a cautionary tale for the robber barons of today. Imagine how much better Gould's historical reputation would be if he had invested his money in naming a great university instead of leaving it to be squandered by his family.

10.4 Explorers and Adventurers

Scientists seek understanding of the world we live in, while explorers search for new places on an increasingly finite globe. The Great Age of Exploration culminated in the discovery of the New World, which has driven much of history ever since. The primary people associated with this quest appeared in Bonnie's textbook, and their exploits were thoroughly discussed in Section 3.5.3.

Here we focus on the final phases of man's geographic exploration: the discovery of the polar regions and the conquest of space. These events are recent enough that our datasets permit us to witness reputations settle into their current historical consensus.

10.4.1 POLAR EXPLORATION

The great era of polar exploration, from roughly 1890 to 1930, is particularly interesting because it occurred during our modern media era. It is difficult to reconstruct how rapidly news of *Christopher Columbus's (1451–1506)* [20] discoveries spread through Europe, but through book Ngram data we can relive the conquest of the poles. Several interesting trends become apparent when we do.

Most Significant Polar Explorers

Sig	Person	Dates	C/G	Description
730	Ernest Shackleton	(1874–1922)	C G	Arctic explorer (*The Endurance*)
808	Robert Falcon Scott	(1868–1912)	C G	British Royal Navy officer, explorer
987	John Franklin	(1786–1847)	C G	Explorer ("Northwest Passage")
1109	Roald Amundsen	(1872–1928)	C G	Norwegian explorer (led first South Pole team)
1550	Fridtjof Nansen	(1861–1930)	C G	Norwegian arctic explorer, Nobel laureate
2026	James Clark Ross	(1800–1862)	C G	British explorer (Ross Ice Shelf)
2199	Douglas Mawson	(1882–1958)	C G	Australian explorer (on first South Pole team)
2227	Edmund Hillary	(1919–2008)	C G	Mountaineer (first to summit Mount Everest)
2255	Robert Peary	(1856–1920)	C G	Explorer (disputedly first to North Pole)
2270	Alexander Mackenzie	(1764–1820)	C G	Scottish explorer (overland crossing of Canada)
2367	Richard Evelyn Byrd	(1888–1957)	C G	American aviator and aerial polar explorer

The three most significant polar explorers are better known for their disasters than their successful missions. *Ernest Shackleton's (1874–1922)* [730] ship *Endurance* was trapped and crushed in pack ice during his 1914–1915 Antarctic expedition. Through his leadership and bravery, all his men were rescued after surviving three years in the region. *John Franklin's (1786–1847)* [987] efforts to navigate the Northwest Passage were less successful: he and his entire crew perished.

The race to discover the South Pole in 1912 pitted a British team led by *Robert Falcon Scott (1868–1912)* [808] against a Norwegian team led by *Roald Amundsen (1872–1928)* [1109]. Amundsen's team was much better prepared for Antarctic conditions, and was first to reach the pole. Scott arrived a month later, only to discover he had been beaten. His team ran out of supplies on the way back and perished, just 11 miles short of their supply depot. The loser of this tragic contest now outranks the winner in historical significance, another demonstration of the harrowing power of martyrdom to capture the popular imagination.

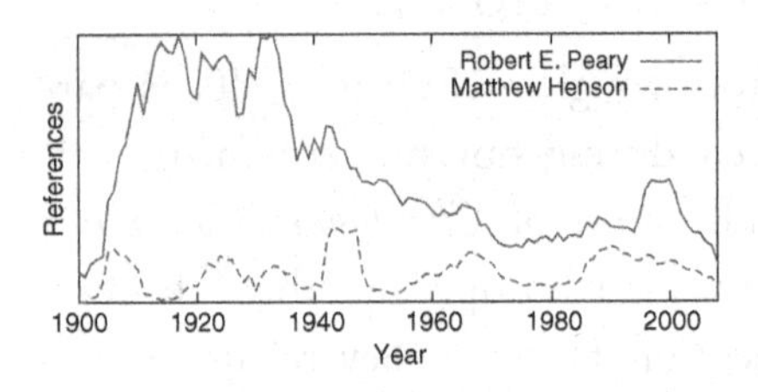

Particularly interesting is the fate of *Robert Peary (1856–1920)* [2255], acclaimed as the discoverer of the North Pole in 1909. The Pole is not a fixed feature of the landscape, but a point in navigation space that looks remarkably like every other spot for hundreds of miles around. Peary returned from his expedition with his partner *Matthew Henson (1866–1955)* [5224] and claimed to have reached the Pole. Many questioned the accuracy of Peary's records and supported *Frederick Cook (1865–1940)* [9333], who claimed to have arrived a year before Peary. A timeplot shows how the controversy has hurt Peary's reputation. As the commander of the expedition, Peary stole much of the glory from the African American Henson. Today, interest in Peary has faded substantially, but Henson has grown to nearly equal stature. South Pole discoverer *Roald Amundsen's (1872–1928)* [1109] claims have held up better than Peary's.

Edmund Hillary (1919–2008) [2227], the first man to reach the peak of Mt. Everest, also successfully traveled to both poles. Hillary's climbing partner was Sherpa *Tenzing Norgay (1914–1986)* [8620]. As was the case with Peary-Henson, Norgay's share of the Ngram data has gradually risen, to approach that of his white partner in adventure.

10.4.2 MAN IN SPACE

Those who have left the planet are a well-remembered group, if less elite than one might suppose. The first man to orbit the earth (*Yuri Gagarin (1934–1968)* [2324]) outranks the first American to do so (*John Glenn*

Most Significant Astronauts

Sig	Person	Dates	C/G	Description
1394	Neil Armstrong	(1930–2012)	C G	Astronaut (first man to walk on the moon)
2324	Yuri Gagarin	(1934–1968)	C G	USSR cosmonaut (first man in space)
2977	John Glenn	(1921–)	C G	Astronaut (first American to orbit earth)
4866	Buzz Aldrin	(1930–)	C G	NASA astronaut (2nd man to walk on moon)
6750	Alan Shepard	(1923–1998)	C G	NASA astronaut (first American in space)
9957	Sally Ride	(1951–2012)	C G	NASA astronaut (first American woman in orbit)
10187	Gus Grissom	(1926–1967)	C G	NASA astronaut (died in *Apollo 1* disaster)
10479	Christa McAuliffe	(1948–1986)	C G	Teacher, astronaut (died in *Challenger* disaster)
10787	V. Tereshkova	(1937–)	C G	USSR cosmonaut (first woman in space)
11590	Jim Lovell	(1928–)	C G	NASA astronaut (survived *Apollo 13* disaster)

(1921–) [2977]), even though Glenn parlayed his astronaut career into a long career in the U.S. Senate. Three female astronauts are highly ranked, with the first American woman in space (*Sally Ride (1951–2012)* [9957]) ahead of the first Russian woman (*Valentina Tereshkova (1937–)* [10787]) even though she flew 20 years earlier. *Christa McAuliffe (1948–1986)* [10479] was the teacher selected to fly into space aboard the doomed Space Shuttle *Challenger*. Her significance outranks all six of her crewmates, most of whom had made previous trips into space.

China is the third nation to achieve manned space flight. The most significant Chinese astronaut is *Yang Liwei (1965–)* [50152], China's first man sent into space.

Two scientists are particularly significant for designing rockets instead of flying on them. American *Robert H. Goddard (1882–1945)* [2413] is credited with building the first liquid-fueled rocket. *Wernher von Braun (1912–1977)* [1939], the German rocket scientist responsible for developing the V-2 rocket for the Nazis, switched loyalty to the United States after World War II and spearheaded spacecraft design for NASA. Von Braun was the inspiration for the *Dr. Strangelove* character in *Stanley Kubrick's (1928–1999)* [1935] classic film of the Cold War.

The long-term significance of these astronauts jibes with that of the first generation of airmen, the pioneering aviators who were first to fly great distances. For example, the first American in orbit (*Alan Shepard (1923–1998)* [6750]) proves less significant than *Chuck Yeager (1923–)* [6294], the pilot who broke the sound barrier. Several were the designers of early gliders and aeroplanes. They generally were their own test pilots and sometimes paid the price, like pre-Wright brothers glider pioneer *Otto Lilienthal (1848–1896)* [2551].

Most Significant Aviators

Sig	Person	Dates	C/G	Description
736	Charles Lindbergh	(1902–1974)	C G	American aviator (New York to Paris)
932	Amelia Earhart	(1897–1937)	C G	American aviation pioneer (disappeared)
2546	A. Santos-Dumont	(1873–1932)	C G	Brazilian aviation pioneer (first dirigible)
2551	Otto Lilienthal	(1848–1896)	C G	German pioneer of aviation (Glider King)
2683	F. von Zeppelin	(1838–1917)	C G	German aircraft manufacturer (Zepplin Airship)
4160	Samuel P. Langley	(1834–1906)	C G	American astronomer, physicist, inventor
4640	Igor Sikorsky	(1889–1972)	C G	Inventor and aviation pioneer (helicopter)
5306	Glenn Curtiss	(1878–1930)	C G	Pioneer of the American aircraft industry
6294	Chuck Yeager	(1923–)	C G	Test pilot (first past the sound barrier)
6937	Octave Chanute	(1832–1910)	C G	Aviation pioneer (worked with Wright brothers)

11

Religion and Philosophy

Religion has shaped man's greatest ideals, from philosophies of ethics and belief to artistic expression in music and the fine arts. Migrations of people seeking freedom to worship sparked the colonization of the New World. Battles between opposing faiths have repeatedly redrawn the political and cultural landscape, from Islamic conquests to Christian crusades. These forces continue to drive the events of today. No historical discussion would be complete without a treatment of religion.

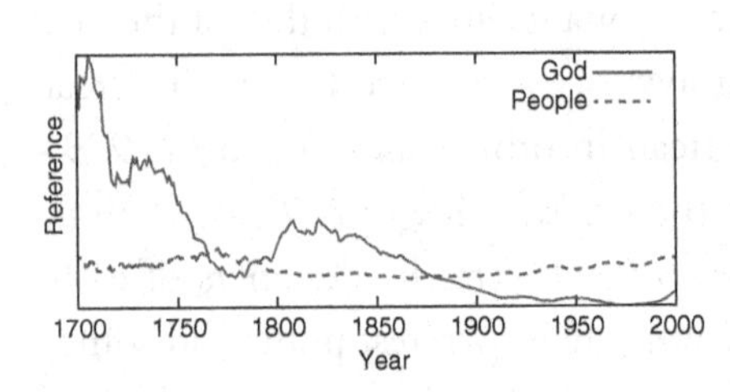

Secularism is an opposing force that has equally driven Western civilization since the Enlightenment. We plot the relative frequency of *God* versus *people* going back to 1700. Granted, the early book data from which this is derived is very thin, and presumably biased to overrepresent religious texts. Still, it shows *God* represented one out of every 400 words appearing in eighteenth-century printed texts. *God*'s relative frequency has steadily declined from this lofty peak, so that it now occurs less than one-tenth as often. Starting about 1880, *people* became a greater preoccupation of the printed word than the Almighty.

This chapter details the changing relationship between man and God, by looking at biblical figures, the saints and popes of the Catholic Church, and the primary leaders of other faiths. We will also measure the stature of the great philosophers, who for many supplement the ideals of religious thought in an increasingly secular world.

It goes without saying that the figures discussed within this chapter are held with reverence by many people, feelings that we respect and admire. Bigger does not mean better. Here, we use our computational methods to measure the historical footprint of religious personalities from a variety of faiths, through the lenses of such admittedly secular and Anglo-centric sources as Wikipedia and Google Ngrams. Our treatment here is intended to be historically but not theologically enlightening.

11.1 Biblical Figures

The Bible stays the same, but how we read it is always changing. In this section, we will identify the most significant Old and New Testament figures, to see what it says about our world today.

11.1.1 OLD TESTAMENT

The historical study of Old Testament figures is crippled by the fact that, for many, there is little independent confirmation that they ever actually existed. *David (1040–970 B.C.)* [57] is regarded as our most significant Old Testament figure primarily because he registers as a historical figure: serving as the second king of Israel and being credited as the founder of Jerusalem. His reign around 1000 B.C.E. makes him a contemporary of the Chinese *King Wen of Zhou (1099–1050 B.C.)* [9618].

David's exploits do not appear in the Pentateuch, the five books of Moses. This renders him less theologically significant than the Patriarchs (*Abraham* [382], *Isaac* [511], and *Jacob* [261]) or *Moses* [203] himself. The lack of documentary or archaeological evidence of their existence undoubtedly impacts their historical ranking, but fifteen Old Testament personalities still manage to rank among our thousand most significant figures.

Being a biblical prophet might seem to be the highest gravitas occupation imaginable, yet the significance of all these figures rests upon both statistical factors. The Old Testament characters whose stories appeal most strongly to children (*Jonah* [914], *Samson* [1027], *Noah* [292], and *Joseph* [691]) all prove lower-gravitas than their peers.

It is popular to give children biblical names. Statistics from the Social Security Administration show that *Jacob* [261] was the most popular boy's

Old Testament Figures

Sig	Person	Ngrams Timeline	C/G	Description
57	David		C G	Biblical king of Israel (Jerusalem)
164	Solomon		C G	Wise biblical king of Israel
203	Moses		C G	Biblical Jewish prophet (parted the Red Sea)
261	Jacob		C G	a.k.a. Israel, third Hebrew patriarch
292	Noah		C G	Final Hebrew Patriarch (Noah's Ark)
336	Adam		C G	Biblical first man created by God (Genesis)
382	Abraham		C G	First Hebrew Patriarch
511	Isaac		C G	Hebrew Patriarch (Binding of Isaac)
522	Aaron		C G	Moses' brother, Jewish high priest
580	Joshua		C G	Figure in the Torah (Book of Joshua)
691	Joseph		C G	"Joseph and the Amazing Technicolor Dreamcoat"
716	Isaiah		C G	Old Testament prophet (Book of Isaiah)
819	Jeremiah		C G	Old Testament prophet ("Weeping prophet")
866	Elijah		C G	Old Testament prophet (Book of Kings)
914	Jonah		C G	Old Testament prophet (swallowed by whale)
1027	Samson		C G	Super-strong until Delilah cut his hair
1168	Ezekiel		C G	Old Testament prophet (Book of Ezekiel)
1731	Hezekiah		C G	King of Judah during Sargon's invasions
1783	Josiah		C G	King of Judah (Dueteronomic reform)

name of the 2000s, with *Joshua (1550–1390 B.C.)* [580] (3), also appearing among the top five. *David (1040–970 B.C.)* [57] (12), *Noah* [292] (20), *Samuel* [2706] (24), and *Benjamin* [2913] (25) round out the top 25. This has not always been the case. In 1900, David ranked thirty-second in frequency, even though it was the most popular Old Testament name of its time. The Ngrams data shows a sharp increase in the number of *Davids* over the past 50 years. Popular biblical names for girls over this period include *Abigail* [7624] (6) and *Sarah* [873] (12).

The book's Ngram data can be used to chart the frequency with which names appear in published texts, as presented in the accompanying capsule time series. This can serve as a proxy for name popularity, although it is not one man, one vote. The frequency of names in the Ngram data depends upon the significance of the individuals with that given name, not just the sheer number of them.

The capsule time series shows sharp declines for several Old Testament figures during this century, including *Moses* [203] and *Hezekiah (?–687 B.C.)* [1731]. In 2011, these names ranked 522th and 877th in popularity according to the Social Security Administration data.

11.1.2 NEW TESTAMENT

The New Testament is only about one-third as long as the Old, yet is equally packed with historically significant figures, including the top man himself. The events of the New Testament occur several hundred years after the events of the Old, into periods of better-documented history. Reasonable birth-death years are known for about half of the important figures.

New Testament Figures

Sig	Person	Dates	C/G	Description
1	Jesus	(7 B.C.–A.D. 30)	C G	Central figure of Christianity
34	Paul the Apostle	(A.D. 5–A.D. 67)	C G	Christian apostle and missionary
65	Saint Peter	(?–?)	C G	Early Christian leader
109	John the Baptist	(5 B.C.–?)	C G	New Testament prophet, baptized Jesus
117	Mary	(?–?)	C G	The Virgin Mary, mother of Jesus
272	Tiberius	(42 B.C.–A.D. 37)	C G	Roman emperor during time of Jesus' death
494	Herod the Great	(73 B.C.–4 B.C.)	C G	Roman client-king of Israel (Herod's Temple)
591	Pontius Pilate	(?–?)	C G	Roman prefect of Judaea, condemned Jesus
848	Saint Matthew	(?–A.D. 34)	C G	One of the 12 apostles (Gospel of Matthew)
861	Mark the Evangelist	(?–A.D. 68)	C G	Gospel of Mark, founded Church of Alexandria
868	Mary Magdalene	(?–?)	C G	Female disciple of Jesus
874	Luke the Evangelist	(?–?)	C G	Author of the Gospel of Luke
882	Saint Joseph	(?–?)	C G	Husband of Mary, Mother of Jesus
991	Saint Stephen	(?–?)	C G	Early Christian martyr (stoned to death)
1141	Judas Iscariot	(?–?)	C G	Apostle and betrayer of Jesus
1377	John the Evangelist	(?–110)	C G	Gospel of John
2247	Barnabas	(?–A.D. 61)	C G	Early Christian disciple and martyr

Family members here include Jesus' mother and father. Among the most prominent figures are the Roman authorities: political leaders *Tiberius (42 B.C.–A.D. 37)* [272] and *Herod the Great (73 B.C.–4 B.C.)* [494], plus the judicial figure *Pontius Pilate* [591].

Much of what is known about Jesus comes from the four gospels in the New Testament. The writers of the gospels, *Saint Matthew (?–A.D. 34)* [848], *Mark the Evangelist (?–A.D. 68)* [861], *Luke the Evangelist* [874], and *John the Evangelist (?–110)* [1377], appear prominently.

The twelve Apostles were early followers of Jesus, often pictured in representations of the Last Supper. Most famously, the guests included *Judas Iscariot* [1141]; however, *Saint Andrew* [524] appears to be the most significant of the Apostles. He was the founder of the Church of Byzantium.

11.2 The Catholic Church

The Roman Catholic Church stands out from other religious denominations in several important ways. This includes its membership of about 1.2 billion people, or almost 20 percent of the earth's population. As an organization, it has a strong central hierarchy, from the pope to the cardinals, and down to parish priests. Its process of canonization provides a formal way to recognize particularly devout and pious members.

The most significant Catholic figures have generally served as pope or have been recognized as saints. But there are exceptions, such as *John Henry Newman (1801–1890)* [534], who was recently beatified: a step toward sainthood, but one that still requires additional performance of miracles to complete the process. We will also look at the ranks of former Catholics, specifically the most famous of those who have been excommunicated by the church.

11.2.1 POPES

The Pope is the spiritual leader of the Roman Catholic Church. The Catholic papacy represents the longest ongoing office in history.[1] As of this writing, 266 men have reigned as pontiff, carrying on a line stretching back almost two thousand years from *Saint Peter* [65] to *Pope Benedict XVI (1927–)* [361].[2]

Figure 11.1 plots the significance of each pope as a function of when they served. It shows that the average significance of the pope has generally risen over time, and currently sits at its highest level in history. This partially reflects the growth of the Catholic Church, but also the increasing volume of historical scholarship over time. Bumps in the curve coincide with major events in history. Surges occurred after the Roman Empire became officially Christian in 380, the advent of the Crusades, and the coming of the Renaissance. Significant dips came with the rise of the Holy Roman Empire and the coming of the Enlightenment.

[1] The pharaohs of Egypt ruled for more than 3,000 years, until A.D. 30.
[2] Pope Francis (1936–) was elected after this book was completed.

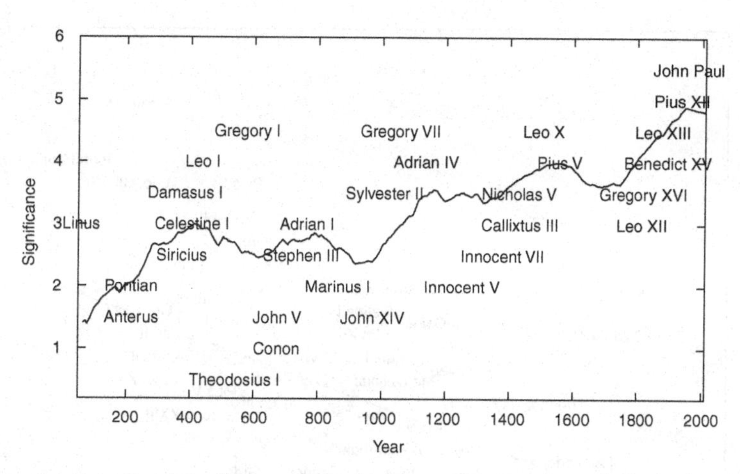

FIGURE 11.1. Papal significance from the Roman Empire through the present day.

One critical factor in the historical significance of popes is longevity: the three longest-serving popes rank among the six most important in history (Figure 11.2). It seems premature to rank the most recent pope (*Pope Benedict XVI (1927–)* [361]) as highly as we have, but much of what otherwise appears to be a recency bias in our analysis can be explained by term length. Just four popes account for more than 60 percent of the time period since 1846. Indeed, we recognize the recent but short-lived *Pope John Paul I (1912–1978)* [3501] as a relatively a minor figure.

Most Significant Popes

Sig	Person	Dates	Reign	C/G	Description
91	Pope John Paul II	(1920–2005)	27	C G	20th-century Polish pope (Solidarity)
294	Pope Pius XII	(1876–1958)	19	C G	Controversial Italian Pope (WWII)
334	Pope Pius IX	(1792–1878)	32	C G	Longest reigning pope (papal infallibility)
361	Pope Benedict XVI	(1927–)	8	C G	Recent pope ("God's Rottweiler")
394	Pope Gregory I	(540–604)	14	C G	6th-century pope ("The last good pope")
424	Pope Leo XIII	(1810–1903)	25	C G	Oldest pope ("Rerum Novarum")
488	Pope Leo X	(1475–1521)	8	C G	Pope during start of Protestant Reformation
571	Pope Pius X	(1835–1914)	11	C G	Traditionalist pope (first Code of Canon Law)
598	Pope Innocent III	(1161–1216)	18	C G	Influential 12th-century pope (4th Crusade)
625	Pope Alexander VI	(1431–1503)	11	C G	Controversial Spanish Renaissance pope
647	Pope Julius II	(1443–1513)	10	C G	Sistine Chapel ("The Fearsome pope")
703	Pope Pius XI	(1857–1939)	16	C G	Interwar Italian Pope
774	Pope Gregory VII	(1020–1085)	12	C G	Reformer pope (Investiture Controversy)
804	Pope Urban II	(1035–1099)	12	C G	French Pope (1st Crusade)
855	Pope John XXIII	(1881–1963)	5	C G	Italian postwar pope (2nd Vatican Council)

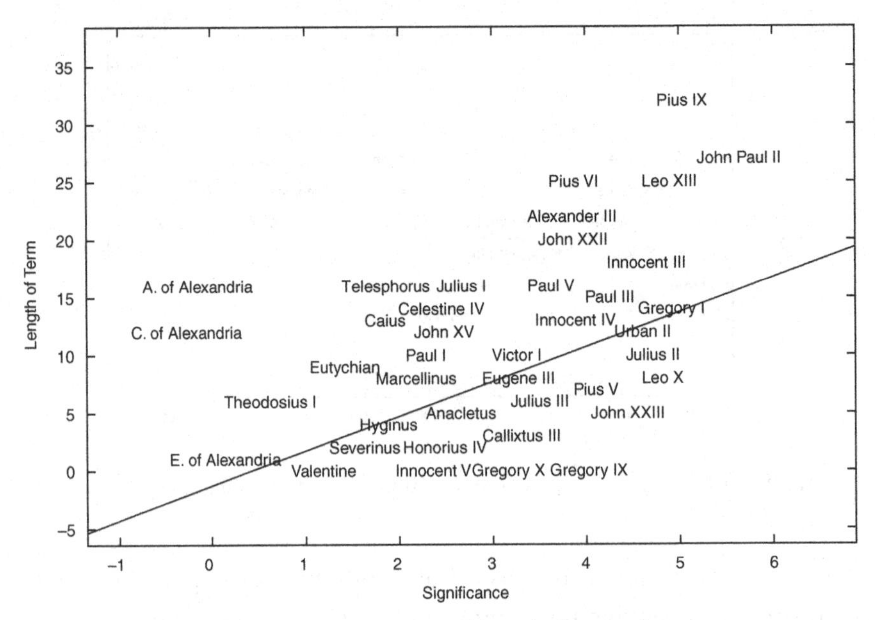

FIGURE 11.2. Significance of popes by length of reign.

Other significant popes may have had short reigns, but they proved very consequential. *Pope John XXIII's (1881–1963)* [855] second Vatican Council modernized many aspects of liturgy and practice. *Pope Gregory I (540–604)* [394] has been called the father of Christian worship and was declared "the last good pope" by Protestant reformer *John Calvin (1509–1564)* [99]. *Pope Leo X (1475–1521)* [488] waged battle against *Martin Luther's (1483–1546)* [17] Protestant Reformation.

11.2.2 SAINTS

The formal procedure employed by the Roman Catholic Church for recognizing saints has varied over the centuries. It begins with an investigation of the merits of their case, which may establish them as "heroic in virtue" or venerable. The next stage is beatification, a declaration that it is "worthy of belief" that the person is in heaven. Canonization requires recognition of an additional miracle on the part of the saint. The office of the devil's advocate, which argued the case against canonization, was abolished in 1983. This is partially responsible for a recent surge of new saints.

Our roster of the most significant saints omits the New Testament figures Paul, Mark, and Luke, whom we discussed in the previous section. Our rankings are striking to the extent that they favor royalty, including Roman, English, and French kings/emperors. Several date back to the early days of the church, when the standards for canonization were less formal.

Most Significant Saints

Sig	Person	Dates	C/G	Description
67	Constantine the Great	(272–337)	C G	Emperor of Rome (first Christian emperor)
72	Augustine of Hippo	(354–430)	C G	Early Christian theologian ("The City of God")
95	Joan of Arc	(1412–1431)	C G	French military leader and saint
169	Thomas More	(1478–1535)	C G	English statesman, philosopher ("Utopia")
197	Saint Patrick	(390–460)	C G	Christian saint, missionary to Ireland
264	Thomas Becket	(1118–1170)	C G	Catholic saint and martyr
275	Francis of Assisi	(1182–1226)	C G	Italian Catholic friar and preacher
331	Saint Nicholas	(270–346)	C G	4th-century Christian saint
505	Saints Cyril and Methodius	(827–869)	C G	Byzantine Christian missionaries
531	Edward the Confessor	(1003–1066)	C G	Last king of the House of Wessex
546	Louis IX	(1214–1270)	C G	King of France
596	Ignatius of Loyola	(1491–1556)	C G	Founder of the Society of Jesus
611	Francis Xavier	(1506–1552)	C G	Roman Catholic missionary (Jesuits)
669	Athanasius of Alexandria	(296–373)	C G	Pope of Alexandria
741	Irenaeus	(130–202)	C G	Early Church Father
874	Luke the Evangelist	(?–?)	C G	Author of the Gospel of Luke
877	Anthony of Padua	(1195–1231)	C G	Portuguese Catholic priest
931	Augustine of Canterbury	(?–604)	C G	Benedictine monk
980	Saint Dominic	(1170–1221)	C G	Founder of the Dominican Order
1044	Basil of Caesarea	(330–379)	C G	Greek bishop and saint

Two early saints have particular connections to popular festivals. *Saint Patrick (390–460)* [197] was the patron saint of Ireland, a missionary largely responsible for converting a pagan land to Christianity. Legend credits Patrick for using the three-leaved shamrock as a visual aid to explain the Holy Trinity, accounting for their ubiquity every March 17. *Saint Nicholas (270–346)* [331] was a Greek bishop who participated in the First Council of Nicaea, and is associated with Orthodox Christianity. His reputation for secret gift giving morphed into the association with Santa Claus. A third, *Pope Sylvester I (?–335)* [4568], died on December 31 in the year 335, and is now associated with the secular new year, often called Sylvester.

Several of the most significant saints were theologians whose teaching shaped the church, such as *Augustine of Hippo (354–430)* [72] and *Thomas Aquinas (1225–1274)* [90]. Others are notable for their martyrdom. *Thomas More (1478–1535)* [169] was a statesman and humanist who refused to support *Henry VIII (1491–1547)* [11] as the head of the English church, and lost his head for it.

11.2.3 THE EXCOMMUNICATED

Excommunication is a formal censure from the Catholic Church that bars the recipient from participation in certain rituals and blessings. A number of historical figures have been excommunicated, which serves as a proxy for the religious and political battles fought by the church throughout its long history.

Political enemies of the church are the most prominent people to be excommunicated. During the Middle Ages, the pope assumed the power to anoint European kings. He used excommunication as a weapon against those who acted against his wishes, like the Holy Roman emperors *Frederick I (1122–1190)* [456] and *Frederick II (1194–1250)* [461]. *Napoleon (1769–1821)* [2] was excommunicated after annexing the Papal States in 1809. In modern times, the excommunicated include the Communist leaders *Fidel Castro (1926–)* [506] and *Josip Broz Tito (1892–1980)* [562].

Most Significant Excommunicated People

Sig	Person	Dates	C/G	Description
2	Napoleon	(1769–1821)	C G	Emperor of France (Battle of Waterloo)
11	Henry VIII	(1491–1547)	C G	King of England (6 Wives)
13	Elizabeth I	(1533–1603)	C G	Queen of England (The Virgin Queen)
17	Martin Luther	(1483–1546)	C G	Protestant Reformation (95 Theses)
95	Joan of Arc	(1412–1431)	C G	French military leader and saint
136	John	(1166–1216)	C G	King of England (Magna Carta)
170	Henry II	(1133–1189)	C G	King of England (House of Plantagenet)
238	Henry IV	(1553–1610)	C G	King of France (Navarre)
455	Thomas Cranmer	(1489–1556)	C G	Archbishop of Canterbury (Church of England)
456	Frederick I	(1122–1190)	C G	Holy Roman Emperor (Frederick Barbarossa)
461	Frederick II	(1194–1250)	C G	German Holy Roman Emperor
470	Robert the Bruce	(1274–1329)	C G	King of Scots
506	Fidel Castro	(1926–)	C G	Cuban revolutionary ruler
562	Josip Broz Tito	(1892–1980)	C G	Yugoslav revolutionary ruler
579	John Wycliffe	(1328–1384)	C G	English theologian / Bible translator

The remaining figures were involved in more overtly theological disputes. Several, such as *Martin Luther (1483–1546)* [17], were leaders of the Protestant Reformation. *Henry VIII's (1491–1547)* [11] marital disputes caused the English split from the Catholic Church. *Joan of Arc (1412–1431)* [95] holds the rare distinction of being both excommunicated and then canonized by the church. Her story is particularly compelling, a young girl whose run of miraculous military victories ended in her burning at the stake, at the age of just 19. She was declared a martyr 25 years later, and finally canonized in 1920.

The person who narrowly missed this table is *Galileo Galilei (1564–1642)* [49], the great scientist whose heretical belief that the earth moved round the sun led to his trial by the Inquisition. He ended his life under house arrest, but was never formally excommunicated.

11.3 Other Religious Denominations

Great leaders found great movements, which then come to perpetuate the memory of their founders. Starting a major religion is perhaps the most effective way to achieve great historical significance. It is a path open only to a chosen few, however.

Founders of Religions

Sig	Person	Dates	C/G	Description
1	Jesus	(7 B.C.–A.D. 30)	C G	Central figure of Christianity
3	Muhammad	(570–632)	C G	Prophet and founder of Islam
52	Gautama Buddha	(563–483 B.C.)	C G	Central figure of Buddhism
55	Joseph Smith	(1805–1844)	C G	American religious leader (Mormonism)
109	John the Baptist	(5 B.C.–?)	C G	New Testament prophet, baptized Jesus
134	Confucius	(551–479 B.C.)	C G	Chinese thinker and philosopher
203	Moses	(?–?)	C G	Biblical Jewish prophet (parted the Red Sea)
382	Abraham	(?–?)	C G	First Hebrew Patriarch
478	Zoroaster	(?–?)	C G	Founder of Zoroastrianism
512	Laozi	(400–?)	C G	Ancient Chinese philosopher ("Tao Te Ching")
1129	Mahavira	(599–527 B.C.)	C G	Indian sage
1208	L. Ron Hubbard	(1911–1986)	C G	Founder of Scientology / science fiction writer
1771	Mary Baker Eddy	(1821–1910)	C G	Founder of Christian Science

In this section, we will briefly review the most significant figures of other major religions.

11.3.1 PROTESTANT CHRISTIANITY

The Protestant Reformation arose from a mix of theological and political forces. In his 95 Theses, *Martin Luther (1483–1546)* [17] inveighed against many church practices, particularly the selling of indulgences remitting punishments for sins. Politically, the rulers of increasingly strong nation-states of Europe chafed under the power of the pope in Rome. Hence, they often welcomed the Reformation and the opportunity to create a local church.

Desiderius Erasmus (1466–1536) [218] was a great scholar and humanist who raised many of the questions leading to the Reformation, yet he remained a member of the Catholic Church.

The most significant Protestant leaders tend to be the founders of denominations that still exist today, such as the Lutherans, Calvinists, Presbyterians (*John Knox (1514–1572)* [425]), and Anglicans (*Thomas Cranmer (1489–1556)* [455]). Other Protestant denominations split off later, such as Methodism, founded by *John Wesley (1703–1791)* [359].

Most Significant Protestant Leaders

Sig	Person	Dates	C/G	Description
17	Martin Luther	(1483–1546)	C G	Protestant Reformation (95 Theses)
99	John Calvin	(1509–1564)	C G	French Protestant theologian (Calvinism)
218	Desiderius Erasmus	(1466–1536)	C G	Dutch Renaissance humanist
359	John Wesley	(1703–1791)	C G	Christian clergyman / founder of Methodism
425	John Knox	(1514–1572)	C G	Scottish clergyman/Presbyterian founder
455	Thomas Cranmer	(1489–1556)	C G	Archbishop of Canterbury (Church of England)
484	Huldrych Zwingli	(1484–1531)	C G	Reformation leader
1187	Philipp Melanchthon	(1497–1560)	C G	Intellectual leader of Protestant Reformation
1290	Martin Bucer	(1491–1551)	C G	German Protestant Reformer (Ecumenism)
1503	George Fox	(?–1691)	C G	English dissident / founder of the Quakers

11.3.2 AMERICAN RELIGIOUS LEADERS

The United States is by certain standards the most religious of Western nations. Colonies such as Massachusetts and Pennsylvania were founded as refuges from religious persecution in Europe. Indeed, Rhode Island was founded as a refuge from religious persecution elsewhere in New England. Important early American theologians include *Roger Williams (1603–1683)* [678], *Jonathan Edwards (1703–1758)* [913], and *Cotton Mather (1663–1728)* [1676].

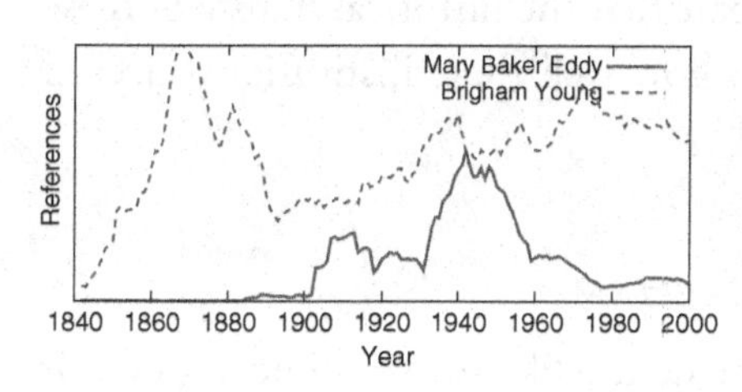

Two faiths that arose from American soil are Mormonism and Christian Science. Both held similar public stature around 1950, according to the Ngrams data. The historical arc of both these faiths can be discerned from the frequency counts of canonical figures. As we can see, the mindshare of Christian Science founder *Mary Baker Eddy (1821–1910)* [1771] has plunged as the church has declined to approximately 85,000 members worldwide. By contrast, the Church of Jesus Christ of Latter-day Saints (the Mormons) now commands a membership exceeding 14 million people. Mormonism, as personified here by *Brigham Young (1801–1877)* [471], has more than held its own in terms of Ngram frequencies.

Evangelical Christian leaders make up most of the remaining figures in our list. *Billy Graham (1918–)* [1780] is the most prominent American Christian evangelist, a pioneer in the use of broadcast media who became the face of Christian worship in America, and the friend of presidents. His son *Franklin Graham (1952–)* [68651] has taken over his ministry, but is of much lower public stature.

American Religious Figures

Sig	Person	Dates	C/G	Description
55	Joseph Smith	(1805–1844)	C G	American religious leader (Mormonism)
471	Brigham Young	(1801–1877)	C G	American Mormonism leader
1208	L. Ron Hubbard	(1911–1986)	C G	Founder of Scientology / science fiction writer
1771	Mary Baker Eddy	(1821–1910)	C G	Founder of Christian Science
1780	Billy Graham	(1918–)	C G	Establishment Christian evangelist
3156	Jerry Falwell	(1933–2007)	C G	Christian televangelist (Moral Majority)
3625	Billy Sunday	(1862–1935)	C G	American baseball player / evangelist
4345	Pat Robertson	(1930–)	C G	Baptist televangelist / media mogul
12124	Fulton J. Sheen	(1895–1979)	C G	American Catholic archbishop
12399	Oral Roberts	(1918–2009)	C G	Pentecostal televangelist / university founder
12579	Jimmy Swaggart	(1935–)	C G	Disgraced Pentecostal televangelist

Two competing evangelists created substantial universities to preserve and expand their ministries: *Jerry Falwell (1933–2007)* [3156] (Liberty University) and *Oral Roberts (1918–2009)* [12399] University. Science fiction writer *L. Ron Hubbard (1911–1986)* [1208] is the founding figure of the Church of Scientology. We anticipate that the historical stature of these individuals will ultimately depend on how well these institutions succeed in preserving their legacy.

11.3.3 JUDAISM

Judaism does not have a hierarchical structure like the Catholic Church. It lost a central religious authority with the destruction of the Second Temple in Jerusalem in A.D. 70, and the subsequent spreading of the Jewish diaspora around the world. Rabbinical Judaism, based on the interpretation of the oral law and written scriptures, arose to replace it. The title of Rabbi means teacher, with the more prominent scholars respected as sages for their profound interpretation of the holy scriptures.

Akiva ben Joseph (A.D. 50–135) [2449] and *Hillel the Elder (A.D. 110–10)* [2701] were major figures in the founding of rabbinical Judaism. Several great scholars of the Middle Ages also rank among the prominent Jewish religious figures, including *Maimonides (1135–1204)* [284], *Rashi (1040–1105)* [658], and *Nahmanides (1194–1270)* [2514].

Most Significant Jewish Leaders

Sig	Person	Dates	C/G	Description
284	Maimonides	(1135–1204)	C G	Medieval Jewish philosopher / physician
658	Rashi	(1040–1105)	C G	French medieval torah commentator
1684	Baal Shem Tov	(1698–1760)	C G	Jewish mystical rabbi
2449	Akiva ben Joseph	(A.D. 50–135)	C G	Early founder of rabbinical Judaism
2514	Nahmanides	(1194–1270)	C G	Medieval Spanish Jewish scholar
2701	Hillel the Elder	(A.D. 110–10)	C G	Jewish religious sage (Mishnah)
2838	Ezra	(?–?)	C G	Jewish priest (the Scribe)
2850	Menachem Mendel Schneerson	(1902–1994)	C G	Prominent Lubavitcher Rebbe
2877	Isaac Luria	(1534–1572)	C G	Jewish mystic (father of Kabbalah)
3123	Judah Halevi	(1075–1141)	C G	Spanish Jewish poet, philosopher

The remaining figures generally concern more modern, mystical strands of Judaism. *Isaac Luria (1534–1572)* [2877] is the father of the Kabbalah, a collection of mystical teachings recently appropriated by celebrities such as *Madonna (1958–)* [121]. The Hasidic movement, founded by the *Baal Shem Tov (1698–1760)* [1684], incorporates elements of mystical thinking. In recent times Hasidism was personified by the late Lubavitcher Rebbe *Menachem Mendel Schneerson (1902–1994)* [2850], so revered by his followers that some thought he might be a candidate to be the Messiah.

11.3.4 ISLAM

Our ranking of the most significant Islamic figures highlights the great schism between the Shia and Sunni sects, following the death of the founding prophet *Muhammad (570–632)* [3]. The Sunni faith traces its leadership to *Abu Bakr (573–634)* [372], Muhammad's father-in-law. He was succeeded as Caliph by *Umar (586–644)* [234].

Muhammad's son-in-law *Ali (598–661)* [89] served as the first leader of the Shia sect of Islam. His son *Husayn ibn Ali (626–680)* [552] was the martyred third leader of the Shia. He was killed by the forces of *Yazid I (645–683)* [1452], the Caliph of the Sunni faith. These two main branches of Islam remain estranged today.

The remaining people are more historical/cultural figures than religious leaders. *Avicenna (980–1037)* [190] was a great medieval physician and philosopher during the Islamic golden age. *Suleiman the Magnificent (1494–1566)* [354] led the Ottoman Empire to great heights of power and civilization during his 46-year reign.

Most Significant Islamic Leaders

Sig	Person	Dates	C/G	Description
3	Muhammad	(570–632)	C G	Prophet and founder of Islam
89	Ali	(598–661)	C G	Early caliph and a central figure of Sufism
190	Avicenna	(980–1037)	C G	Persian physician and philosopher
234	Umar	(586–644)	C G	The second Muslim caliph
316	Rumi	(1207–1273)	C G	Persian Muslim poet, theologian
354	Suleiman the Magnificent	(1494–1566)	C G	10th sultan of the Ottoman Empire
372	Abu Bakr	(573–634)	C G	Father-in-law of Muhammad
437	Ibn Khaldun	(1332–1406)	C G	Arab polymath
552	Husayn ibn Ali	(626–680)	C G	Important figure in Islam
714	Ghazali	(1058–1111)	C G	Persian Islamic philosopher

11.3.5 EASTERN RELIGIONS

The leaders of the great Eastern faiths, such as Buddhism, Hinduism, and Taoism, are little known to Western readers, including the authors of this book. Our analysis of the English Wikipedia here is unlikely to lead to

Enlightenment. Nonetheless, we rank a number of the primary figures from these religions among our top 1000.

Most Significant Eastern Religious Figures

Sig	Person	Dates	C/G	Description
52	Gautama Buddha	(563–483 B.C.)	C G	Central figure of Buddhism
478	Zoroaster	(?–?)	C G	Founder of Zoroastrianism
512	Laozi	(400–?)	C G	Ancient Chinese philosopher ("Tao Te Ching")
601	Ramakrishna	(1836–1886)	C G	Influential Indian Hindu mystic
813	Swami Vivekananda	(1863–1902)	C G	Founder of Ramakrishna Mission
1129	Mahavira	(599–527 B.C.)	C G	Indian sage
1342	Chaitanya Mahaprabhu	(1486–1534)	C G	Hindu saint, social reformer
2949	Karmapa	(–?)	C G	Head of the Karma Kagyu
2980	Swami Dayananda Saraswati	(1824–1883)	C G	Hindu religious scholar, reformer
3545	Padmasambhava	(?–?)	C G	Sage guru (Buddhism)

The most significant living Buddhist figure is the Nobel Peace Prize–winning 14th Dalai Lama *Tenzin Gyatso (1935–)*, the spiritual / political leader of Tibet who has lived in exile from Tibet since 1959. Technical issues prevent us from ranking him with our usual methods, but it appears he should rank in the neighborhood of about 1,500.

11.4 Philosophers

The names of the great philosophers shown in Figure 11.3 may well endure for the rest of human history. Indeed, *Plato (427–347 B.C.)* [25] still ranks among history's most significant people even though he lived nearly 2,500 years ago. Ten other philosophers rank among the hundred most significant figures in history.

To identify a canonical set of philosophers, we started with the selections assembled in the book *The 100 Most Influential Philosophers of All Time* [Duignan, 2009]. We generally agree with who they picked, except for the omission of *Charles Sanders Peirce (1839–1914)* [225], an important nineteenth-century philosopher, who founded pragmatism and whose work in logic anticipated digital computation.

The great philosophers tend to be a historically stable lot. The four founders of twentieth-century analytic philosophy (*Bertrand Russell (1872–1970)* [253], *Ludwig Wittgenstein (1889–1951)* [493], *Gottlob Frege (1848–1925)* [838], and *G. E. Moore (1873–1958)* [3635]) all place respectably in our overall rankings, yet fail to crack the entrenched top 20. Indeed,

Sig	Person	Dates	C/G	Philosophy
8	Aristotle	(384–322 B.C.)	C G	Syllogism
14	Karl Marx	(1818–1883)	C G	Communism
25	Plato	(427–347 B.C.)	C G	Platonic realism
42	Friedrich Nietzsche	(1844–1900)	C G	Nihilism
59	Immanuel Kant	(1724–1804)	C G	Transcendental idealism
68	Socrates	(469–399 B.C.)	C G	Socratic method
80	Jean-Jacques Rousseau	(1712–1778)	C G	General will
81	Francis Bacon	(1561–1626)	C G	Empiricism
90	Thomas Aquinas	(1225–1274)	C G	Thomism
92	René Descartes	(1596–1650)	C G	Method of doubt
100	John Locke	(1632–1704)	C G	Social contract
119	Gottfried Leibniz	(1646–1716)	C G	Principle of sufficient reason
128	David Hume	(1711–1776)	C G	Causality
133	Pythagoras	(570–495 B.C.)	C G	Musica universalis
134	Confucius	(551–479 B.C.)	C G	Moral philosophy
142	Georg Wilhelm Friedrich Hegel	(1770–1831)	C G	Absolute idealism
167	Thomas Hobbes	(1588–1679)	C G	State of nature
168	Niccolò Machiavelli	(1469–1527)	C G	Classical republicanism
190	Avicenna	(980–1037)	C G	Avicennian logic
201	John Stuart Mill	(1806–1873)	C G	Utilitarianism

FIGURE 11.3. The most significant philosophers, with their associated philosophy.

only three of our 20 most significant philosophers lived into the nineteenth century, and none lived into the twentieth.

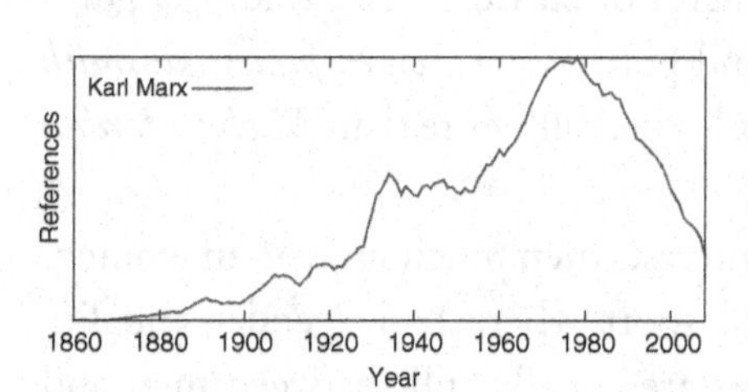

The Ngram dataset traces an interesting philosophical story in the spectacular rise and fall of *Karl Marx (1818–1883)* [14], the most recent figure in the table. Interest in his political philosophy rose steadily until the Cold War, when it stabilized for 10 years before reaching even greater heights in the 1960s. But Marxism as an intellectual force collapsed with the Soviet Union and has since been in steady free fall. We predict that at some point his ideas, once so important to so many, are destined to experience some form of intellectual revival.

Important philosophers continue to think important thoughts, although their work generally commands but a fraction of the public attention accorded to their predecessors. Animal rights ethicist *Peter Singer (1946–)* [8545] is one of the most visible philosophers working today. *Ayn Rand (1905–1982)* [486] and *Noam Chomsky (1928–)* [891] are what pass for popular political philosophers on the right and left, respectively.

12

Sports

Sports captures something important about human culture and society. Literary critic *Wilfrid Sheed (1930–2011)* [294893] observed that "sports communicate a code, a language of the emotions, and a tourist who skips the stadiums will not recoup his losses at Lincoln Center or Grant's Tomb."

Even if you don't give a hoot about sports, skim though this chapter. Think of it as crib notes for a test on cultural literacy. Who are the most significant male and female athletes of all time? The leading figures in boxing, football, hockey, tennis, and golf? What makes *James Naismith (1861–1939)* [847] a more important basketball figure than *Michael Jordan (1963–)* [1115]?

Sports is a domain that generally interests men much more than women. Yet we will identify one professional sport where basic gender equality holds: its ten most significant athletes are equally split between men and women.

12.1 Individual Sports

We start by looking at professional sports where athletes compete as individuals: boxing, tennis, golf, and horse racing. We defer analysis of historically amateur Olympic sports like track and field to Section 12.3.

12.1.1 BOXING

Boxers appear to have been the earliest class of professional athletes to achieve widespread renown. The earliest bare-knuckle champion of

England was *James Figg (1695–1734)* [31351], who won the title in 1719. Boxing was a particularly rough business before the 1860s, when the sport became standardized by the Marquess of Queensberry rules (established by *John Douglas (1844–1900)* [9223]). The shift from bare-knuckle to gloved boxing followed shortly thereafter, with *John L. Sullivan (1858–1918)* [4542] being the major transitional figure.

Top Nineteenth-Century Boxers

Sig	Person	Dates	C/G	Description
4542	John L. Sullivan	(1858–1918)	C G	First gloved heavyweight champion
8002	James J. Corbett	(1866–1933)	C G	Heavyweight boxing champ (Gentleman Jim)
8719	William Poole	(1821–1855)	C G	Leader of the Know Nothing movement
10045	Bob Fitzsimmons	(1863–1917)	C G	British Cornish boxer
11671	Jem Mace	(1831–1910)	C G	English boxing champion

The early boxers were colorful characters. *Jem Mace (1831–1910)* [11671] fought his last fight at the age of 78. *William Poole (1821–1855)* [8719] moonlighted as a gang leader known as "Bill the Butcher," and was murdered at the age of 34 in what appears to have been an act of public service.[1]

Several modern boxing champions resonate as important cultural figures in the history of American race relations. *Jack Johnson (1878–1946)* [2843] was the first black heavyweight champion: a brash, flamboyant figure who established that black men could beat white men at their own game. *Joe Louis's (1914–1981)* [2820] gentlemanly nature made him the first widely popular black champion. His victory over German champion *Max Schmeling (1905–2005)* [9403] in 1938 struck a blow against the master race theories of the Nazis. *Muhammad Ali (1942–)* [612] became a controversial figure in the 1960s, through his involvement with the Black Muslim movement and his refusal to fight in the Vietnam War. But he became arguably the world's most recognizable person in the 1970s, at the height of his popularity. We find him to be the second most significant athlete in history, after *Babe Ruth (1895–1948)* [434].

All but two of these fighters were black, but boxing fans generally discriminate more on the basis of weight than race. The top seven fight-

[1] Poole was also the inspiration for the fictionalized character portrayed by Daniel Day-Lewis in the movie *Gangs of New York*.

Top Modern Boxing Champions

Sig	Person	Dates	C/G	Description
612	Muhammad Ali	(1942–)	C G	Heavyweight boxing champ (The Greatest)
2218	Mike Tyson	(1966–)	C G	Heavyweight boxing champ (baddest man on planet)
2820	Joe Louis	(1914–1981)	C G	Heavyweight boxing champ (Louis vs. Schmeling)
2843	Jack Johnson	(1878–1946)	C G	Heavyweight boxing champ (Great Black Hope)
3164	Jack Dempsey	(1895–1983)	C G	Heavyweight boxing champ / saloon keeper
3377	George Foreman	(1949–)	C G	Heavyweight boxing champ / grill impressario
4838	Rocky Marciano	(1923–1969)	C G	Heavyweight boxing champ (undefeated)
5460	Floyd Mayweather	(1977–)	C G	Welterweight boxing champ (current best pound for pound)
5687	Sugar Ray Leonard	(1956–)	C G	Middleweight boxing champ (popular 5-time champ)
5839	Sugar Ray Robinson	(1921–1989)	C G	Welterweight boxing champ (pound for pound best ever)
6540	Joe Frazier	(1944–2011)	C G	Heavyweight boxing champ (Smokin' Joe)
7511	Sonny Liston	(1932–1970)	C G	Heavyweight boxing champ (lost to Ali)

ers in our rankings were all heavyweights, which is only one of a dozen weight classes recognized by title belts. Three others who won in multiple classes centered around the welterweight and middleweight divisions are tightly bunched in the center: *Floyd Mayweather (1977–)* [5460], *Sugar Ray Leonard (1956–)* [5687], and *Sugar Ray Robinson (1921–1989)* [5839].

Boxing's visibility has declined substantially since the 1980s, when *Mike Tyson (1966–)* [2218] reigned as champion. The sport's reduced state is at least partially due to rival organizations awarding titles, creating a multiplicity of pseudo-champions. As of this writing, two different Klitschkos reign simultaneously as world heavyweight champion, each according to different authorities. *Wladimir Klitschko (1976–)* [16065] is recognized as champion by a somewhat broader consensus, although we more highly regard his brother *Vitali Klitschko (1971–)* [13685], who holds a PhD degree and serves as the leader of a political party in the Ukraine.

12.1.2 TENNIS

The modern sport of tennis emerged in England during the nineteenth century. The men's singles tournament at Wimbledon was first held in 1877, with women's singles following in 1884. These tournaments have been played every year since, except in times of war.

Tennis is unique in that it has been equally accessible to both genders since the sport's beginnings. We present the most significant tennis players of each gender in the table, which shows that women (on right) have achieved a surprising parity with men (on left). The ten tennis players of highest overall significance break evenly into five men and five women.

Top Male Tennis Stars

Sig	Person	Dates
743	Roger Federer	(1981–)
1539	Frederick W. Taylor	(1856–1915)
1946	Rafael Nadal	(1986–)
2622	Andre Agassi	(1970–)
2894	John McEnroe	(1959–)
3882	Pete Sampras	(1971–)
3943	Arthur Ashe	(1943–1993)
4251	Ivan Lendl	(1960–)
4428	Jimmy Connors	(1952–)
4565	Rod Laver	(1938–)

Top Female Tennis Stars

Sig	Person	Dates
2228	Serena Williams	(1981–)
3108	Martina Navratilova	(1956–)
3361	Venus Williams	(1980–)
3374	Steffi Graf	(1969–)
3611	Billie Jean King	(1943–)
4913	Chris Evert	(1954–)
5157	Monica Seles	(1973–)
6754	Margaret Court	(1942–)
7968	Evonne Goolagong	(1951–)
8971	Lindsay Davenport	(1976–)

These rankings generally favor contemporary champions such as *Roger Federer (1981–)* [743] and *Serena Williams (1981–)* [2228] over foundational figures, but this is partially a consequence of the sport's relatively recent professionalism. The Grand Slam tournaments were open only to amateur players until 1968, making it difficult to compare earlier champions with those of the modern (or Open) era. *Frederick Winslow Taylor (1856–1915)* [1539] was an early tennis champion, but is more famous as an industrial efficiency expert, regarded as the father of scientific management. *Helen Wills Moody (1905–1998)* [13461] won 31 Grand Slam tournaments in the 1920s and 1930s, to become perhaps the first widely celebrated woman athlete. *Althea Gibson (1927–2003)* [15789] broke the color barrier as the first African American to win a grand slam tournament in 1956. Yet neither of these women can match the significance of the more recent, Open-era stars.

12.1.3 GOLF

Golf's origins date back to fifteenth-century Scotland, but the modern game emerged much later. The oldest major tournament, the British Open, began in 1860. Many of the early stars were amateurs, most notably *Bobby Jones (1902–1971)* [5105], who dominated the sport in the 1920s while earning his living as a lawyer. He was legendary for his sportsmanship, and for his role in building the Augusta National Golf Club, now the premiere golf course in the United States.

Golfers enjoy longer careers than competitors in other sports. *Tom Watson (1949–)* [9450] came within an eyelash of winning the 2009 U.S. Open, at just shy of 60 years old and 26 years after his last major championship victory. The top stars of the 1960s and 1970s (Nicklaus, Palmer, Player) still dominate the top of our rankings, a tribute to their long rivalries and the extent to which they remain a popular presence at tournaments. We rank the most significant men's and women's players.

Top Male Golfers

Sig	Person	Dates
2101	Tiger Woods	(1975–)
2607	Jack Nicklaus	(1940–)
4144	Arnold Palmer	(1929–)
5105	Bobby Jones	(1902–1971)
5399	Greg Norman	(1955–)
5897	Gary Player	(1935–)
6504	Ben Hogan	(1912–1997)
9194	Sam Snead	(1912–2002)
9450	Tom Watson	(1949–)
10623	Nick Faldo	(1957–)

Top Female Golfers

Sig	Person	Dates
7563	Babe Zaharias	(1911–1956)
20986	Annika Sörenstam	(1970–)
21064	Nancy Lopez	(1957–)
21498	Michelle Wie	(1989–)
30589	Patty Berg	(1918–2006)
40384	Mickey Wright	(1935–)
41672	Kathy Whitworth	(1939–)
44694	Louise Suggs	(1923–)
46928	Paula Creamer	(1986–)
48860	Betsy King	(1955–)

Tiger Woods (1975–) [2101] is the lone active player at the top of our rankings. He has dominated professional golf for more than 15 years. His public persona has changed wildly with time: first a supernaturally talented youth, then a high-character symbol of business excellence, and now a fallen celebrity struggling to regain what once seemed so effortless. One imagines that his story will continue to resonate, long after older champions pass from the scene.

Women's professional golf has a much shorter history than the men's game. The Ladies Professional Golf Association (LPGA) tour only commenced play in 1950, which contributes to the large disparity between the highest-ranked players of each gender. We count more than 30 male golfers ranked higher than the second significant female, *Annika Sörenstam (1970–)* [20986]. In general, the early champions of women's golf (*Patty Berg (1918–2006)* [30589], *Mickey Wright (1935–)* [40384], and *Kathy Whitworth (1939–)* [41672]) are not ranked as major cultural figures.

The exception is *Babe Zaharias (1911–1956)* [7563], often regarded as the outstanding female athlete of the twentieth century. A three-time Olympic medalist in track and field and an All-American basketball player, she achieved her greatest success in golf. Despite the limited sports opportunities for women athletes of her time, she dominated the young LPGA tour and even competed respectably against men on the PGA tour.

12.1.4 HORSE RACING

Strictly speaking, horse racing isn't an individual sport, because each race is a partnership between the horse and its jockey. It isn't immediately obvious whether human or animal should be the bigger star.

Successful race horses have short but brilliant careers, usually lasting no more than two or three years before retirement to stud. The Kentucky Derby, America's most famous horse race, is only open to three-year-olds, thus ensuring that no horse basks in its spotlight more than once. By comparison, the greatest jockeys can ride professionally for 30 years or more, and appear annually in the showcase Triple Crown races (the Kentucky Derby, Preakness, and Belmont Stakes). Who's bigger?

This issue is one we can resolve easily, because the great racehorse *Secretariat* [1444] was mistakenly classified as a person by our algorithms. He was thus graded head-to-head against all the people in this book. He acquits himself quite well, edging Palestinian leader *Yasser Arafat (1929–2004)* [1486] by a nose, and beating Supreme Court justices *Thurgood Marshall (1908–1993)* [1707] and *Charles Evans Hughes (1862–1948)* [1760] by three lengths. With this significance score, Secretariat leaves even the greatest jockeys eating his dust.

Most Significant Jockeys

Sig	Person	Dates	C/G	Description
35824	Bill Shoemaker	(1931–2003)	C G	American jockey (rode Ferdinand)
39698	Eddie Arcaro	(1916–1997)	C G	American jockey (rode Citation)
69155	Russell Baze	(1958–)	C G	Winningest Canadian jockey
72014	Angel Cordero	(1942–)	C G	American jockey (rode Bold Forbes)
75735	Steve Cauthen	(1960–)	C G	American jockey (rode Affirmed)

Russell Baze (1958–) [69155] has the most wins of any jockey in North America, and remains an active rider, as of this writing. Yet his public visibility has been limited because he does not race in the Triple Crown races. By comparison, *Bill Shoemaker (1931–2003)* [35824] won 11 Triple Crown races and *Eddie Arcaro (1916–1997)* [39698], 16 more. Both raced at a time when thoroughbred racing had a much greater hold on the popular imagination.

12.2 Team Sports

We focus here on the major American professional team sports – football, basketball, and hockey – baseball was thoroughly covered in Chapter 6. Soccer is the world's most popular team sport, so we also identify the most significant international players.

The cultural magnitude of the most popular American sports (baseball, football, and basketball) is revealed by the frequency with which each sport's athletes appear among highly significant people. Figure 12.1a presents a cumulative distribution plot of the number of athletes by sport among the 50,000 most highly ranked people. Baseball quickly comes to dominate these totals, with as many prominent figures as the other two sports combined.

But the picture changes when we normalize for the number of players for each sport. Figure 12.1b presents the total percentage of each sport's

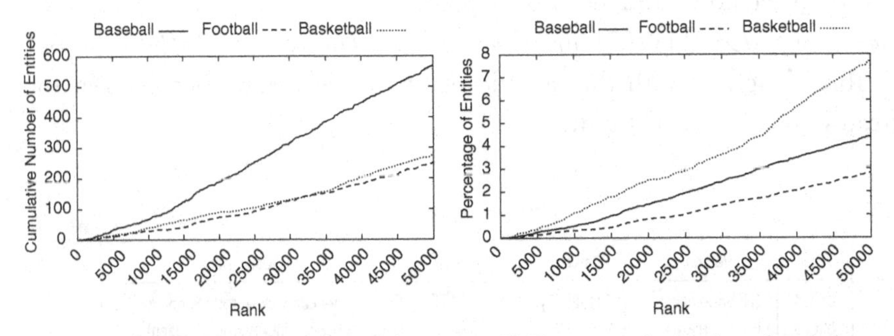

(a) Team sports by total number of players. (b) Team sports by fraction of total players.

FIGURE 12.1. Cumulative significance distribution of all major league baseball, football, and basketball players.

stars among our 50,000 most significant people. The National Basketball Association (NBA) has proven remarkably adept at building its players into historical-class figures. Fully 7 percent of NBA players to date appear in this top ranked cohort, compared with only 4 percent of the baseball players and 2.5 percent of NFL players.

12.2.1 FOOTBALL

Football is the most popular sport in the United States, as measured by television ratings. Although more spectators attend professional baseball games each year, this is attributable to a much longer schedule: 162 games per baseball team versus 16 per National Football League team.

We limit attention here to the members of the Professional Football Hall of Fame in Canton, Ohio. Players must have been retired for at least five years to be eligible for inauguration.

Top Football Hall of Famers

Sig	Person	Dates	C/G	Description
1356	Jim Thorpe	(1888–1953)	C G	Native American athlete (Olympics, football)
2904	Vince Lombardi	(1913–1970)	C G	NFL football coach (Packers)
2932	Joe Montana	(1956–)	C G	NFL football quarterback (49ers)
3307	Jerry Rice	(1962–)	C G	NFL football wide receiver (49ers)
3464	O. J. Simpson	(1947–)	C G	NFL running back, convicted felon
3825	Dan Marino	(1961–)	C G	NFL football quarterback (Dolphins)
4768	Joe Namath	(1943–)	C G	NFL football quarterback (NY Jets)
5184	Troy Aikman	(1966–)	C G	NFL football quarterback (Cowboys)
5358	George Halas	(1895–1983)	C G	NFL football pioneer player / coach
6274	John Elway	(1960–)	C G	NFL football quarterback (Broncos)
6541	John Madden	(1936–)	C G	NFL football coach, video game namesake
6643	Terry Bradshaw	(1948–)	C G	NFL football quarterback (Steelers)
6677	Mike Ditka	(1939–)	C G	NFL football player / coach (Bears)
6960	Tom Landry	(1924–2000)	C G	NFL football coach (Cowboys)
7126	Steve Young	(1961–)	C G	American professional football player

The most significant football player was *Jim Thorpe (1888–1953)* [1356], who played in the NFL during its 1920 inaugural season while serving as the league's first president. His greatest fame accrued from his performance in the 1912 Olympic games, where he won gold medals for the pentathlon and decathlon and was proclaimed the best athlete of the first half of the twentieth century. His medals from the 1912 Olympics were later revoked, for a trivial infraction.

The breakdown of highly significant players by position is revealing. This table includes seven quarterbacks and five coaches, but no one on defense. Although defense wins championships, there are several reasons why the best players do not strongly resonate with the public. First, it is difficult to measure performance quality statistically. Defensive play is best analyzed as a unit, aggregated over all the players on the team. Indeed, a truly dominating player may be active on fewer plays, because opposing teams avoid attacking his side of the field. Their helmets and other protective gear also mask their identities from the public.

Coach *Vince Lombardi (1913–1970)* [2904] led his Green Bay Packers to five league championships, but his image as a leader who was closely focused on excellence turned him into an important management icon of his time. Inspirational sayings like "Winners never quit and quitters never win" and "Winning isn't everything; its the only thing" are popularly attributed to him. His early death, near the peak of his career, left his reputation untrammeled by memories of decline. However, this win-at-all-costs message does not register quite as intensely today as it did in his time.

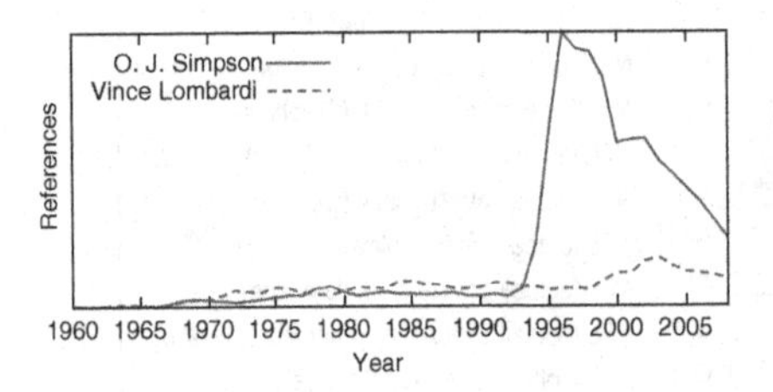

The final case of interest is *O. J. Simpson (1947–)* [3464], a great running back turned popular movie actor, who apparently murdered his former wife in 1994. The case and subsequent trial riveted the nation, culminating in a controversial not guilty verdict. The Ngrams dataset shows how Vince Lombardi possessed roughly twice Simpson's mindshare prior to the murder, before interest in the case came to dominate the affairs of the day. His rapid Ngram decline after the trial reflects the lack of genuine mystery as to the killer's identity. Fully 78 percent of respondents to a June 2004 Gallup poll believed that Simpson was probably or definitely guilty of murder. These numbers have only increased following his convictions for wrongful death and kidnapping, combined with greater public acceptance of DNA evidence.

12.2.2 BASKETBALL

Basketball has a distinct moment of creation, December 1891, when *James Naismith (1861–1939)* [847] set up peach baskets at the Springfield, Massachusetts YMCA and defined the rules to "Basket Ball." His sport spread quickly within the YMCA movement. The first college team was founded in 1898, and it became an Olympic demonstration sport in 1904.

Basketball entered the modern era in 1954, with the invention of the 24-second shot clock that dramatically increased scoring. Naismith is the only historically significant figure to punch out before the clock. *George Mikan (1924–2005)* [11221] was the most significant early professional basketball player. Seven foot, one inch *Wilt Chamberlain (1936–1999)* [1786] became the dominant player of the post-clock era, exploiting the higher-speed game to average more than 50 points per game over a full season. His massive size and outsized personality resonated beyond the sport.

Most Significant Basketball Figures

Sig	Person	Dates	C/G	Description
847	James Naismith	(1861–1939)	C G	Inventor of basketball
1115	Michael Jordan	(1963–)	C G	American basketball player (Air Jordan)
1786	Wilt Chamberlain	(1936–1999)	C G	American basketball center (Wilt the Stilt)
2126	Magic Johnson	(1959–)	C G	American basketball point guard (Lakers)
2252	Kareem Abdul-Jabbar	(1947–)	C G	American basketball center (skyhook)
2280	Larry Bird	(1956–)	C G	American basketball forward (Celtics)
3563	Bill Russell	(1934–)	C G	American basketball center (Celtics)
3839	Amos Alonzo Stagg	(1862–1965)	C G	Longtime college football coach
4092	John Wooden	(1910–2010)	C G	American college basketball coach (UCLA)
4802	Julius Erving	(1950–)	C G	American basketball player (Dr. J)
6341	Charles Barkley	(1963–)	C G	American basketball player (Round Mound)
6936	Jerry West	(1938–)	C G	American basketball player (Lakers)
7417	Adolph Rupp	(1901–1977)	C G	American college basketball coach (Kentucky)
7537	Bill Bradley	(1943–)	C G	Basketball player / senator (Dollar Bill)
7564	Pete Maravich	(1947–1988)	C G	American basketball player (Pistol Pete)

As we can see from our ranking of the top basketball figures, *Michael Jordan (1963–)* [1115] is the most significant basketball player ever, a man who was the most recognizable athlete of his time and a cultural/marketing phenomenon. He also stands as vivid testimony to the difficulty of human selection processes. Recently graduated college basketball players get assigned to professional teams via the NBA draft, which is structured so that weaker teams have better opportunities to select the players of their choice.

Houston used its first selection in the 1984 draft to pick *Hakeem Olajuwon (1963–)* [9590], who became a Hall of Fame player himself. More notoriously, Portland then passed on the chance to pick Jordan, grabbing center *Sam Bowie (1961–)* [35887] instead. This decision appears foolish in hindsight, which is always 20-20. Injuries limited Bowie's career, while Jordan developed into a much greater player than could reasonably be foreseen at the time.

Our earlier studies of the accuracy of human selection procedures (Hall of Fame memberships) focused on domains where complete information about the candidates existed at the time they were evaluated. The poor performance we have established in these better-understood domains should increase our sympathy for those faced with sport draft decisions. It is vastly harder to predict the future than recognize the past, and even the latter proves much harder than is generally believed.

A curious founding figure here is *Amos Alonzo Stagg (1862–1965)* [3839], best known as a longtime college football coach. But he also played in the very first basketball game as Naismith's associate. He is credited with a seemingly impossible series of innovations that shaped football (the huddle, tackling dummy, and lateral pass), basketball (five-man teams), and baseball (the batting cage). He is a member of both the Basketball and College Football Halls of Fame.

12.2.3 HOCKEY

Ice hockey is the most popular winter sport in Canada and several European countries. The first organized game was played in 1875, and the premier professional league (the National Hockey League, or NHL) founded in 1917. Our rankings of the most significant figures reflect players from most of this long history, although with some bias consistent with the league's expansion from six teams to 30 since 1967.

Five of our top ranked players were born in a narrow interval from 1961 to 1965, including the most significant one of all. *Wayne Gretzky (1961–)*

Most Significant Hockey Figures

Sig	Person	Dates	C/G	Description
1949	Wayne Gretzky	(1961–)	C G	Professional hockey star (The Great Gretzky)
3128	Mario Lemieux	(1965–)	C G	Professional hockey star (Super Mario)
3747	Frederick Stanley	(1841–1908)	C G	6th Governor General of Canada
5071	Maurice Richard	(1921–2000)	C G	Professional ice hockey player (the Rocket)
5765	Patrick Roy	(1965–)	C G	Canadian ice hockey goaltender
5855	Gordie Howe	(1928–)	C G	Longtime professional ice hockey star
9020	Brett Hull	(1964–)	C G	American NHL player
9065	Conn Smythe	(1895–1980)	C G	Canadian soldier and sportsman (ice hockey)
9119	Mark Messier	(1961–)	C G	Canadian ice hockey center
10588	Jacques Plante	(1929–1986)	C G	Canadian ice hockey goaltender

[1949] scored at unprecedented rates, raising the season record for goals from 76 to 92 and assists from 102 to 163. He changed the sense of what was possible. Indeed, his contemporaries approached several of his standards only a few seasons later, after watching him do it. His leadership explains the concentration of his peers in our rankings.

The two highest gravitas figures in professional hockey are founder-figures who lent their names to hockey's most important prizes. The *Frederick Stanley (1841–1908)* [3747] Cup goes to the championship team, while the *Conn Smythe (1895–1980)* [9065] Trophy is awarded to the most valuable player during these playoffs.

12.2.4 SOCCER

FIFA, the international football association, was founded in 1904. Dozens of nations have their own professional leagues, with several capable of making coherent claims to being the world's strongest. The World Cup tournament, held every four years, identifies the world's best national team.

The players we consider here are drawn from the *FIFA 100* list, which was compiled in 2004 by the great soccer player *Pelé (1940–)* [1418] in recognition of the 100th anniversary of the foundation of the international governing body of football. He refused to be constrained to the numerical target, so his Top 100 contains 125 people. The only two players from the United States appearing in the list were also the only two women: *Mia Hamm (1972–)* [10420] and *Michelle Akers (1966–)* [62368].

Most Significant Soccer Figures

Sig	Person	Dates	C/G	Description
1264	Diego Maradona	(1960–)	C G	Argentine soccer player / coach
1418	Pelé	(1940–)	C G	Brazilian soccer star
1462	David Beckham	(1975–)	C G	English soccer player
1863	Zinedine Zidane	(1972–)	C G	French soccer player
2040	Thierry Henry	(1977–)	C G	French soccer player
2128	Ronaldo	(1976–)	C G	Brazilian soccer player
2907	Johan Cruyff	(1947–)	C G	Dutch soccer player
2951	George Best	(1946–2005)	C G	Irish professional footballer
3090	Alan Shearer	(1970–)	C G	English soccer player
3365	Ronaldinho	(1980–)	C G	Brazilian soccer player
3961	Franz Beckenbauer	(1945–)	C G	German soccer player / coach

What is notable is the international diversity of the top ranked players. The eleven men presented here represent six different countries: Argentina, Brazil, England, France, Germany, and the Netherlands. International football is administrated by FIFA, an international organization long ruled by its president, *Sepp Blatter (1936–)* [8430]. Although widely controversial, he has a significance rank comparable to the leaders of another international organization, the United Nations.

12.3 Olympic Sports Stars

The Summer Olympic games dominate the world's attention for exactly one month every four years. And then they are gone. This gap between games looms large for the athletes themselves, who seldom can maintain peak skills long enough to compete in multiple Olympiads. These athletes provide a laboratory to study the effect of brief, intense fame on historical significance.

The accompanying table presents the most significant summer Olympians of each gender. Male champions are generally more highly regarded than females, but this varies by sport. Olympic champions do not score particularly well according to our historical significance measures, with only a dozen or so ranked among the top 10,000 figures. One explanation is that each new champion replaces her predecessors, by filling the specific niche each once held in the public consciousness. Only exceptional figures have achievements that remain indelible years later, or parlay their

Top Male Olympic Stars

Sig	Person	Dates
1356	Jim Thorpe	(1888–1953)
2116	Jesse Owens	(1913–1980)
4887	Carl Lewis	(1961–)
5510	Michael Phelps	(1985–)
6096	Mark Spitz	(1950–)
6587	Johnny Weissmuller	(1904–1984)
6812	Ben Johnson	(1961–)
7139	Sebastian Coe	(1956–)
7260	Steve Prefontaine	(1951–1975)
7459	Avery Brundage	(1887–1975)
7519	Ian Thorpe	(1982–)
7568	Paavo Nurmi	(1897–1973)
8095	Eric Liddell	(1902–1945)
8225	Usain Bolt	(1986–)
9739	Duke Kahanamoku	(1890–1968)

Top Female Olympic Stars

Sig	Person	Dates
7563	Babe Zaharias	(1911–1956)
9148	Sonja Henie	(1912–1969)
9705	F. Griffith-Joyner	(1959–1998)
10257	Nadia Comneci	(1961–)
13560	Dawn Fraser	(1937–)
13770	Wilma Rudolph	(1940–1994)
13861	J. Joyner-Kersee	(1962–)
16628	Mary Lou Retton	(1968–)
18119	Lisa Leslie	(1972–)
19711	Fanny Blankers-Koen	(1918–2004)
25335	Larissa Latynina	(1934–)
25478	Nastia Liukin	(1989–)
26044	Olga Korbut	(1955–)
26057	Betty Cuthbert	(1938–)
26945	Natalie Coughlin	(1982–)

athletic success into accomplishments in other fields. We will review the stars of several prominent sports separately.

The significance rankings we report here employ data from before the 2012 Olympics in London, and hence inadequately represent the achievements of these athletes. Still, two of the biggest stars of the 2012 games do quite well in our rankings: swimmer *Michael Phelps (1985–)* [5510] and sprinter *Usain Bolt (1986–)* [8225].

Most Significant Track-and-Field Stars

Sig	Person	Dates	C/G	Description
1356	Jim Thorpe	(1888–1953)	C G	Native American athlete (Olympics, football)
2116	Jesse Owens	(1913–1980)	C G	American track star (4 Olympic gold, 1936)
4887	Carl Lewis	(1961–)	C G	American track star (4 Olympic gold, 1984)
6812	Ben Johnson	(1961–)	C G	Drug-enhanced Canadian sprinter
7139	Sebastian Coe	(1956–)	C G	English athlete (middle distance runner)
7260	Steve Prefontaine	(1951–1975)	C G	American middle/long-distance runner
7459	Avery Brundage	(1887–1975)	C G	American Olympic athlete and official
7563	Babe Zaharias	(1911–1956)	C G	Greatest woman athlete (golf, track, basketball)
7568	Paavo Nurmi	(1897–1973)	C G	Finnish long distance runner
8095	Eric Liddell	(1902–1945)	C G	Scottish athlete (*Chariots of Fire*)

12.3.1 TRACK AND FIELD

Track-and-field events represent the popular ideal of Olympic sport: elementary, unchanging, traditionally amateur events. We count seven of the ten most prominent Olympians as track-and-field champions. *Jesse Owens (1913–1980)* [2116] won four gold medals in the 1936 Berlin Olympics; the symbolism of a black man triumphing over the master race in front of the Nazis explains his enduring fame. *Carl Lewis (1961–)* [4887] and *Ben Johnson (1961–)* [6812] were rival sprinters over several Olympiads, with Johnson ultimately disgraced for illegal use of steroids.

Two of these athletes are best known because of movies. Because of religious convictions, sprinter *Eric Liddell (1902–1945)* [8095] declined to run the finals of the 1924 Olympic 100-meter dash on a Sunday, a tale told in the film *Chariots of Fire*. Long-distance runner *Steve Prefontaine (1951–1975)* [7260] is a more perplexing case: an American (but never Olympic) champion, he was killed in an automobile accident at age 24. This cult figure among runners became the subject of two competing movies, which came out in 1997 and 1998, respectively.

12.3.2 SWIMMING

Much of the attention Olympic swimming receives is due to the large number of closely related events, which enable champions to accumulate eye-popping totals of medals. Four of the most significant swimmers are known for their gaudy Olympic medal totals: seven gold for Spitz in 1972, eight gold for Phelps in 2008, ten overall for Thorpe in 2000 and 2004, and eight for Fraser over three Olympiads.

Most Significant Swimmers

Sig	Person	Dates	C/G	Description
5510	Michael Phelps	(1985–)	C G	American swimmer (8 Olympic gold, 2008)
6096	Mark Spitz	(1950–)	C G	American swimmer (7 Olympic gold, 1972)
6587	Johnny Weissmuller	(1904–1984)	C G	American swimmer and actor (Tarzan)
7519	Ian Thorpe	(1982–)	C G	American Olympic swimmer (the Thorpedo)
9739	Duke Kahanamoku	(1890–1968)	C G	Hawaiian swimmer and surfing pioneer
13560	Dawn Fraser	(1937–)	C G	Australian Olympic swimmer
17720	Buster Crabbe	(1908–1983)	C G	American swimmer and actor (Flash Gordon)

In the years before World War II, Hollywood regularly fished the Olympic pools for stars, catching both Tarzan (*Johnny Weissmuller (1904–1984)* [6587]) and Flash Gordon (*Buster Crabbe (1908–1983)* [17720]). Special mention goes to aquamusical star *Esther Williams (1921–2013)* [12359], who lost her moment with the cancellation of the 1940

Olympic games due to the outbreak of war. *Duke Kahanamoku (1890–1968)* [9739] was another swimming star who turned to acting, but today is best known for popularizing surfing.

12.3.3 GYMNASTICS

Women completely dominate the rankings of the most significant gymnasts, filling nine of the top ten positions. The lone male, *Li Ning (1963–)* [33106], won six medals at the 1984 games and is one of China's most famous athletes.

Most Significant Gymnasts

Sig	Person	Dates	C/G	Description
10257	Nadia Comneci	(1961–)	C G	Romanian gymnast (1976 Olympic gold)
16628	Mary Lou Retton	(1968–)	C G	American gymnast (1984 Olympic gold)
25335	Larissa Latynina	(1934–)	C G	Soviet gymnast (1956/60/64 Olympic gold)
25478	Nastia Liukin	(1989–)	C G	American gymnast (2008 Olympic gold)
26044	Olga Korbut	(1955–)	C G	Soviet gymnast (1972 Olympic gold)
33106	Li Ning	(1963–)	C G	Chinese male gymnast / entrepreneur
35532	Shannon Miller	(1977–)	C G	American gymnast (1996 Olympic gold)
36809	Nellie Kim	(1957–)	C G	Soviet gymnast (1980 Olympic gold)
40885	Ludmilla Tourischeva	(1952–)	C G	Soviet gymnast (1968 Olympic gold)
41882	Svetlana Khorkina	(1979–)	C G	Russian gymnast (2000 Olympic gold)

The strong performance of women's gymnastics here reflects its broad popularity among American girls. Gymnastics is a young person's sport, reflected by champions *Nadia Comaneci (1961–)* [10257] (age 14), *Mary Lou Retton (1968–)* [16628] (16), *Olga Korbut (1955–)* [26044] (17), and *Nastia Liukin (1989–)* [25478] (18). Concern about injuries and abusive coaching has raised minimum age limits; since 1997, gymnasts must be at least 16 years old to participate in the Olympics.

12.3.4 WINTER SPORTS

The Winter Olympics have been held since 1924, and played second fiddle to the summer games ever since. The London summer Olympics of 2012 featured 204 nations, but only 82 assembled for the Vancouver winter games of 2010. Climate restricts the number of nations that can meaningfully participate in winter sports; the mere existence of a Jamaican bobsled team was enough to inspire the movie *Cool Runnings*.

Top Figure Skaters

Sig	Person	Dates
9148	Sonja Henie	(1912–1969)
15458	Kristi Yamaguchi	(1971–)
16118	Michelle Kwan	(1980–)
17356	Brian Boitano	(1963–)
19402	Tonya Harding	(1970–)
22197	Nancy Kerrigan	(1969–)
24619	Dorothy Hamill	(1956–)
24922	Dick Button	(1929–)

Top Speed Skaters

Sig	Person	Dates
15132	Apolo Ohno	(1982–)
20089	Eric Heiden	(1958–)
46995	Shani Davis	(1982–)
70624	Chad Hedrick	(1977–)
80666	Bonnie Blair	(1964–)
113968	Beth Heiden	(1959–)
122314	Sheila Young	(1950–)
125149	Joey Cheek	(1979–)

Skating is an interesting Olympic sport because it comes in two distinct flavors. Figure skating is a test of grace and beauty, while speed skating is an aerobic race against the clock. We present the most significant champions in each of these sports, with figure skaters on the left and speed skaters on the right of the table.

The figure skaters prove much higher ranked than their faster brethren, even though they compete in only one event per Olympiad. Champion figure skaters generally go on to professional careers giving exhibitions, increasing their visibility. *Sonja Henie (1912–1969)* [9148] was a champion skater who became famous for her later Hollywood career. It is revealing that the most significant speed skater (*Apolo Ohno (1982–)* [15132]) won eight Olympic medals, yet made his biggest splash as the winner of *Dancing with the Stars.*

Skiing is another popular sport at the Winter Olympics. The two most significant champions, *Bode Miller (1977–)* [26961] and *Jean-Claude Killy (1943–)* [32918], are also substantially outranked by the most prominent figure skaters.

13

The Arts

Most historical figures are accorded a one-way ride down the road from glory. Long-dead figures rarely regain prominence, except perhaps in response to external events, such as the release of a movie about their life. But the most dramatic swings in historical reputation we have encountered occur in the arts, reflecting long-term fluctuations in taste. Authors and their books can rise in response to fresh critical attention, and obscure painters may suddenly find their work back in style.

In this chapter, we will identify the most significant figures in literature and the fine arts. Along the way we will address certain larger-scale questions: In which media do contemporary artists have enough stature to compete with the classical masters? How well do best-seller lists and literary awards identify the significant voices of their times? Have modern architects usurped the role traditionally occupied by sculptors?

13.1 Literature

Ranking the world's greatest literary figures is a parlor game of comparable popularity to the ranking of presidents. It provides free rein to express the biases inherent in everyone's worldview. Which is better: classical or contemporary, domestic or international, or prose vs. verses?

The *Literary 100* [Burt, 2000] is a representative ranking, which tries to identify "the most influential novelists, playwrights, and poets of all time." Burt provides an alternate perspective to evaluate our "Literary 50" rankings, presented in Figure 13.1.

Sig	Writer	Dates	Lit100	Field	Well-known Work
4	William Shakespeare	(1564–1616)	1	plays	*Hamlet*
33	Charles Dickens	(1812–1870)	6	novels	*A Tale of Two Cities*
53	Mark Twain	(1835–1910)	63	novels	*The Adventures of Tom Sawyer*
54	Edgar Allan Poe	(1809–1849)	55	poems	"The Raven"
64	Voltaire	(1694–1778)	69	poems	*Candide*
77	Oscar Wilde	(1854–1900)	100	plays	*The Importance of Being Earnest*
88	Johann Wolfgang von Goethe	(1749–1832)	10	novels	*The Sorrows of Young Werther*
96	Dante Alighieri	(1265–1321)	2	poems	*Divine Comedy*
118	Lewis Carroll	(1832–1898)		novels	*Alice's Adventures in Wonderland*
131	Henry David Thoreau	(1817–1862)		essays	*Walden*
139	Jane Austen	(1775–1817)	18	novels	*Pride and Prejudice*
141	Samuel Johnson	(1709–1784)		essays	*Dictionary of the English Language*
152	Homer	(?–?)	3	poems	*The Odyssey*
158	Lord Byron	(1788–1824)	75	poems	*Don Juan*
160	Walt Whitman	(1819–1892)	40	poems	*Leaves of Grass*
165	John Milton	(1608–1674)	8	poems	*Paradise Lost*
173	Geoffrey Chaucer	(1343–1400)	5	poems	*The Canterbury Tales*
177	Virgil	(?–?)	9	poems	*The Aeneid*
182	William Wordsworth	(1770–1850)	33	poems	*Lyrical Ballads*
191	Stephen King	(1947–)		novels	*The Shining*
194	Emily Dickinson	(1830–1886)	54	poems	I taste a liquor never brewed
196	Leo Tolstoy	(1828–1910)	4	novels	*War and Peace*
208	Victor Hugo	(1802–1885)	89	poems	*Les Miserables*
213	George Bernard Shaw	(1856–1950)	45	plays	*Man and Superman*
227	Nathaniel Hawthorne	(1804–1864)	84	novels	*The House of the Seven Gables*
244	Fyodor Dostoyevsky	(1821–1881)	15	novels	*Crime and Punishment*
246	Miguel de Cervantes	(1547–1616)	11	novels	*Don Quixote*
248	Ernest Hemingway	(1899–1961)	46	novels	*The Old Man and the Sea*
249	H. G. Wells	(1866–1946)		novels	*The Time Machine*
251	Herman Melville	(1819–1891)	24	novels	*Moby-Dick*
259	Rudyard Kipling	(1865–1936)		novels	*The Jungle Book*
274	Sophocles	(496–406 B.C.)	13	plays	*Oedipus the King*
280	Samuel Taylor Coleridge	(1772–1834)		poems	"The Rime of the Ancient Mariner"
305	John Keats	(1795–1821)	25	poems	*The Complete Poems*
317	Robert Burns	(1759–1796)		poems	"Auld Lang Syne"
326	Petrarch	(1304–1374)	53	poems	"Il Canzoniere"
329	Percy Bysshe Shelley	(1792–1822)	72	poems	"Ozymandias"
342	George Orwell	(1903–1950)		novels	*1984*
374	Christopher Marlowe	(1564–1593)		poems	"Doctor Faustus"
378	Thomas Hardy	(1840–1928)	44	novels	*Tess of the D'Urbervilles*
386	Aeschylus	(525–456 B.C.)	29	plays	*Oresteia*
391	Jonathan Swift	(1667–1745)	42	essays	*Gulliver's Travels*
397	Rabindranath Tagore	(1861–1941)	90	novels	*The Home and the World*
403	Henrik Ibsen	(1828–1906)	36	poems	"A Doll's House"
406	James Joyce	(1882–1941)	7	novels	*Ulysses*
408	Henry James	(1843–1916)	38	novels	*The Turn of the Screw*
418	Aristophanes	(446–386 B.C.)	34	plays	*The Knights*
420	Alexander Pushkin	(1799–1837)	21	novels	*Eugene Onegin*
421	Ben Jonson	(1572–1637)		plays	*The Alchemist*
436	T. S. Eliot	(1888–1965)	16	poems	*The Waste Land*

FIGURE 13.1. The most significant literary figures, with *Literary 100* rankings [Burt, 2000].

There is good basic agreement between his ranking and ours. Our top 50 contains 39 members of the *Literary 100*, including Burt's 11 highest-ranked figures. There is a strong correlation between our significance measure and the *Literary 100* ranking, indicating broad general agreement.

Mark Twain (1835–1910) [53] famously defined a classic as a book that "everybody wants to have read and nobody wants to read." Our rankings respect this distinction. We generally score popular writers such as *Oscar Wilde (1854–1900)* [77], *Lewis Carroll (1832–1898)* [118], and Twain himself higher than Burt does in his *Literary 100*.

We particularly expect the literary establishment to blanch at our significance rank for best-selling horror novelist *Stephen King (1947–)* [191]. He is the only contemporary writer to come close to a spot in our Literary 50.[1] We interpret King as the *Charles Dickens (1812–1870)* [33] of our time, both sharing immense popularity, mind-boggling productivity, and even the serial novel genre. We feel confident that a hundred years from now King will be read more than most of the *Literary 100*'s tail.

Least Significant Members of the *Literary 100*

Sig	Person	Dates	Lit100	Field	Work
29330	Günter Grass	(1927–)	92	Novels	*The Tin Drum*
17736	Zeami Motokiyo	(1363–1443)	99	Plays	*Noh drama*
14490	Cao Xueqin	(1715–1763)	67	Novels	*Dream of the Red Chamber*
9104	Robert Musil	(1880–1942)	79	Novels	*The Man Without Qualities*
7819	Ralph Ellison	(1913–1994)	86	Novels	*Invisible Man*
5197	Isaac Bashevis Singer	(1902–1991)	95	Novels	*Gimple the Fool*
4885	Theodore Dreiser	(1871–1945)	85	Novels	*An American Tragedy*
4702	Richard Wright	(1908–1960)	97	Novels	*Uncle Tom's Children*
4254	Lu Xun	(1881–1936)	93	Novels	*The True Story of Ah Q*
3534	Federico García Lorca	(1898–1936)	83	Poems	*Gypsy Ballads*

Indeed, the least significant members of the Literary 100 reveal biases in both Burt's rankings and our own. The literature of China and Japan is not adequately covered in the English-language Wikipedia, so we presumably underrate certain writers like *Murasaki Shikibu (973–?)* [2256], the author of the early Japanese novel *The Tale of Genji*. She was Burt's #12, but did

[1] I (Charles) can't bear to hear what my father will say when he reads this.

not crack our Literary 100. However, these Asian literatures are so vast that it is debatable whether the *Literary 100* really found the right individuals, either. We rank the classical Chinese writers *Li Bai (701–762)* [1552], *Du Fu (712–770)* [2192], and *Su Shi (1037–1101)* [3162] all as stronger than his picks of *Zeami Motokiyo (1363–1443)* [17736] and *Cao Xueqin (1715–1763)* [14490].

Burt's rankings show a weakness for several twentieth-century writers who we do not anticipate will age very well. The recognition accorded to Nobel Prize winners Singer and Grass largely reflects the cultural aftermath of World War II. The novels of Ellison and Wright reflect the black experience before the successes of the Civil Rights movement erased the worst vestiges of this era. This will limit these books' future resonance. Others reflect idiosyncratic choices. The Wikipedia article for *Robert Musil (1880–1942)* [9104] apologizes that his work "has not been widely read because of its delayed publication and also because of the lengthy and intricate plot."

Time Trends

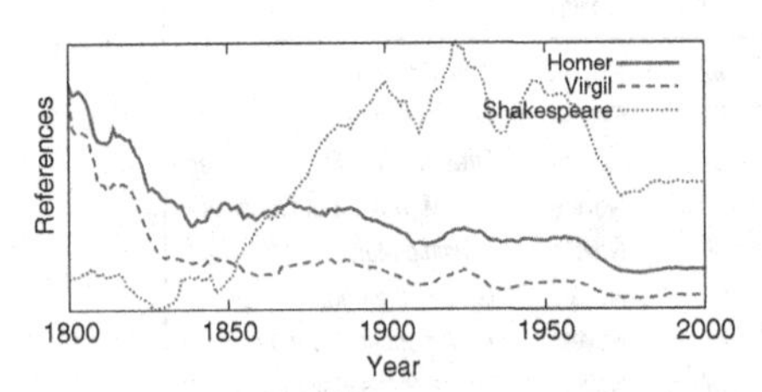

The Google Ngram data enables us to track changes in authors' reputations over time. Two interesting trends emerge with respect to classical writers. First, *William Shakespeare's (1564–1616)* [4] mindshare exploded after 1850, from a position roughly comparable to *Geoffrey Chaucer (1343–1400)* [173]. This increase mirrors the growth of the theater in America. The Roman poet *Virgil* [177] has declined from a peak comparable to *Homer* [152] to a mindshare now half as large. His flagging status probably correlates with the decline in the importance of Latin in a liberal arts education.

The career of novelist *Jane Austen (1775–1817)* [139] exhibits a particularly dramatic reputational arc. Her first novel was published only five years before her early death, and then went out of print for twelve years. Since then, however, her reputation has exhibited sustained growth.

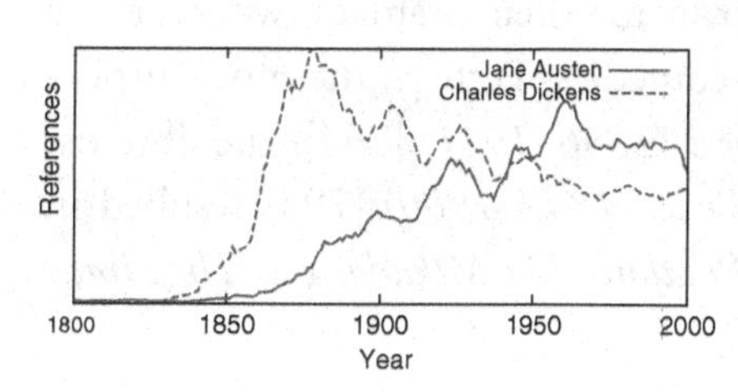

This remarkable trend shows no sign of abating. Since about 1950, she has been beating the dickens out of *Charles Dickens (1812–1870)* [33], and is still increasing in mind-share long after he began to fade. Recent surges reflect increased interest in gender, successful motion picture adaptations of her work, and even its ability to withstand a zombie invasion [Grahame-Smith and Austen, 2009].

13.1.1 PLAYWRIGHTS

Plays may be the literary form that has best survived the passage of time. The works of ancient Greek playwrights continue to be performed fully 2,500 years after they were written. Indeed, four of our seven most significant playwrights were active during the peak of classical Greek civilization. And of course, the most significant writer in our rankings, *William Shakespeare (1564–1616)* [4], is known primarily for his plays.

Modern writers also contribute to this ancient form, with half of our top 15 playwrights living in the twentieth century.

Most Significant Playwrights

Sig	Person	Dates	C/G	Description
4	William Shakespeare	(1564–1616)	C G	English playwright (*Hamlet*)
213	George Bernard Shaw	(1856–1950)	C G	Irish playwright (*Pygmalion*)
274	Sophocles	(496–406 B.C.)	C G	Ancient Greek tragedian (*Oedipus Rex*)
348	Euripides	(480–406 B.C.)	C G	Ancient Greek tragedian
386	Aeschylus	(525–456 B.C.)	C G	Ancient Greek tragedian
403	Henrik Ibsen	(1828–1906)	C G	Norwegian playwright (*Enemy of the People*)
418	Aristophanes	(446–386 B.C.)	C G	Greek comic playwright (*The Frogs*)
421	Ben Jonson	(1572–1637)	C G	English Renaissance dramatist (*Volpone*)
523	Anton Chekhov	(1860–1904)	C G	Russian playright (*The Cherry Orchard*)
744	W. S. Gilbert	(1836–1911)	C G	English dramatist (Gilbert and Sullivan)
876	August Strindberg	(1849–1912)	C G	Swedish playwright (*Miss Julie*)
933	Bertolt Brecht	(1898–1956)	C G	German playwright (*The Threepenny Opera*)
1003	Jean-Paul Sartre	(1905–1980)	C G	French existentialist philosopher
1081	Arthur Miller	(1915–2005)	C G	American playwright (*Death of a Salesman*)
1213	Tennessee Williams	(1911–1983)	C G	American playwright (*Cat on a Hot Tin Roof*)

Most of these playwrights are known for their dramatic works, but a few earned their reputations through comedies. Perhaps the most surprising member of the club is *W. S. Gilbert (1836–1911)* [744], the librettist whose collaboration with *Arthur Sullivan (1842–1900)* [795] resulted in such famous comic operas as *H.M.S. Pinafore*, *The Mikado*, and *The Pirates of Penzance*.

13.1.2 POETS

The statement "More people write poetry than read it" is attributed to *George Carlin (1937–2008)* [5466], and it has the ring of truth. Consider the most prominent living English-language poets: *Billy Collins (1941–)* [26048], *Robert Pinsky (1940–)* [16364], *W. S. Merwin (1927–)* [41425], and *Rita Dove (1952–)* [47391]. All rank as fairly minor cultural figures, particularly by the standards of their illustrious predecessors.

Most Significant Poets

Sig	Person	Dates	C/G	Description
102	William Blake	(1757–1827)	C G	English poet and painter ("The Tyger")
158	Lord Byron	(1788–1824)	C G	British poet ("Don Juan")
160	Walt Whitman	(1819–1892)	C G	American poet ("Leaves of Grass")
182	William Wordsworth	(1770–1850)	C G	English poet ("Lyrical Ballads")
194	Emily Dickinson	(1830–1886)	C G	American female poet (The Belle of Amherst)
280	Samuel Taylor Coleridge	(1772–1834)	C G	English poet ("Kubla Khan")
305	John Keats	(1795–1821)	C G	English Romantic poet ("Ode to a Grecian Urn")
349	Henry Wadsworth Longfellow	(1807–1882)	C G	American poet ("Paul Revere's Ride")
378	Thomas Hardy	(1840–1928)	C G	British novelist (*Tess of the d'Urbervilles*)
412	William Butler Yeats	(1865–1939)	C G	Irish poet and dramatist
436	T. S. Eliot	(1888–1965)	C G	English poet ("The Waste Land")
507	Ezra Pound	(1885–1972)	C G	American expatriate poet (*The Cantos*)
564	Friedrich Schiller	(1759–1805)	C G	German poet and philosopher
568	John Donne	(1572–1631)	C G	English poet ("Death Be not Proud")
644	Robert Frost	(1874–1963)	C G	American poet ("The Road Less Traveled")
806	John Dryden	(1631–1700)	C G	English Restoration poet
893	Dr. Seuss	(1904–1991)	C G	American children's writer (Cat in the Hat)
905	Matthew Arnold	(1822–1888)	C G	British poet (Dover Beach)
1050	Francis Scott Key	(1777–1843)	C G	Poet (lyrics of National Anthem)
1211	Langston Hughes	(1902–1967)	C G	Black poet (Harlem Renaissance)

We have taken the liberty of removing ancient and medieval lyric poets such as *Homer* [152], *Geoffrey Chaucer (1343–1400)* [173], and *Dante Alighieri (1265–1321)* [96]. These writers might be better thought of as storytellers than wordsmiths, and who might have been novelists had the genre been invented in their times.

Truly contemporary poets are nowhere to be seen in our rankings, but we are not so far removed from a time when major poets walked the earth. Four members of our top twenty were active after World War II: *T. S. Eliot (1888–1965)* [436], *Robert Frost (1874–1963)* [644], *Ezra Pound (1885–1972)* [507], and *Langston Hughes (1902–1967)* [1211]. The Beat poets of the late 1950s, including *Allen Ginsberg (1926–1997)* [1775] and *William S. Burroughs (1914–1997)* [1561], were major influences on the counterculture of the 1960s. But perhaps the poet who best speaks to today's sensibilities is the children's author *Dr. Seuss (1904–1991)* [893].

Popular successors to these poets live on in rock-and-roll bands, not in academia. Music lyrics have been published as poems since at least the *Poetry of Rock* [Astor, 2010; Goldstein, 1969] in 1969, and today even respected poetry anthologies [Axelrod et al., 2012; Lehman, 2006] manage to include works by such "poets" as *Bob Dylan (1941–)* [130], *Queen Latifah (1970–)* [6367], and *Patti Smith (1946–)* [5825]. Certainly an ambitious poet can reach a greater audience writing lyrics instead of verse, and singing their words instead of printing them. This harkens back to ancient times, when pre-literate audiences had their poetry sung to them by balladeers rather than of reading it.

13.1.3 NOVELISTS

The novel is today's major literary form, but its history is short relative to that of poetry or plays. Some call *Don Quixote* by *Miguel de Cervantes (1547–1616)* [246] the first major novel, while others cast their lot with the Japanese *The Tale of Genji* by *Murasaki Shikibu (973–?)* [2256].

We will tell the story of the novel in two chapters, beginning with the pre-twentieth century novelists presented in Figure 13.2.

Our highest ranked novelists wrote mostly in English, but include writers in German (Goethe), French (Hugo and Verne), Spanish (Cervantes), and Russian (Tolstoy and Dostoevsky). All remain widely read today, and are popular choices for high school and college literature classes. We probably rank *Jules Verne (1828–1905)* [229] and *Mary Shelley (1797–1851)* [223] higher than most English professors would, but they had important roles in developing the genre of science fiction.

The capsule time series in Figure 13.2 permits us to distinguish authors who were posthumously rediscovered (such as *Herman Melville (1819–1891)* [251]) from those whose reputations peaked in their lifetimes and are now declining (for instance, *Rudyard Kipling (1865–1936)* [259]).

Sig	Person	Ngrams Timeline	C/G	Description
33	Charles Dickens		C G	English novelist (*David Copperfield*)
53	Mark Twain		C G	American author (*Huckleberry Finn*)
77	Oscar Wilde		C G	Irish author and poet (*Dorian Gray*)
88	J. W. von Goethe		C G	German writer and polymath (*Faust*)
118	Lewis Carroll		C G	Author (*Alice's Adventures in Wonderland*)
139	Jane Austen		C G	English author (*Pride and Prejudice*)
196	Leo Tolstoy		C G	Russian novelist (*War and Peace*)
208	Victor Hugo		C G	French poet (*Les Misérables*)
223	Mary Shelley		C G	English novelist (*Frankenstein*)
227	Nathaniel Hawthorne		C G	Novelist (*The Scarlet Letter*)
229	Jules Verne		C G	French author (*Around the World in 80 Days*)
244	Fyodor Dostoyevsky		C G	Russian novelist (*Crime and Punishment*)
246	Miguel de Cervantes		C G	Spanish author of first novel (*Don Quixote*)
251	Herman Melville		C G	American novelist (*Moby-Dick*)
259	Rudyard Kipling		C G	English author of Empire (*The Jungle Book*)
378	Thomas Hardy		C G	British novelist (*Tess of the d'Urbervilles*)
391	Jonathan Swift		C G	Anglo-Irish satirist (*Gulliver's Travels*)
408	Henry James		C G	American literary novelist (*Washington Square*)
420	Alexander Pushkin		C G	Russian author (*Eugene Onegin*)
433	Washington Irving		C G	Author (*Legend of Sleepy Hollow*)

FIGURE 13.2. The most significant novelists before 1900.

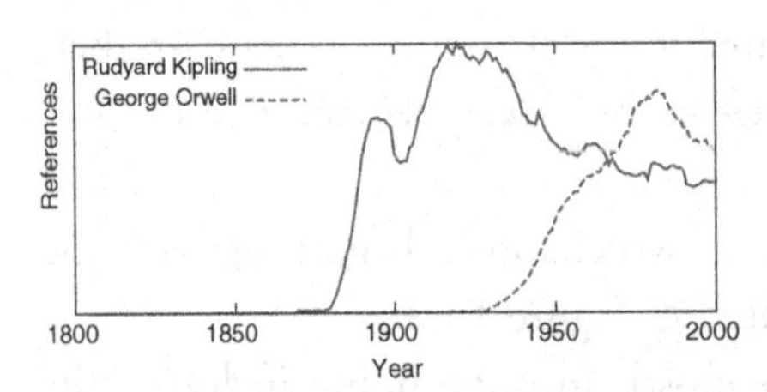

Kipling and *George Orwell* [342] are an interesting matched pair, as reflected in their Ngrams trends. Kipling writes from a nineteenth-century viewpoint, that of "The White Man's Burden." Orwell himself called Kipling a "prophet of British Imperialism." But as the British Empire declined, so did the draw of Kipling's narratives, replaced by the growing disillusionment in Empire, and the active distrust in authority reflected in Orwell's work.

We make a fairly arbitrary distinction between these early novelists and those most active during the twentieth century.

They include representatives from several genres and wildly differing levels of literary ambition. Some live on today primarily as "young adult" writers, such as *H. G. Wells* [249] and *Jack London* [578]. Others pushed the boundaries of narrative to create demanding (at times, almost unreadable) works, including *James Joyce* [406], *William Faulkner* [907], and *Virginia Woolf* [650].[2] More popular genres include mystery, particularly the authors who created the detectives Sherlock Holmes (*Arthur Conan Doyle* [247]) and Hercule Poirot (*Agatha Christie* [708]), and science fiction, represented here by the astonishingly prolific *Isaac Asimov* [916].

Most Significant Twentieth- Century Novelists

Sig	Person	Dates	C/G	Description
191	Stephen King	(1947–)	C G	American popular novelist (*Carrie*)
247	Arthur Conan Doyle	(1859–1930)	C G	Scottish creator of Sherlock Holmes
248	Ernest Hemingway	(1899–1961)	C G	Author (*The Sun Also Rises*)
249	H. G. Wells	(1866–1946)	C G	Science fiction writer (*The Time Machine*)
342	George Orwell	(1903–1950)	C G	English author (*1984*)
406	James Joyce	(1882–1941)	C G	Stream of consciousness novelist (*Ulysses*)
486	Ayn Rand	(1905–1982)	C G	Anti-altruist author (*The Fountainhead*)
544	Joseph Conrad	(1857–1924)	C G	Polish novelist (*Heart of Darkness*)
576	John Steinbeck	(1902–1968)	C G	Author / Nobelist (*The Grapes of Wrath*)
578	Jack London	(1876–1916)	C G	American author (*White Fang*)
650	Virginia Woolf	(1882–1941)	C G	English author (*Mrs Dalloway*)
708	Agatha Christie	(1890–1976)	C G	British crime writer (Hercule Poirot)
724	Franz Kafka	(1883–1924)	C G	German novelist (*The Metamorphosis*)
747	F. Scott Fitzgerald	(1896–1940)	C G	Novelist (*This Side of Paradise*)
907	William Faulkner	(1897–1962)	C G	Author / Nobelist (*The Sound and the Fury*)
916	Isaac Asimov	(1919–1992)	C G	American science fiction author (*I, Robot*)

13.1.4 THE NOBEL PRIZE IN LITERATURE

The Nobel Prize is the world's most prestigious literary award, but its roster of honorees includes several puzzling choices. Two of the most prominent laureates are best known for other accomplishments. *Winston Churchill's (1874–1965)* [37] six-volume history of World War II was less important than his efforts in winning that war. Philosopher *Bertrand Russell (1872–1970)* [253] was recognized for his humanitarian writings, not the symbolic

[2] Others manage to be unreadable without explicitly trying, like *Ayn Rand (1905–1982)* [486].

logic of his most famous book, *Principia Mathematica*. The least significant recipients will be unknown to all but the most dedicated students of world literature.

Top Nobel Prize Authors

Sig	Person	Dates
37	Winston Churchill	(1874–1965)
213	George Bernard Shaw	(1856–1950)
248	Ernest Hemingway	(1899–1961)
253	Bertrand Russell	(1872–1970)
259	Rudyard Kipling	(1865–1936)
397	Rabindranath Tagore	(1861–1941)
412	William B. Yeats	(1865–1939)
436	T. S. Eliot	(1888–1965)
576	John Steinbeck	(1902–1968)
907	William Faulkner	(1897–1962)

Bottom Nobel Prize Authors

Sig	Person	Dates
89721	Tomas Tranströmer	(1931–)
86978	José Echegaray	(1832–1916)
86674	Imre Kertész	(1929–)
85832	Wisława Szymborska	(1923–2012)
79311	Roger M. d. Gard	(1881–1958)
59642	Claude Simon	(1913–2005)
57617	Frans E. Sillanpää	(1888–1964)
56185	Jacinto Benavente	(1866–1954)
51146	Vicente Aleixandre	(1898–1984)
44002	Camilo José Cela	(1916–2002)

By contrast, many great writers never received the Nobel Prize. A few, such as *Anton Chekhov (1860–1904)* [523] and *Herbert Spencer (1820–1903)* [619], died early enough in the history of the award to absolve the committee of responsibility here. Because the prize is awarded to only one writer each year, it should have taken several years to clear the backlog of great available authors. However, the judges did not take advantage of this deep bench, for only three of the first 20 winners are familiar to us today: *Rudyard Kipling (1865–1936)* [259] (1907), *Rabindranath Tagore (1861–1941)* [397] (1913), and *Anatole France (1844–1924)* [2891] (1921).

Top Authors Without a Nobel Prize

Sig	Person	Dates	C/G	Description
53	Mark Twain	(1835–1910)	C G	American author (*Huckleberry Finn*)
196	Leo Tolstoy	(1828–1910)	C G	Russian novelist (*War and Peace*)
249	H. G. Wells	(1866–1946)	C G	Science fiction writer (*The Time Machine*)
378	Thomas Hardy	(1840–1928)	C G	British novelist (*Tess of the d'Urbervilles*)
406	James Joyce	(1882–1941)	C G	Stream of consciousness novelist (*Ulysses*)
523	Anton Chekhov	(1860–1904)	C G	Russian playright (*The Cherry Orchard*)
544	Joseph Conrad	(1857–1924)	C G	Polish novelist (*Heart of Darkness*)
619	Herbert Spencer	(1820–1903)	C G	English philosopher (survival of the fittest)
644	Robert Frost	(1874–1963)	C G	American poet ("The Road Less Traveled")
650	Virginia Woolf	(1882–1941)	C G	English author (*Mrs Dalloway*)

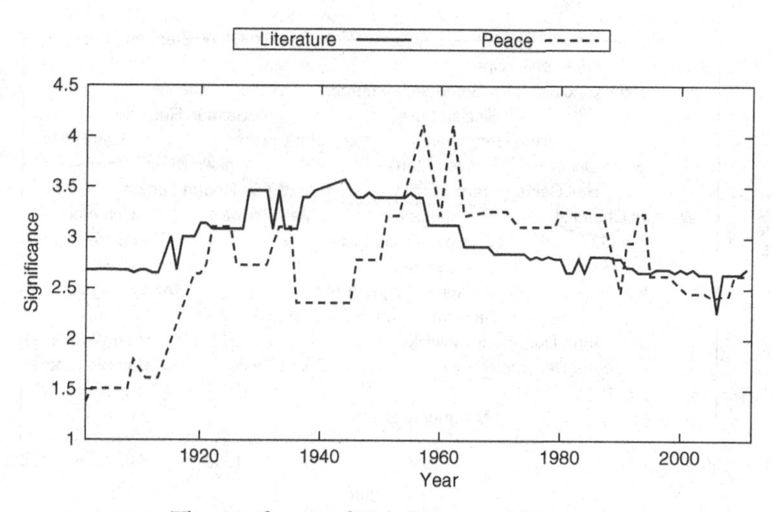

FIGURE 13.3. The significance of Nobel Peace and Literature prize winners.

These generally lackluster selections reflect the early reluctance of Scandinavian judges to consider English-language writers for the award. The significance timeline in Figure 13.3 shows that the stature of literary prize winners rose steadily from 1920 until about 1960, when it began a gentle slide back to the early standards. This decline reflects the Swedish academy's decision to spread the award geographically, recognizing literatures around the world. This is a noble sentiment and worthy objective. However, it diminishes the stature of the prize to honor writers who have not yet been translated outside their native language. Indeed, it calls into question exactly how the academy makes their decisions, since many judges could never have read the writers under consideration.

13.2 Popular Writers

There is a cultural divide between authors with literary ambitions, and the writers who strive to reach a broad popular audience. We have assembled the authors of the 10 best-selling books each year from 1900 to 2010, so we can monitor how the reputations of popular authors fare.

Figure 13.4 presents a timeline measuring the significance of best-selling authors over the past century. Writers of similar significance sit at the same

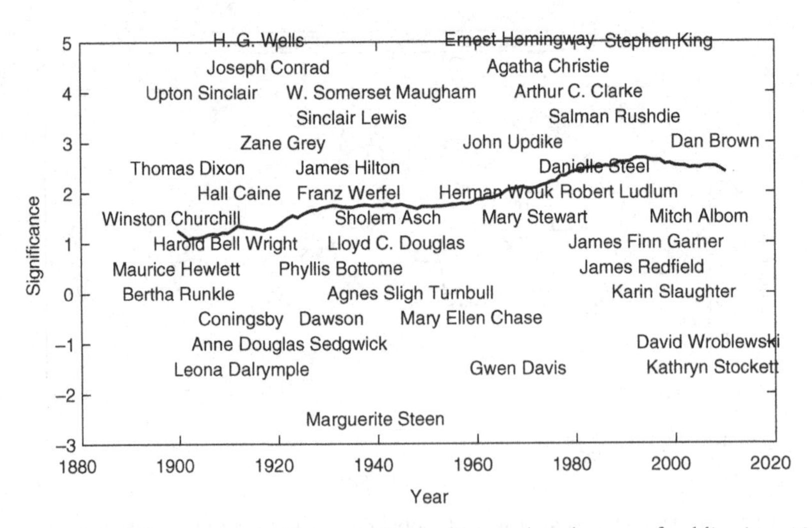

FIGURE 13.4. The average significance of best-selling authors by year of publication, 1900 to 2010. The trend is for more recent authors to have increasing significance, yet there appear to be fewer high-profile literary figures today than in the past.

level from left to right. The stars of recent best-seller lists generally seem to lack the gravitas of those of earlier times.

The plot shows a gradual increase in average significance from 1900 up to the present day. This trend reflects the fact that even the weakest best-selling writers from the last 40 years remain preserved in the living memory of the Wikipedia community. But as each generation passes, only the best of its books survive to find new readers. The great have been treated with deference, while the lesser figures have essentially been forgotten.

To get a better handle on this phenomenon, we compared the most significant authors of best sellers from the first half of our data (1900–1955) with those appearing since 1956. Recognizable literary authors have regularly made the best-seller lists, although the frequency appears to have declined in recent years:

Generally speaking, the names on the right score as more significant, but these post-1955 best sellers are a surprisingly mixed bag. Five are Nobel Prize winners (Hemingway, Steinbeck, Faulkner, Solzhenitsyn, and Pasternak), but two others are better known as television personalities (astronomer *Carl Sagan (1934–1996)* [2687] and political commentator *Glenn Beck (1964–)* [3727]. Particularly curious here is *D. H. Lawrence*

Top Best-selling Authors before 1955			Top Best-selling authors after 1955		
Sig	Person	Dates	Sig	Person	Dates
37	Winston Churchill	(1874–1965)	191	Stephen King	(1947–)
247	Arthur Conan Doyle	(1859–1930)	248	Ernest Hemingway	(1899–1961)
248	Ernest Hemingway	(1899–1961)	486	Ayn Rand	(1905–1982)
249	H. G. Wells	(1866–1946)	576	John Steinbeck	(1902–1968)
544	Joseph Conrad	(1857–1924)	708	Agatha Christie	(1890–1976)
576	John Steinbeck	(1902–1968)	893	Dr. Seuss	(1904–1991)
650	Virginia Woolf	(1882–1941)	907	William Faulkner	(1897–1962)
1150	Upton Sinclair	(1878–1968)	1166	D. H. Lawrence	(1885–1930)
1179	Aldous Huxley	(1894–1963)	1440	Vladimir Nabokov	(1899–1977)
1482	W. Somerset Maugham	(1874–1965)	1522	Arthur C. Clarke	(1917–2008)
1562	Edith Wharton	(1862–1937)	1974	Ian Fleming	(1908–1964)
1798	George Santayana	(1863–1952)	2019	Graham Greene	(1904–1991)
2619	Willa Cather	(1873–1947)	2056	Kurt Vonnegut	(1922–2007)
2925	Frances H. Burnett	(1849–1924)	2197	A. Solzhenitsyn	(1918–2008)
3106	Norman Mailer	(1923–2007)	2687	Carl Sagan	(1934–1996)
3360	Sinclair Lewis	(1885–1951)	3105	Salman Rushdie	(1947–)
3990	Pearl S. Buck	(1892–1973)	3302	Simone de Beauvoir	(1908–1986)
4043	J. B. Priestley	(1894–1984)	3522	Elia Kazan	(1909–2003)
4066	Zane Grey	(1872–1939)	3727	Glenn Beck	(1964–)
4748	Thornton Wilder	(1897–1975)	4083	Boris Pasternak	(1890–1960)

(1885–1930) [1166], whose *Lady Chatterley's Lover* was a posthumously published best seller in 1959, after emerging victorious in a landmark obscenity trial. It is relatively little read today.

There is a sense that the figures on the left are more "important" than those on the right, but how much of this is real and how much is a temporal illusion? Our celebrity/gravitas scale provides a tool to investigate. Generally speaking, modern writers hold up surprisingly well by this measure, with *Stephen King (1947–)* [191] of similar gravitas to *Ernest Hemingway (1899–1961)* [248], James Bond's *Ian Fleming (1908–1964)* [1974] comparable to Sherlock Holmes's *Arthur Conan Doyle (1859–1930)* [247], and *Kurt Vonnegut (1922–2007)* [2056] on par with *Upton Sinclair (1878–1968)* [1150]. Books of past eras endure largely through the endorsement of schools and other authorities, more than through their inherent attractiveness to lay readers.

Best-seller lists have generally not recognized the authors of children's books. *Dr. Seuss (1904–1991)* [893] hit the best-seller list for *"Oh, the Places You'll Go!"* which was aimed at adults. Indeed, the most popular author of our time, Harry Potter creator *J. K. Rowling (1965–)* [2094], did not qualify here despite having sold more than 400 million books.

By contrast, contemporary literary novelists do not get much respect from our rankings, with *Salman Rushdie (1947–)* [3105] being the most obvious active presence. Much of his reputation is a legacy of years in hiding, to escape a fatwa in response to his novel *The Satanic Verses.* The best-selling books of *Philip Roth (1933–)* [11135] and *Toni Morrison (1931–)* [5588] have won honors from the National Book Award to the Nobel Prize, yet we do not recognize either as cultural figures of historical magnitude.

13.2.1 THE EFFECT OF COPYRIGHT LAWS

Without the benefit of copyright laws to protect rights to intellectual property, authors like us would have little financial incentive to write books. But perpetual copyright ties up access to knowledge well beyond the point where substantial economic value remains. The Google book Ngrams are a great example of new uses found for old texts, although copyright concerns and ambiguities still restrict access to millions of these scanned texts.

The length of U.S. copyright protection has greatly increased in contemporary times, from an original term of 14 years to today's 70 years past the life of the author. The changes in copyright laws yield a sharp divide. Books published before 1923 have all fallen into the public domain, free to be used for any purpose. Those published after this cutoff date are generally still protected.

Controversy exists as to whether such strong protection promotes creativity, by incentivizing authors and marshaling professional stewardship of their works, or stifles it by restricting access [Pollock et al., 2010]. Certainly the book you are reading was made possible only because of the open nature of Wikipedia, which provided access to text for our analysis.

Our significance rankings provide an approach to thinking about this problem. Do the authors of books published before 1923 maintain greater mindshare because their works are in the public domain?

We compared the best-selling authors for the five years before the divide (1918–1922) versus the five years after the wall fell (1923–1927). We deleted all writers of post-1922 books who also had an earlier best seller from before the cutoff date. These include both popular authors like *Edith*

Wharton (1862–1937) [1562], *Sinclair Lewis (1885–1951)* [3360], and *Zane Grey (1872–1939)* [4066] as well as obscurities like *Gertrude Atherton (1857–1948)* [33072] and *Elizabeth von Arnim (1866–1941)* [40622]. The top authors among the remaining older (left) and younger (right) cohorts are presented in the accompanying table.

1918–1922 Authors			1923–1927 Authors		
Sig	Person	Dates	Sig	Person	Dates
544	Joseph Conrad	(1857–1924)	6778	John Galsworthy	(1867–1933)
2925	Frances H. Burnett	(1849–1924)	12055	Edna Ferber	(1885–1968)
20695	Robert W. Chambers	(1865–1933)	13897	Rafael Sabatini	(1875–1950)
45627	Eleanor H. Porter	(1868–1920)	14215	Anita Loos	(1888–1981)
73331	E. P. Oppenheim	(1866–1946)	41702	Dorothy C. Fisher	(1879–1958)

The results are suggestive, although not conclusive. The top two authors who lost the rights to their work (Conrad and Burnett) remain broadly in print today, in cheap editions from several publishers. The top two authors on the right are also accomplished, respected writers. *John Galsworthy (1867–1933)* [6778] won the Nobel Prize for *The Forsyte Saga* (which was published before 1923), while *Edna Ferber's (1885–1968)* [12055] books have been the source of several popular films. Yet neither have the significance of the top figures before the divide. On the other hand, the authors of many other freely available works have been forgotten. As many a blogger knows, putting a work before the public does not necessarily mean that the public will care about it.

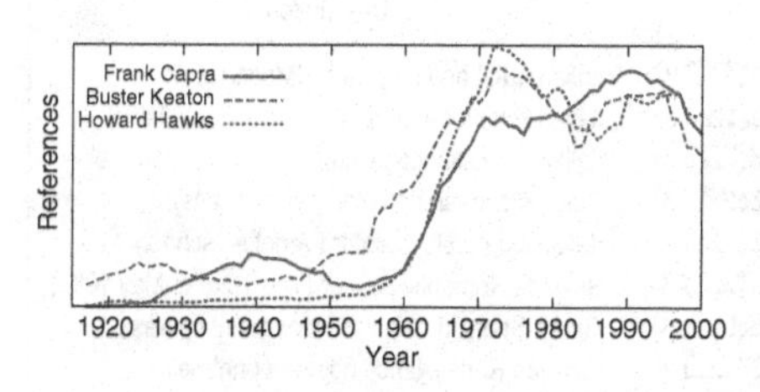

The effects of early copyright expiration become more obvious in the movies, because television stations had substantial economic incentive to identify films whose protection lapsed when copyright was not renewed. Several motion pictures that had fallen into complete obscurity became popular again when revived in the public domain. *It's a Wonderful Life* is the canonical tale: a 1946 film directed by *Frank Capra (1897–1991)* [3113] entered the public domain in 1975. The Ngram data on the left shows the impact for Capra and two other artists whose masterworks became freely available: *Buster*

Keaton's *(1895–1966)* [1857] *The General* in 1955 and *Howard Hawks*'s *(1896–1977)* [6178] *His Girl Friday* in 1967. All showed sharp, durable increases in mindshare as soon as their works became freely available.

13.3 Painters and Sculptors

Creating art has been a uniquely human activity for more than 40,000 years. But the names of the artists went unrecorded over most of this period. The ancient Greeks represent the beginning of Western artistic tradition, and the identity of several prominent Greek sculptors survived through contemporary written accounts and Roman copies of their work, including *Phidias (490–430 B.C.)* [1994], *Myron* [6168], and *Praxiteles* [12331].

Painting was equally esteemed by the Greeks, yet these works have not endured because of the perishable nature of their materials. The large collections of painted vases appearing in major art museums reflect their extreme durability more than their artistic importance to the people of the era. The most significant Greek painter appears to be *Apelles* [9774], although none of his works have survived.

But the Romans were content with copies of Greek sculpture. Hence the notion of artists with distinct identities faded, not to be revived until the late Middle Ages. *Cimabue (1240–1302)* [4245] and his student *Giotto di Bondone (1267–1337)* [610] were among the first Italians to break away from the Byzantine style. The great painters of the Renaissance dominate our rankings of the most significant pre-twentieth-century artists.

Top Pre-Twentieth-Century Artists

Sig	Person	Dates	C/G	Description
29	Leonardo da Vinci	(1452–1519)	C G	Italian artist and polymath ("Mona Lisa")
86	Michelangelo	(1475–1564)	C G	Italian sculptor and Renaissance man (David)
140	Raphael	(1483–1520)	C G	Italian renaissance painter
189	Rembrandt	(1606–1669)	C G	Dutch old master painter (self-portraits)
319	Titian	(1485–1576)	C G	Italian old master painter (Venetian school)
366	Francisco Goya	(1746–1828)	C G	Spanish Romantic painter (The Third of May 1808)
465	El Greco	(1541–1614)	C G	Greek-Spanish painter of elongated figures
503	Albrecht Dürer	(1471–1528)	C G	German Renaissance painter / printmaker
555	Hans Holbein the Younger	(1497–1543)	C G	German painter (portrait of Sir Thomas More)
567	Johannes Vermeer	(1632–1675)	C G	Dutch painter ("Girl with a Pearl Earring")
607	Jacques-Louis David	(1748–1825)	C G	French neoclassical painter ("Death of Socrates")
610	Giotto di Bondone	(1267–1337)	C G	Italian artist (first Renaissance painter)
693	Diego Velázquez	(1599–1660)	C G	Spanish painter ("Portraits of Philip IV")
965	Gustave Courbet	(1819–1877)	C G	French realist painter
983	Hieronymus Bosch	(1450–1516)	C G	Painter ("The Garden of Earthly Delights")

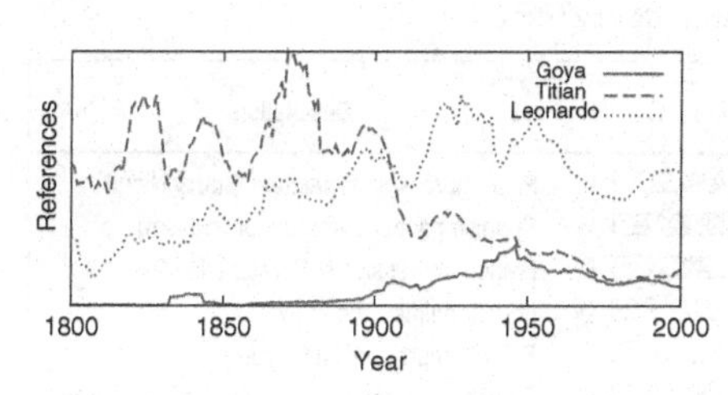

The book Ngram data provides a mirror on how certain classical artists have risen or fallen in reputation over the past 200 years. *Titian's (1485–1576)* [319] stock has declined substantially, while interest in *Leonardo da Vinci (1452–1519)* [29] grew throughout the nineteenth century into the great esteem he enjoys today. *Francisco Goya's (1746–1828)* [366] mindshare leaps forward starting around 1900, after he was posthumously recognized as a forerunner of modern painting.

13.3.1 MODERN PAINTING

The emergence of modern painting in the late nineteenth century marked a major cultural breakthrough. From that point on, painting would be about ideas, instead of reproductive fidelity. A dozen artists from this period rank among the thousand most significant figures of history.

Vincent van Gogh (1853–1890) [73] is our most significant figure in modern painting. He fits the popular image of the starving artist, and the Bohemian lifestyles of the avant garde. However, the most significant artists (presented in Figure 13.5) tended to live long lives dedicated to creating and promoting their work. Oil painting was their preferred medium, and producing a sizable body of work required resources and time.

The Impressionist painters and their successors fill our table of the most significant modern artists. Later movements like surrealism (*Salvador Dalí (1904–1989)* [1021]) and abstract expressionism (*Jackson Pollock (1912–1956)* [1013]) are represented, but by relatively few artists at this rarefied level.

The table makes two things clear: how many of the most prominent artists are modern, and yet how few important modern artists are contemporary. *Andy Warhol (1928–1987)* [485] is the only painter of great significance to emerge over the last 60 years.

Some of Warhol's success resulted from his perceptive sense of Who's Bigger. Warhol's iconic portraits feature personalities such as *Marilyn Monroe (1926–1962)* [347], *Mao Zedong (1893–1976)* [151], *Elvis Presley (1935–1977)* [69], and *Jacqueline Kennedy Onassis (1929–1994)* [1387];

Most Significant Twentieth-Century Artists

Sig	Person	Dates	C/G	Description
73	Vincent van Gogh	(1853–1890)	C G	Post-Impressionist painter ("Starry Night")
171	Pablo Picasso	(1881–1973)	C G	Spanish painter and sculptor (cubism)
178	Claude Monet	(1840–1926)	C G	French Impressionist ("Water Lilies")
376	Henri Matisse	(1869–1954)	C G	French painter (the Fauves)
389	Paul Cézanne	(1839–1906)	C G	French Impressionist (apples)
422	Edgar Degas	(1834–1917)	C G	French Impressionist (dancers and bathers)
485	Andy Warhol	(1928–1987)	C G	American Pop artist (Brillo boxes)
540	Paul Gauguin	(1848–1903)	C G	French post-Impressionist artist
549	Pierre-Auguste Renoir	(1841–1919)	C G	French artist (Impressionist style)
574	Auguste Rodin	(1840–1917)	C G	French bronze sculptor ("The Thinker")
618	Wassily Kandinsky	(1866–1944)	C G	Russian painter and art theorist
640	Édouard Manet	(1832–1883)	C G	French pre-Impressionist painter
815	Camille Pissarro	(1830–1903)	C G	French painter (post-Impressionism)
915	Diego Rivera	(1886–1957)	C G	Mexican painter and muralist
944	Edvard Munch	(1863–1944)	C G	Norwegian painter ("The Scream")
1002	James Abbott McNeill Whistler	(1834–1903)	C G	Artist ("Whistler's Mother")
1013	Jackson Pollock	(1912–1956)	C G	American abstract expressionist (drip painting)
1021	Salvador Dalí	(1904–1989)	C G	Spanish surrealist painter
1051	Piet Mondrian	(1872–1944)	C G	Dutch painter (De Stijl)
1178	Georgia O'Keeffe	(1887–1986)	C G	American painter (flowers and bones)

FIGURE 13.5. Most significant twentieth-century artists.

celebrities whose reputations burn as brightly today as they did when he painted them.

Today's primary visual arts involve moving images, and one senses that the most imaginative artists have left traditional media for greener pastures. Even today's most prominent painters/sculptors like *Damien Hirst (1965–)* [7808], *Jeff Koons (1955–)* [23874], and *Cindy Sherman (1954–)* [31600] prove relatively small potatoes, despite the vast sums being spent on their work. So-called popular contemporary artists like *Thomas Kinkade (1958–2012)* [39334], *LeRoy Neiman (1921–2012)* [55656], and *Peter Max (1937–)* [24126] rank no higher, despite their ubiquity on living room mantles and in cruise ship lounges.

We speculate that the age of the major painter or sculptor has now permanently ended, just as it has for classical composers. Technology now favors other art forms, and the media of hands, brush, and orchestra cannot compete with the camera and the rock-and-roll band.

13.3.2 SCULPTORS

Unlocking living figures from stone is perhaps the greatest technical challenge in the fine arts. Sculpture and painting seem intrinsically different media, although prominent artists in one field often dabble in the other. *Michelangelo (1475–1564)* [86] is probably more celebrated for his sculptures, particularly the *Pietà* and the *David*, than he is even for his frescos in the Sistine Chapel.

However, the greatest painters prove to be of higher significance than the greatest sculptors. This is largely a function of how their works are disseminated. Marble sculptures are massive works, expensive to produce and hard to reproduce. They seldom physically travel. But famous painted images are ubiquitous through prints in books, posters, and interactive media. The majesty and power of sculpture are not properly conveyed through two-dimensional photographs.

Most Significant Sculptors

Sig	Person	Dates	C/G	Description
86	Michelangelo	(1475–1564)	C G	Italian sculptor and Renaissance man (*David*)
443	Donatello	(1386–1466)	C G	Renaissance Italian sculptor
574	Auguste Rodin	(1840–1917)	C G	French bronze sculptor (*The Thinker*)
680	Gian Lorenzo Bernini	(1598–1680)	C G	Italian sculptor and architect
1245	Marcel Duchamp	(1887–1968)	C G	French Dadaist artist (readymades)
1485	Henry Moore	(1898–1986)	C G	English abstract sculptor
1994	Phidias	(490–430 B.C.)	C G	Ancient Greek sculptor
2016	Benvenuto Cellini	(1500–1571)	C G	Italian artist (Mannerism)
2509	Lorenzo Ghiberti	(1378–1455)	C G	Italian artist (*Gates of Paradise*)
2806	Bertel Thorvaldsen	(1770–1844)	C G	Danish neoclassicist sculptor
2870	Alexander Calder	(1898–1976)	C G	American sculptor (mobiles)
2897	Daniel Chester French	(1850–1931)	C G	American sculptor (Lincoln Memorial)
3411	Antonio Canova	(1757–1822)	C G	Venician neoclassicist sculptor
3497	Frederic Remington	(1861–1909)	C G	Painter and sculptor of the Old West
3659	Andrea del Verrocchio	(1435–1488)	C G	Italian artist (taught da Vinci)

The most significant modern sculptor, *Auguste Rodin (1840–1917)* [574] of *The Thinker* fame, achieved renown through bronze casts of his plaster models. Repeated casting of these models multiplied his work, so that it could be widely represented in the world's art museums and public gardens. It is this ubiquitousness that most distinguishes Rodin from his rivals.

Perhaps new developments in three-dimensional printing technologies will revive the prestige of classical sculpture. High-resolution scanning efforts like the Digital Michelangelo Project [Levoy et al., 2000] now make available data for producing perfect reproductions of the *David* and other masterpieces at any scale, in a variety of colors and materials. If you could download a bust by Rodin as easily as a *Bob Dylan (1941–)* [130] recording, perhaps the proper prestige of sculpture could be restored.

13.3.3 THE PRICE OF ART

The global art market puts exorbitant prices on the work of master artists. Obviously some of this price reflects the beauty and quality of the work, but the artist's name accounts for much of the value. Our significance scores provide a way to measure the price of fame. By training on a database of auction prices, we can build a simple model for the value of an artist's work.

In particular, we extracted the record price paid at auction for 200 of the most expensive U.S. and international artists from the database at `www.findartinfo.com`. The vast majority of these high-priced artists prove to be modern: the best Old Masters material now resides in museums, permanently off the auction market.

Figure 13.6 shows that artist significance correlates quite well (0.52) with the logarithm of the maximum auction price. The biggest outlier here is *Li Keran (1907–1989)* [663492], whose sales price of $17 million[3] would seem at odds with his meager -2.13 significance score. He barely has an English Wikipedia page, yet his paintings are very valuable to the new class of super-wealthy Chinese, eager to pay exorbitant sums of money for national artists.

Under our model, we would project a painting by *Leonardo da Vinci (1452–1519)* [29] to be worth in the neighborhood of $450 million. None have come on the market in recent years, but a recent find has been valued in the $150 million range, if its authenticity can be established [Brooks, 2012]. The *Mona Lisa* would seem certain to fetch our price, if the Louvre decided to sell. Our model values work by the least significant painter we identified in Wikipedia, *Dora Holzhandler (1928–)* [840677], at around

[3] Indeed, recent press reports have raised his top price even higher, to $46 million [Qi, 2012].

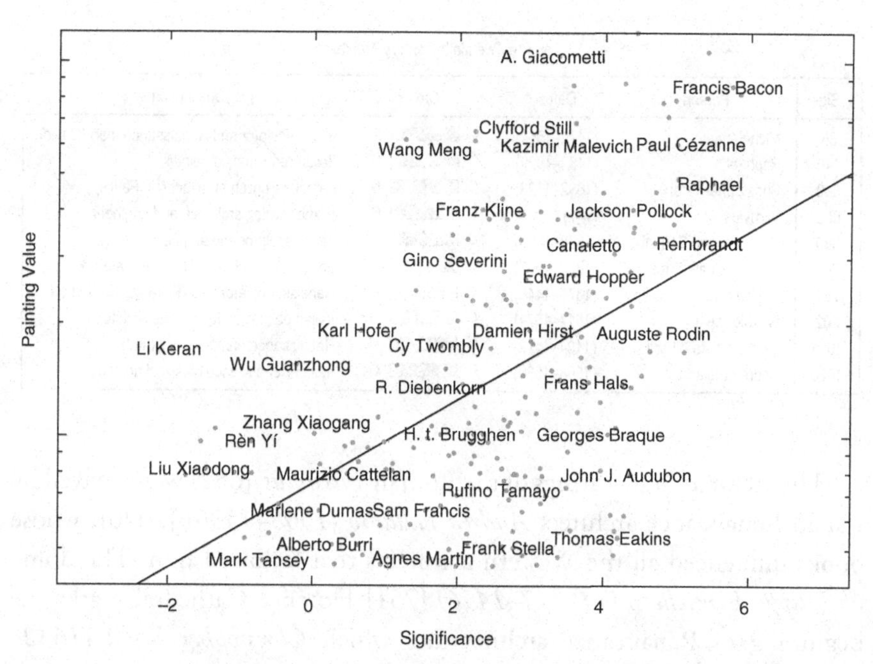

FIGURE 13.6. Dotplot of logged painting prices against the historical significance of the artists. Price and significance exhibit a 0.52 correlation.

$30,000. This isn't crazy. Indeed, she has previously sold for $5,898 at auction.

13.4 Architects

For much of history, architects were jacks-of-all-trades, people for whom designing buildings was only part of their portfolio. The leading court artists ran substantial studios, and got pressed into service during all the major projects of their day. Thus, rosters of important Renaissance architects include those with more renown as painters and sculptors, such as *Leonardo da Vinci (1452–1519)* [29], *Michelangelo (1475–1564)* [86], and *Raphael (1483–1520)* [140].

Architects remained polymaths until recent times. President *Thomas Jefferson (1743–1826)* [10] can be credited as America's first great architect, on the strength of his design of the University of Virginia. The most significant classical (pre-twentieth century) figures in architecture are given in the accompanying table.

Top Pre-Twentieth-Century Architects

Sig	Person	Dates	C/G	Description
86	Michelangelo	(1475–1564)	C G	Italian sculptor and Renaissance man (*David*)
140	Raphael	(1483–1520)	C G	Italian renaissance painter
600	Christopher Wren	(1632–1723)	C G	Famous English architect (St. Paul's)
652	Vitruvius	(?–?)	C G	Roman writer, architect, and engineer
680	Gian Lorenzo Bernini	(1598–1680)	C G	Italian sculptor and architect
740	Frederick Law Olmsted	(1822–1903)	C G	Landscape architect (NYC Central Park)
751	Filippo Brunelleschi	(1377–1446)	C G	Renaissance architect (Florence Cathedral)
902	Giorgio Vasari	(1511–1574)	C G	Italian painter, historian, and architect
1011	Leon Battista Alberti	(1404–1472)	C G	Renaissance humanist polymath
1026	Andrea Palladio	(1508–1580)	C G	Italian Renaissance classical architect

The architectural writings of the Roman *Vitruvius* [652] were revived by Italian Renaissance architect *Andrea Palladio (1508–1580)* [1026], whose books influenced all the Western architects that followed him. The dome of *Filippo Brunelleschi's (1377–1446)* [751] Florence Cathedral marks the beginning of Renaissance architecture, which *Christopher Wren (1632–1723)* [600] brought to London's St. Paul's Cathedral. *Giorgio Vasari (1511–1574)* [902] is most famous for his book on the lives of Renaissance painters, but he worked primarily as an architect.

Architecture really emerged as a distinct discipline only in modern times, when the complexities of modern materials and structural engineering required greater training and sophistication to master. Modern architects turn concrete, glass, and metal into sculpture: works large and distinct enough to register with the broader public.

Today's architects have become public figures, whose fame towers over that of technical professionals in fields such as engineering. Star architects are brands, the public faces of the small number of large firms that attract the vast majority of public commissions. Their buildings are essentially ghostwritten by large teams, but it is the architect whose name is forever engraved on the marble cornerstone.

The greatest modern architects have been credited with creating sculptural forms since at least the times of *Antoni Gaudí (1852–1926)* [558] and *Frank Lloyd Wright (1867–1959)* [245]. New building technologies and computerized drafting methods have opened the door to ever wilder forms, epitomized by *Frank Gehry's (1929–)* [2438] Guggenheim Museum in Bilbao, Spain.

Top Modern Architects

Sig	Person	Dates	C/G	Description
245	Frank Lloyd Wright	(1867–1959)	C G	American architect ("Fallingwater")
558	Antoni Gaudí	(1852–1926)	C G	Spanish Catalan architect
775	Rudolf Steiner	(1861–1925)	C G	Austrian philosopher, social thinker
1101	Le Corbusier	(1887–1965)	C G	French architect (chapel at Ronchamp)
1544	Ludwig Mies van der Rohe	(1886–1969)	C G	German American architect (less is more)
1581	Albert Speer	(1905–1981)	C G	German architect / Nazi minister of production
1719	Louis Sullivan	(1856–1924)	C G	American architect (first skyscraper)
1960	Daniel Burnham	(1846–1912)	C G	American architect (D.C. Union Station)
1990	Buckminster Fuller	(1895–1983)	C G	Polymath, author, inventor (geodesic dome)
2086	Henry Hobson Richardson	(1838–1886)	C G	American architect (Trinity Church, Boston)
2272	Walter Gropius	(1883–1969)	C G	German architect (Bauhaus school)
2438	Frank Gehry	(1929–)	C G	Modern architect (Guggenheim Bilbao)

These architects have become the important sculptors of today, creating works on a scale that reshapes the perceptions of cities. The functionality of these newfangled buildings often seems to be a secondary concern compared to their sheer visual impact.

14

The Performing Arts

For thousands of years, man has watched stories told through flickering images, be they in front of a campfire or displayed on a television. Performance modes and styles have changed, but the human need to entertain and be entertained has remained a constant of our species.

The performing arts are both the oldest forms of human expression and the greatest contemporary source of new cultural memes. Primitive man and the hippest club-goers both loved music and dance, although it is hard to tell whether they would appreciate each other's art.

In this chapter, we will review the most significant people in the performing arts. We start from the era before recorded media and continue through the present day. We will analyze trends in classical and popular music, for clues as to which of these cultural streams is likely to dominate historically. We do the same for broadcast and visual media, studying the history of film, radio, and television as a guide to the future of communications.

14.1 Before Recording Technology

Sunday preachers can be thought of as early America's most prominent class of performing artist. The rigors of the six-day workweek and the strictures of the Christian Sabbath left little time for secular entertainment, but spellbinding ministers drew large crowds for Sunday worship. In his autobiography [Franklin, 1818], *Benjamin Franklin (1706–1790)* [35] estimated that more than 30,000 people attended a sermon by *George Whitefield (1714–1770)* [1258] in Philadelphia.

Professional theater in the United States began with the arrival of the *Lewis Hallam (1714–1756)* [145007] troupe from England in 1752. The Lyceum movement of the mid-nineteenth century made stars out of popular lecturers, including *Ralph Waldo Emerson (1803–1882)* [122], *Henry David Thoreau (1817–1862)* [131], *Susan B. Anthony (1820–1906)* [432], and *Nathaniel Hawthorne (1804–1864)* [227].

Before *Thomas Edison's (1847–1931)* [40] invention of the phonograph in 1877, all performances were ephemeral. They existed for a moment, only to be enjoyed by those within earshot. No effective trace of the skills of any singer or instrumentalist remains from before this time, and there is no hope of historical resurrection.

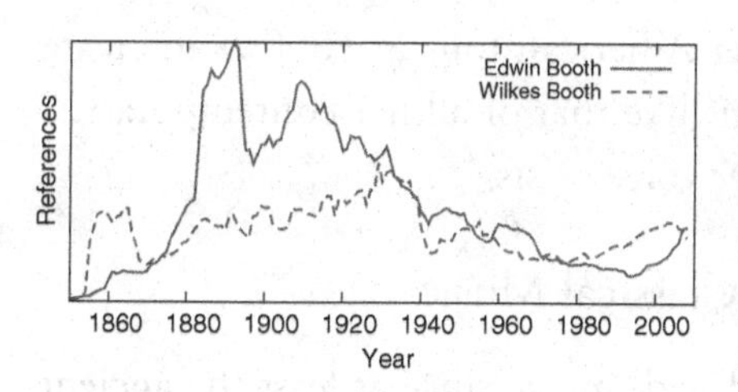

Several of the most historically significant pre-Edison performing artists are best remembered today for other things. *John Wilkes Booth (1838–1865)* [451] is indeed famous for what he did in a theater, but it was shooting *Abraham Lincoln (1809–1865)* [5], rather than his acting. *Edwin Booth (1833–1893)* [3686] was, by consensus, the greatest actor of his generation. But he died in the year Edison invented motion pictures, so his performances were completely ephemeral. Ngrams data of him and his actor/assassin brother *John Wilkes Booth (1838–1865)* [451] are revealing. The slayer of Lincoln dominated the great actor in the immediate aftermath of the killing, but Edwin Booth remained more prominent in the book Ngrams data from the 1880s until roughly 1960.

Pre-1900 Performing Artists

Sig	Person	Dates	C/G	Description
451	John Wilkes Booth	(1838–1865)	C G	Assassin (Abraham Lincoln)
664	Buffalo Bill	(1846–1917)	C G	Bison hunter / Wild West Show
1112	Constantin Stanislavski	(1863–1938)	C G	Russian acting coach (Stanislavski Method)
1465	Sarah Bernhardt	(1844–1923)	C G	French stage actress (Divine Sarah)
1995	David Garrick	(1717–1779)	C G	English playwright and theater manager
2121	Jenny Lind	(1820–1887)	C G	Swedish opera singer (Swedish Nightingale)
2720	Annie Oakley	(1860–1926)	C G	American exhibition sharpshooter
2946	Colley Cibber	(1671–1757)	C G	English playwright and Poet Laureate
3196	Henry Irving	(1838–1905)	C G	English stage actor
3349	Ellen Terry	(1847–1928)	C G	English stage actress

Constantin Stanislavski (1863–1938) [1112] is more famous for his system of acting, outlined in several books, than his performances on stage. *Sarah Bernhardt (1844–1923)* [1465] was the great actress of her generation, whose popularity endured into the early silent film era. *David Garrick (1717–1779)* [1995] was a great Shakespearian actor who had a major impact on raising the profile of the theater. According to his friend *Samuel Johnson (1709–1784)* [141], "his profession made him rich and he made his profession respectable."

Fame generally accrued to the impresarios of theatrical productions, like *P. T. Barnum (1810–1891)* [767] and *Buffalo Bill (1846–1917)* [664], more so than the stars themselves. Probably the most eminent performing artist of the era was singer *Jenny Lind (1820–1887)* [2121], the "Swedish Nightingale." Barnum so avidly promoted her American tour in 1850 as to create the first modern celebrity. But her art, like that of all her contemporaries, has been silenced. No recording of her voice exists.

14.2 Western Classical Music

Civilized societies have been suffused with music since at least the ancient Greeks. Enough evidence of Greek musical notation survives to surmise that their music would sound quite alien to our ears. Greek musical instruments ranged from strummed stringed devices to wind instruments and even the hydraulis, an hydraulically powered pipe organ attributed to *Ctesibius* [9757] of the third century B.C.E. The musical critiques of *Plato (427–347 B.C.)* [25], disconcerted by the then–avant garde, sound familiar today [Plato]:

> Our music was once divided into its proper forms... But later, an unmusical anarchy was led by poets who had natural talent, but were ignorant of the laws of music.

Roman music was heavily influenced by the peoples that were absorbed into their expanding empire. The Romans likely took their music from the Greek tradition, as they did in other cultural matters. After Christianity took hold in Europe, this rich and diverse musical tradition passed away, replaced with the sounds of the ascendant Catholic Church.

Gregorian chant is the best known form of early music, the formalization of which is frequently credited to *Pope Gregory I (540–604)* [394].

Gregorian chant is quite simple, consisting of a single melodic line, and lacking harmony, rhythm, and any other adornment. Reflecting its austere religious aesthetic, it marks the beginnings of Western classical music.

Today's classical music world comprises several types of musician: composers, conductors, and performers. Their relative rankings, discussed in the following sections, tells us something about the historical processes of making music and how they have changed with time.

14.2.1 COMPOSERS

Early composers are frequently anonymous, due in part to the lack of surviving attributions. One of the earliest identifiable music composers was *Hildegard of Bingen (1098–1179)* [1721], a female polymath and Catholic saint. By the late medieval period, other composers began to move Western music away from plainchant and into polyphony. However, the polyphonic music of the early Renaissance would not be recognizable to a modern listener as classical music. And unless you can sing it yourself, you would have a hard time finding a performance of music from the late Renaissance. Indeed, despite my training as a classical pianist, I (Charles) have only a nodding familiarity with one composer from this time, *Giovanni Pierluigi da Palestrina (1525–1594)* [955].

But classical music takes on its modern shape once we enter the seventeenth century. The history of classical music is usually divided into the Baroque (1600–1760), Classical (1730–1820), Romantic (1815–1910), and Modern (1900–present) eras. Of classical composers since the Baroque era, 19 rank among the 500 most significant figures in history.

Most Significant Classical Composers

Sig	Person	Dates	C/G	Description
24	Wolfgang Amadeus Mozart	(1756–1791)	C G	Austrian composer (*Don Giovanni*)
27	Ludwig van Beethoven	(1770–1827)	C G	German composer (*Ode to Joy*)
48	Johann Sebastian Bach	(1685–1750)	C G	Classical composer (*Well-Tempered Clavier*)
114	George Frideric Handel	(1685–1759)	C G	Baroque composer (*Messiah*)
116	Franz Liszt	(1811–1886)	C G	Hungarian Romantic composer and pianist
129	Joseph Haydn	(1732–1809)	C G	Austrian composer (Classical period)
195	Frédéric Chopin	(1810–1849)	C G	Polish Romantic composer and pianist

(continued)

Sig	Person	Dates	C/G	Description
198	Franz Schubert	(1797–1828)	C G	Austrian composer (Romantic era)
199	Felix Mendelssohn	(1809–1847)	C G	German composer (*Midsummer Night's Dream*)
204	Johannes Brahms	(1833–1897)	C G	German composer (The third "B")
228	Giuseppe Verdi	(1813–1901)	C G	Italian composer (*Aida*)
262	Antonio Vivaldi	(1678–1741)	C G	Italian Baroque composer (*Four Seasons*)
321	Robert Schumann	(1810–1856)	C G	German composer (Kinderszenen)
364	Gustav Mahler	(1860–1911)	C G	Late-Romantic Austrian composer
368	Igor Stravinsky	(1882–1971)	C G	American composer (*The Rite of Spring*)
380	Claude Debussy	(1862–1918)	C G	French composer (Impressionist music)
390	Gioachino Rossini	(1792–1868)	C G	Italian composer (*The Barber of Seville*)
419	Hector Berlioz	(1803–1869)	C G	French Romantic composer (*Symphonie Fantastique*)
445	Edward Elgar	(1857–1934)	C G	English composer (*Pomp and Circumstance*)

Half of these composers date from the Romantic era. Indeed, the second half of the nineteenth century was the most productive and popular period in Western classical music. The popular imagination, however, is still disproportionately fascinated with the earlier big three of Bach, Beethoven, and Mozart.

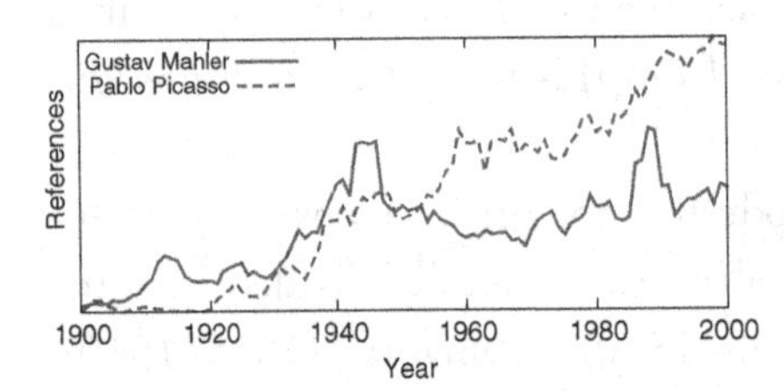

It is easy to underestimate the degree to which classical music dominated the fine arts at the turn of the twentieth century. *Gustav Mahler (1860–1911)* [364] and *Pablo Picasso (1881–1973)* [171] were roughly contemporaries. Picasso is the dominant figure in modern art, and outlived the composer by 50 years. Yet the Ngram data show that Mahler remained the more talked-about cultural figure through World War II.

14.2.2 CONDUCTORS

The symphony has long been considered the preeminent compositional form in classical music. But performing a symphony is the collective effort of dozens of musicians. For this reason, great conductors have become the faces of great orchestras. The image of a classical conductor is of a stern perfectionist demanding complete obedience from his stable of a hundred-plus musicians.

The cult of the great maestro appears to have emerged during the broadcast era, when their work left the concert hall and went out to the public at large. The conductors we rank as most significant, *Arturo Toscanini (1867–1957)* [1190] and *Leonard Bernstein (1918–1990)* [1209], both had extensive broadcasting careers in radio and television. *Leopold Stokowski's (1882–1977)* [2939] enduring fame rests largely on his gig with Mickey Mouse in *Fantasia*.

The middle of the twentieth century proved the heyday for classical conductors. The current generation of conductors have not yet risen to the same stature as their predecessors, with *Gustavo Dudamel (1981–)* [79745] perhaps the brightest star on the horizon. This reflects popular music eclipsing the contemporary classical scene, as well as the increasing attention individual soloists are receiving at the expense of the symphony itself.

Most Significant Conductors

Sig	Person	Dates	C/G	Description
1190	Arturo Toscanini	(1867–1957)	C G	Italian conductor (NBC Orchestra)
1209	Leonard Bernstein	(1918–1990)	C G	American conductor (*West Side Story*)
2939	Leopold Stokowski	(1882–1977)	C G	Orchestral conductor (*Fantasia*)
3356	Herbert von Karajan	(1908–1989)	C G	Austrian orchestra conductor
6293	Serge Koussevitzky	(1874–1951)	C G	Russian-born Jewish conductor
7091	Otto Klemperer	(1885–1973)	C G	German conductor
7582	Georg Solti	(1912–1997)	C G	Orchestral and operatic conductor
9266	Eugene Ormandy	(1899–1985)	C G	Hungarian-born conductor

14.2.3 PERFORMERS

Many of the great classical composers were also star performers in their day. By contrast, the pre-recording era musicians that didn't leave compositions behind are generally forgotten. *Niccolò Paganini (1782–1840)* [1034] is perhaps the best known exception: although he was a composer, he is far better remembered as one of the outstanding violin virtuosi of all time. Unfortunately, he died before recording technology could preserve his work. "The last romantic" pianist, *Vladimir Horowitz (1903–1989)* [3896], is a similar modern figure, who never committed his compositions to paper but was fortunate to live in the age of recordings.

We divide performers into two groups: instrumentalists and opera singers. The great instrumentalists are dominated by those with long careers: our three most significant artists each performed for more than 60 years. Today, the most prominent instrumentalists have similar stature to conductors. Indeed, artists like *Yo-Yo Ma (1955–)* [6410] and *Itzhak Perlman (1945–)* [10067] register as more significant than any contemporary conductor.

Most Significant Classical Performers

Sig	Person	Dates	C/G	Description
3112	Pablo Casals	(1876–1973)	C G	Spanish cellist and conductor
3896	Vladimir Horowitz	(1903–1989)	C G	American classical pianist
3934	Ignacy Jan Paderewski	(1860–1941)	C G	Pianist / prime minister of Poland
4252	Arthur Rubinstein	(1887–1982)	C G	Polish American classical pianist
4989	Glenn Gould	(1932–1982)	C G	Canadian pianist / Bach interpreter
5017	Yehudi Menuhin	(1916–1999)	C G	Jewish American violinist
6410	Yo-Yo Ma	(1955–)	C G	Chinese American cellist
6837	Jascha Heifetz	(1901–1987)	C G	Twentieth-century violinist
7381	Fritz Kreisler	(1875–1962)	C G	Austrian-born violinist
10067	Itzhak Perlman	(1945–)	C G	Israeli-born violinist

The rankings for opera singers are remarkable to the extent that they favor figures from the very early years of recording. The opera was the most popular music form of the late nineteenth century, and the phonograph gave the mass public a way to hear the world's greatest singers. Although *Luciano Pavarotti (1935–2007)* [3032] is perhaps more recognizable to contemporary readers, the outsized personalities of *Enrico Caruso (1873–1921)* [1249] and *Maria Callas (1923–1977)* [2986] still register today, divas demanding our attention from beyond the grave.

Most Significant Opera Singers

Sig	Person	Dates	C/G	Description
1249	Enrico Caruso	(1873–1921)	C G	Great Italian opera tenor
2121	Jenny Lind	(1820–1887)	C G	Swedish opera singer (Swedish Nightingale)
2375	Nellie Melba	(1861–1931)	C G	Australian operatic soprano (Melba toast)
2986	Maria Callas	(1923–1977)	C G	Greek opera soprano / diva
3032	Luciano Pavarotti	(1935–2007)	C G	Recent Italian operatic tenor
5035	Francesco Tamagno	(1850–1905)	C G	Early recorded operatic tenor
5493	Joan Sutherland	(1926–2010)	C G	Australian dramatic coloratura soprano
5739	Adelina Patti	(1843–1919)	C G	Nineteenth-century opera singer
7474	Leontyne Price	(1927–)	C G	Black American soprano
7733	Andrea Bocelli	(1958–)	C G	Blind Italian tenor

14.2.4 DANCE

The most significant figures in dance prove to be a mix of classical ballet, modern dance, and stars from the era of movie musicals. *Fred Astaire (1899–1987)* [1506] significantly outranks his dance partner *Ginger Rogers (1911–1995)* [3504], even though (as Rogers said) she did everything he did but "backwards . . . and in high heels."

Ballet is represented more strongly by choreographers and ballet masters than by the dancers themselves. The father of the French ballet, *Marius Petipa (1818–1910)* [1981], outranks the more notorious Russian master *Sergei Diaghilev (1872–1929)* [2943]. *Isadora Duncan (1877–1927)* [1734] is the seminal figure in modern dance: bringing to it natural forms of movement that were lost in the rigors of ballet, along with great attention through her scandalous lifestyle.

Most Significant Dance Figures

Sig	Person	Dates	C/G	Description
1506	Fred Astaire	(1899–1987)	C G	American Broadway stage dancer
1734	Isadora Duncan	(1877–1927)	C G	Creator of modern dance
1981	Marius Petipa	(1818–1910)	C G	Father of the French ballet
2200	Martha Graham	(1894–1991)	C G	Modern dancer choreographer
2943	Sergei Diaghilev	(1872–1929)	C G	Russian ballet impresario
3254	Gene Kelly	(1912–1996)	C G	American dancer (*Singing in the Rain*)
3402	George Balanchine	(1904–1983)	C G	American ballet choreographer
3504	Ginger Rogers	(1911–1995)	C G	American actress (*Kitty Foyle*)
3912	Anna Pavlova	(1881–1931)	C G	Russian ballerina
5717	Vaslav Nijinsky	(1890–1950)	C G	Tragic male ballet dancer

14.3 Popular Music

No area of popular culture cycles more rapidly and unpredictably than popular music. Each year new stars blaze across the heavens, then quickly fade to black. *Frankie Avalon (1940–)* [17639] was the *Justin Bieber (1994–)* [8633] of the late 1950s. Between them this teen idol role was filled by *David Cassidy (1950–)* [11154] in the 1970s, and then *Debbie Gibson (1970–)* [6577] in the late 1980s. "Those who cannot remember the past are condemned to repeat it" (*George Santayana (1863–1952)* [1798]).

But some artists remain in the collective memory forever. *Elvis Presley (1935–1977)* [69] was the genuine article: a music pathbreaker whose influence will endure far into the future. We anticipate that, 100 years from now, educated people will be more familiar with "Hound Dog" than they are today with the "Grand March" from *Aida* (composed by *Giuseppe Verdi (1813–1901)* [228]).

The persistence of genre seems to be the most important factor affecting the historical fame of popular musicians. It is an amazing fact that early rock-n-roll recordings like *Bill Haley's (1925–1981)* [2452] "Rock Around the Clock" remain omnipresent in the public realm, nearly 60 years after their creation. By contrast, the popular music from just before this transition is now hardly listened to, and stars like *Patti Page (1927–2013)* [12640], who sold hundreds of millions of records in the 1950s hold little cultural resonance today.

14.3.1 JAZZ AND THE BIG BAND ERA

The ranks of the leading jazz stars cross several different eras, ranging from multigenerational stars like *Louis Armstrong (1901–1971)* [441] and *Ella Fitzgerald (1917–1996)* [1327], to tragic figures like *John Coltrane (1926–1967)* [1827] and *Billie Holiday (1915–1959)* [1218]. The most significant figures in jazz are black, with the exception of *Bing Crosby (1903–1977)* [947], whose image as a jazz singer was eclipsed by his subsequent movie career.

Top Early Pop Singers

Sig	Person	Dates	C/G	Description
357	Frank Sinatra	(1915–1998)	C G	Singer ("My Way," "New York, New York")
441	Louis Armstrong	(1901–1971)	C G	Jazz singer and trumpeter (Satchmo)
495	Duke Ellington	(1899–1974)	C G	American composer, pianist
947	Bing Crosby	(1903–1977)	C G	American singer, actor (*Going My Way*)
973	Miles Davis	(1926–1991)	C G	American cool jazz musician
1194	Irving Berlin	(1888–1989)	C G	Broadway songwriter ("America the Beautiful")
1218	Billie Holiday	(1915–1959)	C G	Jazz singer (*Lady Sings the Blues*)
1327	Ella Fitzgerald	(1917–1996)	C G	American jazz and song vocalist
1508	Scott Joplin	(1867–1917)	C G	Ragtime composer and pianist
1578	Glenn Miller	(1904–1944)	C G	American Big Band musician

Frank Sinatra (1915–1998) [357] ranks as the most historically significant singer, at least before Elvis. The Big Band era of the 1930s is represented here by *Glenn Miller (1904–1944)* [1578], the popular band leader who disappeared in a plane crash while entertaining American troops during World War II. Songwriter *Irving Berlin (1888–1989)* [1194] created many popular American standards, from "White Christmas" to "God Bless America."

14.3.2 ROCK AND ROLL

The boundaries of a genre are difficult to pin down. Which artists should be considered as rock and roll stars, as opposed to jazz, country, pop music, or some other style? We use as our defining characteristic membership in the Rock and Roll Hall of Fame, an institution dedicated to honor the leaders of the genre.

This still presents several complexities. First, the Rock and Roll Hall of Fame often admits groups instead of individuals. Thus the Rolling Stones are in, but not lead singer *Mick Jagger (1943–)* [1055]. Also, they have made a point of recognizing foundational figures, who often turn out to be strongly associated with other genres: such as jazz legends *Louis Armstrong (1901–1971)* [441] and *Miles Davis (1926–1991)* [973], and country singer *Johnny Cash (1932–2003)* [900]. Several of these rank among the most significant people in the building.

Top Rock and Roll HoF Stars

Sig	Person	Dates	C/G	Description
69	Elvis Presley	(1935–1977)	C G	The "king of rock and roll"
121	Madonna	(1958–)	C G	Singer and songwriter ("Like a Virgin")
130	Bob Dylan	(1941–)	C G	Folk / rock singer ("Blowin' in the Wind")
162	John Lennon	(1940–1980)	C G	Beatle ("Imagine")
180	Michael Jackson	(1958–2009)	C G	American recording artist ("Thriller")
338	Jimi Hendrix	(1942–1970)	C G	American guitarist ("Purple Haze")
441	Louis Armstrong	(1901–1971)	C G	Jazz singer and trumpeter (Satchmo)
447	Paul McCartney	(1942–)	C G	English musician (The Beatles)
657	George Harrison	(1943–2001)	C G	Lead guitarist of the Beatles
829	David Bowie	(1947–)	C G	Androgynous rock musician (Ziggy Stardust)
900	Johnny Cash	(1932–2003)	C G	Country singer ("Folsam Prison Blues")
961	Otis Redding	(1941–1967)	C G	Soul singer ("Sitting on the Dock of the Bay")
966	Bob Marley	(1945–1981)	C G	Jamaican reggae musician
973	Miles Davis	(1926–1991)	C G	American cool jazz musician
1042	Elton John	(1947–)	C G	1970s English singer-songwriter

Three of the four Beatles rank among the top 1,000 people in history, with *Ringo Starr (1940–)* [1875] being the odd man out. *Madonna (1958–)* [121] is the highest ranking female music artist, in any genre. That we rate her significance ahead of *Bob Dylan (1941–)* [130] will likely upset many readers, but our rankings among the top figures in music are generally consistent with the book Ngram data.

About half of these artists died premature deaths: some in their twenties, others in their thirties, forties, and fifties. The brevity of certain careers is astonishing. *Jimi Hendrix (1942–1970)* [338] lived only three years after first becoming a star. *Otis Redding (1941–1967)* [961] died in a plane crash at the age of 26. The fact that they lived fast and died young is an important part of their legacy and mystique.

One popular singer who deserves discussion is *Britney Spears (1981–)* [689]. It's difficult to guess where her long-term historical significance will lie. She was the leading teenage star at the beginnings of the Internet era, and was the most popular search term on Yahoo! for seven years, from 2000 to 2008. One can question whether her music will hold up over the coming years, yet she is only 12 years older than *Justin Bieber (1994–)* [8633]. We project her as a figure of greater or comparable stature to the opera diva *Maria Callas (1923–1977)* [2986]: both major musical talents, with back stories that endure well beyond the memory of their actual performances.

14.4 Motion Pictures

The question of who deserves the credit for inventing motion pictures touches on murky issues of priority. Different people are recognized as the inventor of image projection, the movie camera, and celluloid film. *Thomas Edison's (1847–1931)* [40] Kinetoscope was basically a movie projector, first demonstrated in 1893, and his company's early films started the motion picture industry. Edison's precursors include *Eadweard Muybridge (1830–1904)* [1015], whose pioneering high-speed photo series unlocked many secrets of human and animal motion.

Motion pictures rapidly progressed from a curiosity to an experimental medium of short films. The modern industry emerged roughly a century ago, with the first full-length films of *D. W. Griffith (1875–1948)* [1276]. It has survived threats from radio, television, and video games, yet seems destined to remain the most prestigious and popular form of mass entertainment. Hollywood has reigned long enough to be analyzed on historical timescales, making it meaningful to compare stars from different eras. The accompanying table presents the most significant actors and actresses whose fame is drawn primarily from motion pictures.

Actors			Actresses		
Sig	Person	Dates	Sig	Person	Dates
295	Charlie Chaplin	(1889–1977)	347	Marilyn Monroe	(1926–1962)
459	John Wayne	(1907–1979)	1031	Judy Garland	(1922–1969)
729	Orson Welles	(1915–1985)	1396	Katharine Hepburn	(1907–2003)
1077	Sean Connery	(1930–)	1490	Meryl Streep	(1949–)
1118	Laurence Olivier	(1907–1989)	1556	Marlene Dietrich	(1901–1992)
1254	Clint Eastwood	(1930–)	1605	Bette Davis	(1908–1989)
1291	Humphrey Bogart	(1899–1957)	1632	Audrey Hepburn	(1929–1993)
1292	Robert De Niro	(1943–)	1689	Joan Crawford	(1905–1977)
1393	Cary Grant	(1904–1986)	1743	Mary Pickford	(1892–1979)
1401	James Stewart	(1908–1997)	1768	Greta Garbo	(1905–1990)
1506	Fred Astaire	(1899–1987)	2366	Doris Day	(1924–)
1664	Jack Nicholson	(1937–)	2443	Barbra Streisand	(1942–)
1788	Groucho Marx	(1890–1977)	2460	Jodie Foster	(1962–)
1851	Clark Gable	(1901–1960)	2572	Grace Kelly	(1929–1982)
1857	Buster Keaton	(1895–1966)	2651	Mae West	(1893–1980)

In this section, we will consider three largely distinct groups of figures associated with the motion picture industry: (1) the stars of the silent film era, (2) actors honored by Academy Awards, and (3) the directors most responsible for creating the films we love.

14.4.1 THE SILENT FILM ERA

The earliest film stars were those of the silent era, whose images remain vibrant today. The "Little Tramp" character created by *Charlie Chaplin (1889–1977)* [295] is still recognizable by children, and even served as the focus of IBM's advertising campaign that brought personal computers to the general public. The physical comedies of *Stan Laurel (1890–1965)* [3649] and *Oliver Hardy (1892–1957)* [4129] remain fresh despite the passage of time. We highlight the most significant figures of the silent film era.

Top Silent Film–Era Stars

Sig	Person	Dates	C/G	Description
295	Charlie Chaplin	(1889–1977)	C G	Silent film actor (The Little Tramp)
621	Harry Houdini	(1874–1926)	C G	American magician, escapologist
1276	D. W. Griffith	(1875–1948)	C G	Film director (*The Birth of a Nation*)

(continued)

Sig	Person	Dates	C/G	Description
1743	Mary Pickford	(1892–1979)	C G	Silent film actress (America's Sweetheart)
1857	Buster Keaton	(1895–1966)	C G	Silent film comic ("The General")
1921	Will Rogers	(1879–1935)	C G	American cowboy, comedian
2486	Rudolph Valentino	(1895–1926)	C G	Italian actor (the Latin lover)
2737	John Barrymore	(1882–1942)	C G	American actor of stage and screen
2788	Roscoe Arbuckle	(1887–1933)	C G	American silent film actor (Fatty)
2816	Sergei Eisenstein	(1898–1948)	C G	Russian film director (*Battleship Potemkin*)
3223	W. C. Fields	(1880–1946)	C G	American comedian, juggler, and drinker
3276	Harold Lloyd	(1893–1971)	C G	Silent film star (*Safety Last*)
3393	Lon Chaney	(1883–1930)	C G	Silent film actor (*Man of a Thousand Faces*)
3649	Stan Laurel	(1890–1965)	C G	English comic actor (Laurel and Hardy)
4001	Gloria Swanson	(1899–1983)	C G	Silent film actress (*Sunset Boulevard*)
4036	Clara Bow	(1905–1965)	C G	American actress (silent film era's "It girl")
4129	Oliver Hardy	(1892–1957)	C G	American comic actor (Laurel and Hardy)
4190	Lillian Gish	(1893–1993)	C G	American actress (*The Birth of a Nation*)
5047	Mack Sennett	(1880–1961)	C G	American silent film director
5726	Carole Lombard	(1908–1942)	C G	American film comedian (*My Man Godfrey*)

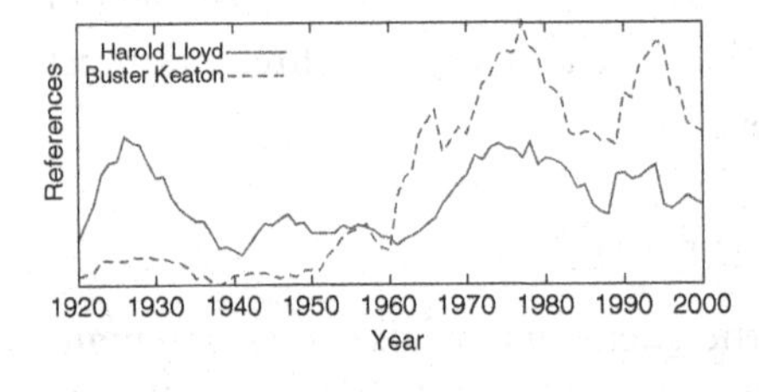

Today's perception of certain silent film stars differs from how they were received in their heyday. Recall *Harold Lloyd (1893–1971)* [3276], dangling off a clock in his 1923 film *Safety Last.* The frequency data on the left in the table shows that he was a substantially larger figure than his fellow silent comedian *Buster Keaton (1895–1966)* [1857] for most of this time period. Keaton's star only began to exceed Lloyd's during the 1960s. This is in part because Keaton continued to appear as a character actor late in life, while Lloyd enjoyed a more prosperous retirement. But a much bigger factor was the availability of their films. Lloyd retained ownership, but so restricted access to his films that he allowed his star to dim. Keaton's greatest films fell into the public domain, allowing them to be rediscovered by a new audience.

Preservation and dissemination only partially explain why several of the greatest silent-era stars do not cut recognizable figures today. *Mary Pickford (1892–1979)* [1743] co-founded United Artists. Yet she could not make the transition to talking pictures, and her characterization as "America's Sweetheart" soon became hopelessly dated. *Rudolph Valentino (1895–1926)*

[2486] still personifies the "Latin lover," but only because few have actually seen his films.

Silent-era actors seldom successfully transitioned to the sound era. Those that did generally had earlier careers in vaudeville (like *W. C. Fields (1880–1946)* [3223]) or the legitimate stage (like *John Barrymore (1882–1942)* [2737]). Certain popular silent stars did their best work in other venues. *Harry Houdini (1874–1926)* [621] was a great magician and escape artist, while *Will Rogers (1879–1935)* [1921] rose above his beginnings as a cowboy vaudeville star to become a major social commentator and celebrity.

14.4.2 THE ACADEMY AWARDS

The motion picture industry honors itself with the Academy Awards (or Oscars), in a ceremony watched by more than 40 million Americans each year. The Oscars have been presented annually since 1929, the end of the silent movie era. This makes them an excellent laboratory to study trends in American cinema. Are today's movie stars larger or smaller than they were in years past? How has the balance shifted with time between actors and actresses, or stars and supporting actors?

The three most significant movies stars of all time were never even nominated for an Oscar: *Ronald Reagan (1911–2004)* [32], *Elvis Presley (1935–1977)* [69], and *Marilyn Monroe (1926–1962)* [347]. This tells you something important about the motion picture industry: it makes heros (Reagan), imports them (Presley), and then often destroys them (Monroe). Yet Academy Award winners serve as a reasonable proxy for the industry. Of the thirty greatest movie stars we identified earlier, about half received either best actor or supporting actor awards, and several others garnered career awards or recognition in different categories.

Best Actor Winners

Figure 14.1 presents the most significant winners of the best actor and best actress awards. The top names prove interesting. *John Wayne (1907–1979)* [459] and *Bing Crosby (1903–1977)* [947] were major box office draws for more than 30 years, and each sold in excess of 1 billion movie tickets. Yet both of their images seem to be fading in popular culture. The genres where Wayne built his reputation (westerns and war pictures) are of declining relevance to younger audiences. It seems difficult to believe his pictures will

Sig	Person	Film	Year
459	John Wayne	*True Grit*	1970
947	Bing Crosby	*Going My Way*	1945
1118	Laurence Olivier	*Hamlet*	1949
1291	Humphrey Bogart	*The African Queen*	1952
1292	Robert De Niro	*Raging Bull*	1981
1401	James Stewart	*The Philadelphia Story*	1941
1664	Jack Nicholson	*As Good As It Gets*	1998
1851	Clark Gable	*It Happened One Night*	1935
2266	Marlon Brando	*The Godfather*	1973
2284	Tom Hanks	*Forrest Gump*	1995
2464	Gary Cooper	*High Noon*	1953
2571	Henry Fonda	*On Golden Pond*	1982
2613	James Cagney	*Yankee Doodle Dandy*	1943
2658	Dustin Hoffman	*Rain Man*	1989
2673	Al Pacino	*Scent of a Woman*	1993

(a)

Sig	Person	Film	Year
1373	Cher	*Moonstruck*	1988
1396	Katharine Hepburn	*On Golden Pond*	1982
1490	Meryl Streep	*Sophie's Choice*	1983
1605	Bette Davis	*Jezebel*	1939
1632	Audrey Hepburn	*Roman Holiday*	1954
1689	Joan Crawford	*Mildred Pierce*	1946
1743	Mary Pickford	*Coquette*	1930
2443	Barbra Streisand	*Funny Girl*	1969
2460	Jodie Foster	*The Silence of the Lambs*	1992
2572	Grace Kelly	*The Country Girl*	1955
2892	Vivien Leigh	*A Streetcar Named Desire*	1952
2965	Claudette Colbert	*It Happened One Night*	1935
2968	Ingrid Bergman	*Anastasia*	1957
3205	Nicole Kidman	*The Hours*	2003
3462	Sophia Loren	*Two Women*	1962

(b)

FIGURE 14.1. The most significant Academy Award winners for (a) Best Actor and (b) Best Actress.

hold up as well as, say, those of *Humphrey Bogart (1899–1957)* [1291]. *Bing Crosby (1903–1977)* [947] achieved even greater success as a singer than as a motion picture star, but little of his legacy remains popular in the rock-and-roll era except for the annual invasion of his record *White Christmas.*

We identify *Robert De Niro (1943–)* [1292], *Jack Nicholson (1937–)* [1664], and *Tom Hanks (1956–)* [2284] as the contemporary actors of greatest significance, but their reputations will ultimately depend upon whether their films remain popular into the future. We venture to guess that Hanks will emerge as the one with the greatest staying power, retaining a similar status as, say, *James Stewart (1908–1997)* [1401].

Among the women, the most significant Best Actress winner turns out to be *Cher (1946–)* [1373]. This is a surprise, and is more attributable to her singing and television work than her successful but limited film career. *Barbra Streisand (1942–)* [2443] is also more recognizable as a singer than an actress: she received her Best Actress award in a musical comedy. The juxtaposition of *Katharine Hepburn (1907–2003)* [1396] and *Meryl Streep (1949–)* [1490] at the top of the list is evocative, because both have similar personas as intelligent, sophisticated actresses.

Best Supporting Actors

Next we highlight the most significant winners of the Best Supporting Actor awards. Several won Oscars as both lead and supporting actors, including *Robert De Niro (1943–)* [1292], *Meryl Streep (1949–)* [1490], and *Ingrid Bergman (1915–1982)* [2968]. The most prominent recipients of the supporting actor awards would generally be considered as stars: their supporting actor awards often implicitly recognize a broader body of work. Indeed, only one of the actors in Figure 14.2 made her reputation on the strength of a single supporting role.

That exception would be *Hattie McDaniel (1895–1952)* [7163], the black woman who played Scarlett O'Hara's maid in *Gone with the Wind.* She was the first black person to win an Academy Award, in 1939. *Sidney Poitier (1927–)* [3803] was the first black to win Best Actor, in 1963.

Trends by Time and Gender

Figure 14.3 shows trend lines measuring the significance of the award-winning actors of each gender. It reveals several interesting things. The Best Actor/Actress winners were of roughly equal significance until the mid-1980s, when the male winners came to dominate best actresses for the next 15 years.

This largely reflects the dearth of significant roles for women outside certain prescribed categories. Four women who received their Oscars in this period rank among the eleven youngest ever (*Marlee Matlin (1965–)* [34956], *Hilary Swank (1974–)* [12415], *Jodie Foster (1962–)* [2460], and *Gwyneth Paltrow (1972–)* [5144]). Only Foster matured into a major star.

Sig	Person	Film	Year
357	Frank Sinatra	*From Here to Eternity*	1954
1077	Sean Connery	*The Untouchables*	1988
1292	Robert De Niro	*The Godfather Part II*	1975
1664	Jack Nicholson	*Terms of Endearment*	1984
2777	George Clooney	*Syriana*	2006
3004	John Gielgud	*Arthur*	1982
3076	Robin Williams	*Good Will Hunting*	1998
3432	Heath Ledger	*The Dark Knight*	2009
3502	Denzel Washington	*Glory*	1990
3509	Michael Caine	*The Cider House Rules*	2000
3780	George Burns	*The Sunshine Boys*	1976
4141	Jack Lemmon	*Mister Roberts*	1956
4648	Anthony Quinn	*Lust for Life*	1957
4693	Morgan Freeman	*Million Dollar Baby*	2005
4704	Gene Hackman	*Unforgiven*	1993

(a)

Sig	Person	Film	Year
1490	Meryl Streep	*Kramer vs. Kramer*	1980
2968	Ingrid Bergman	*Murder on the Orient Express*	1975
3728	Whoopi Goldberg	*Ghost*	1991
4293	Angelina Jolie	*Girl, Interrupted*	2000
5729	Goldie Hawn	*Cactus Flower*	1970
6179	Maggie Smith	*California Suite*	1979
6189	Ethel Barrymore	*None but the Lonely Heart*	1945
6935	Cate Blanchett	*The Aviator*	2005
7070	Judi Dench	*Shakespeare in Love*	1999
7163	Hattie McDaniel	*Gone with the Wind*	1940
7179	Helen Hayes	*Airport*	1971
7395	Jennifer Connelly	*A Beautiful Mind*	2002
7397	Kim Basinger	*L.A. Confidential*	1998
7424	Jennifer Hudson	*Dreamgirls*	2007
8193	Shelley Winters	*A Patch of Blue*	1966

(b)

FIGURE 14.2. The most significant Academy Award winners for (a) Best Supporting Actor and (b) Best Supporting Actress.

By contrast, the Best Actor winners tend to be older, more experienced actors. The youngest best actor winner ever, *Adrien Brody (1973–)* [15319], received his Oscar just shy of 30: but he was older than 29 Best Actress winners. Among actors, nineteen men were more than 50 years old when they received their awards, compared to only nine women.

Similar misogyny appears among the Best Supporting winners, with a gender separation beginning about the same year and lasting about the same duration. Fortunately these trends appear to have since reverted, with women regaining rough parity with men.

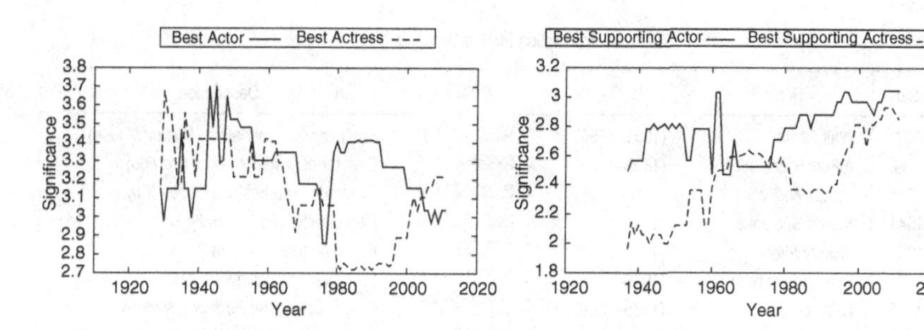

FIGURE 14.3. Significance of Oscar winners by year of award: Best Actors/Actresses (left) and Best Supporting Actors/Actresses (right), highlighting the differences between male and female winners.

A second important trend is the increasing significance of the supporting actor winners, whose statures have grown steadily since 1980. This seems somewhat counterintuitive. The studio film era was marked by distinctive character actors: people like *Peter Lorre (1904–1964)* [4132], *Edward Everett Horton (1886–1970)* [20056], and *Walter Brennan (1894–1974)* [8006] (who won best supporting actor three times). Yet there is now an increased willingness of major stars to take on interesting supporting roles, in search of the prestige that is absent from formulaic big-money vehicles.

14.4.3 FILM DIRECTORS

The director is the person with the greatest responsibility for a film's quality and integrity. Sitting behind the camera, directors are far more immune to the effects of aging than actors. Successful directors can have very long careers. *Alfred Hitchcock (1899–1980)* [439] directed more than 50 major films from 1927 to 1976. *Cecil B. DeMille (1881–1959)* [2107] followed his 1923 silent film of *The Ten Commandments* with the better-known 1956 version, nominated for seven Academy Awards.

The most significant Hollywood directors generally have distinct styles, which remains obvious even though they themselves do not appear on screen. It would seem impossible to mistake a *Woody Allen (1935–)* [1069] film for one by *Martin Scorsese (1942–)* [1041], even though both are contemporaries intimately associated with New York. Many are synonymous with specific genres: animation (Disney), suspense thrillers (Hitchcock), westerns (Ford), special-effect blockbusters (Spielberg and Lucas), and Woody Allen films (Allen).

Top Motion Picture Directors

Sig	Person	Dates	C/G	Description
337	Walt Disney	(1901–1966)	C G	American film animation (*Mickey Mouse*)
439	Alfred Hitchcock	(1899–1980)	C G	Director of suspense films (*Vertigo*)
729	Orson Welles	(1915–1985)	C G	Film/radio wunderkind (*Citizen Kane*)
1041	Martin Scorsese	(1942–)	C G	Film director (*Raging Bull*)
1069	Woody Allen	(1935–)	C G	Film director (*Annie Hall*)
1079	Steven Spielberg	(1946–)	C G	Film director (*Schindler's List*)
1276	D. W. Griffith	(1875–1948)	C G	Film director (*The Birth of a Nation*)
1855	John Ford	(1894–1973)	C G	Western film director (*High Noon*)
1935	Stanley Kubrick	(1928–1999)	C G	American film director (*2001*)
1941	Francis Ford Coppola	(1939–)	C G	American film director (*The Godfather*)
2107	Cecil B. DeMille	(1881–1959)	C G	American film director (*Ten Commandments*)
2149	Fritz Lang	(1890–1976)	C G	Austrian American filmmaker (*Metropolis*)
2428	Quentin Tarantino	(1963–)	C G	American film director (Pulp Fiction)
2531	George Lucas	(1944–)	C G	Lucasfilm producer (*Star Wars*)
3052	Billy Wilder	(1906–2002)	C G	American filmmaker (*The Apartment*)

Walt Disney (1901–1966) [337] should probably be regarded more as a producer and studio head than a director. *Orson Welles (1915–1985)* [729] directed only 13 full-length films in his career, but his significance rests on *Citizen Kane*, considered by many to be the greatest film of all time. It is notable that almost half of the most significant directors are alive as of this writing, and most of these remain active.

14.5 Radio and Television

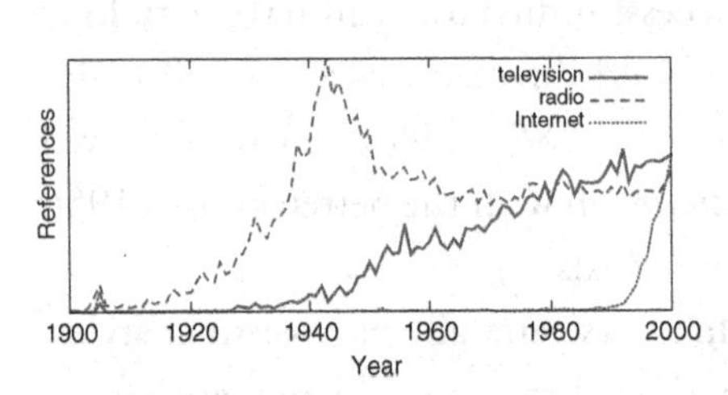

Most experienced actors favor one of the three major performance venues: television, movies, or the live stage. Which one they prefer is usually determined by where they made their initial splash. Although actors often try to transition to the movies (to maximize their income), or to the stage (for honor and prestige), true ubiquity is rare. Successful actors can be readily identified as either movie stars, TV stars, or Broadway stars. We are comfortable seeing our favorites in one particular medium.

Contemporary stage stars are rare, a victim of Broadway's decline in the face of recorded media. It is possible that traditional television stars will also soon disappear, a victim of the exploding number of cable channels and

Internet media outlets. Broadcasting was once *broad*casting, where a small number of national networks controlled what 100 million people watched each night. Today we have a larger number of smaller magnitude television stars, whose long-term staying power remains unknown.

14.5.1 RADIO

For insight, we can look to the history of radio, which dominated popular entertainment from the first commercial stations in 1920 until the advent of television after World War II. Broadcast radio still thrives, but it has become the ambient sound we drive to in our cars, eat to in restaurants, and type to while in the office.

A National Radio Hall of Fame based in Chicago (`http://www.radiohof.org/`) has inducted more than 150 of the most significant figures in radio. *Franklin D. Roosevelt (1882–1945)* [43] was an inspired choice: his thirty "fireside chats" narrowed the gap between the White House and the public, and his spirit lives on in the weekly radio addresses given by all U. S. presidents since 1982.

Among the members are inventors whose technologies made radio broadcasting possible: *Guglielmo Marconi (1874–1937)* [777], *Lee De Forest (1873–1961)* [2525], and *Edwin Howard Armstrong (1890–1954)* [4562]. Others were popular music stars of the Big Band era: *Benny Goodman (1909–1986)* [1800], *Tommy Dorsey (1905–1956)* [3879], and *Gene Autry (1907–1998)* [3973]. Finally, there are several comedians who starred on radio but achieved bigger impact in film and television: *Bob Hope (1903–2003)* [1098], *Groucho Marx (1890–1977)* [1788], and *Jack Benny (1894–1974)* [2568].

The most significant contemporary radio stars are conservative broadcaster *Rush Limbaugh (1951–)* [2837] and shock jock *Howard Stern (1954–)* [2130]. No other active radio figures approach their popularity and stature.

Top Members of the Radio Hall of Fame

Sig	Person	Dates	C/G	Description
43	Franklin D. Roosevelt	(1882–1945)	C G	32nd U.S. president (New Deal, WWII)
729	Orson Welles	(1915–1985)	C G	Film / radio wunderkind (*Citizen Kane*)
777	Guglielmo Marconi	(1874–1937)	C G	Italian inventor (radio broadcasting)

(continued)

Sig	Person	Dates	C/G	Description
947	Bing Crosby	(1903–1977)	C G	American singer, actor ("Going My Way")
1098	Bob Hope	(1903–2003)	C G	Longtime radio / film comedian
1788	Groucho Marx	(1890–1977)	C G	American comedian (Marx Brothers)
1800	Benny Goodman	(1909–1986)	C G	American jazz clarinetist
2130	Howard Stern	(1954–)	C G	American radio personality (shock jock)
2525	Lee De Forest	(1873–1961)	C G	American inventor (radio tube)
2568	Jack Benny	(1894–1974)	C G	American comedian and violinist
2837	Rush Limbaugh	(1951–)	C G	Right-wing radio talk show host
3236	Edward R. Murrow	(1908–1965)	C G	American broadcast journalist
3879	Tommy Dorsey	(1905–1956)	C G	American jazz trombonist
3973	Gene Autry	(1907–1998)	C G	American performer (singing cowboy)
4562	Edwin Howard Armstrong	(1890–1954)	C G	American electrical engineer

The biggest threat facing radio today is not television, but personal music players and customized music services such as Pandora, which provides an individualized soundtrack to everyone. Each of us gets to hear what we want in our own bubble, but at the loss of communal experiences and exposure to new sounds and ideas.

14.5.2 TELEVISION

Topping our ranking of most significant television stars is *Lucille Ball (1911–1989)* [785], whose *I Love Lucy* show and popular successors ran from 1951 to 1974, and live on in syndication today.

However, the television stars of the 1950s generally have not held up quite as strongly as the movie stars of that era. *Honeymooners'* star *Jackie Gleason (1916–1987)* [3881] was the very biggest male television star of the 1950s. Yet today his significance is comparable to contemporaries *Lee Marvin (1924–1987)* [6452] and *Jack Lemmon (1925–2001)* [4141]: stars, but not stars of the scale of *Paul Newman (1925–2008)* [2970] or *Marlon Brando (1924–2004)* [2266]. Elvis and the Beatles were introduced to America by variety host *Ed Sullivan (1901–1974)* [4908], and *Tonight Show* host *Johnny Carson (1925–2005)* [4834] put America to bed for 30 years. Yet they register more as middle-rank celebrities than historical figures.

The significance rankings here are dominated by more contemporary celebrities, such as Disney star *Hilary Duff (1987–)* [1626] and *Stephen Colbert (1964–)* [2700], whose followings aggressively use social media.

Most Significant Television Stars

Sig	Person	Dates	C/G	Description
1391	Lucille Ball	(1911–1989)	C G	American comedian (*I Love Lucy*)
1626	Hilary Duff	(1987–)	C G	American teen actress, singer
2700	Stephen Colbert	(1964–)	C G	American political satirist (*Colbert Report*)
2709	Roger Ebert	(1942–2013)	C G	TV film critic (*Siskel & Ebert*)
2947	Jennifer Aniston	(1969–)	C G	American actress (*Friends*)
2969	Jon Stewart	(1962–)	C G	American political satirist
3104	Bill Cosby	(1937–)	C G	American comedian (*The Cosby Show*)
3236	Edward R. Murrow	(1908–1965)	C G	American broadcast journalist
3593	William Shatner	(1931–)	C G	TV actor (Captain Kirk)
3739	Jay Leno	(1950–)	C G	Longtime *Tonight Show* host
3746	Ron Howard	(1954–)	C G	Child actor turned film director
3881	Jackie Gleason	(1916–1987)	C G	TV comedian (*The Honeymooners*)
3903	Julia Child	(1912–2004)	C G	American TV's French chef
3937	David Letterman	(1947–)	C G	Late night television host
4241	John Goodman	(1952–)	C G	American TV / film actor (*Roseanne*)
4477	Conan O'Brien	(1963–)	C G	Very late night television host
4834	Johnny Carson	(1925–2005)	C G	*Tonight Show* television host
4908	Ed Sullivan	(1901–1974)	C G	No-necked TV variety show host
4924	Alan Alda	(1936–)	C G	American actor (*M*A*S*H*)
5000	Jerry Seinfeld	(1954–)	C G	Stand-up / TV comedian (*Seinfeld*)

Colbert is particularly notorious in this regard, having directed his viewers to win a NASA Internet poll to name part of the International Space Station after him. We anticipate that their reputations will decline with the passage of time. Place not your trust in the present world.

15

Devils and Angels

The battle between good and evil has been fought since the beginning of time. Which is the surer path to historical recognition? There seems to be conflicting evidence wherever you look:

- For every *Jesus (7 B.C.–A.D. 30)* [1], there is an *Adolf Hitler (1889–1945)* [7].
- For every charitable missionary like *Mother Teresa (1910–1997)* [820] lurks a gangster like *Al Capone (1899–1947)* [646].
- For every social worker like *Jane Addams (1860–1935)* [1256] exists an assassin like *Lee Harvey Oswald (1939–1963)* [1435].
- For every civil rights leader like *Martin Luther King (1929–1968)* [221], we find a serial killer like *Jack the Ripper* [166].

In this chapter, we will consider a range of figures from both ends of the morality scale, to open a discussion on which is the more enduring path to glory. Our conclusion is that true virtue is generally rewarded, but spectacular acts of infamy endure longer than similar acts of heroism.

15.1 The Killers

Hammurabi (1796–1750 B.C.) [899], the lawmaker, has been followed by three millennia of law breakers. But several criminals richly deserving of their punishment endure as cultural figures because of the imagination and magnitude of their crimes. In this section we will identify the most prominent outlaws, assassins, and other killers, to better understand the source of this fascination.

15.1.1 OUTLAWS IN THE WEST

Yes, there were ranchers and farmers and businessmen and preachers who helped settle the American West. But the mythology of the West is of the outlaw, and the lawmen who brought them to justice. How much of what we think we know about the settling of the American West comes from movies?

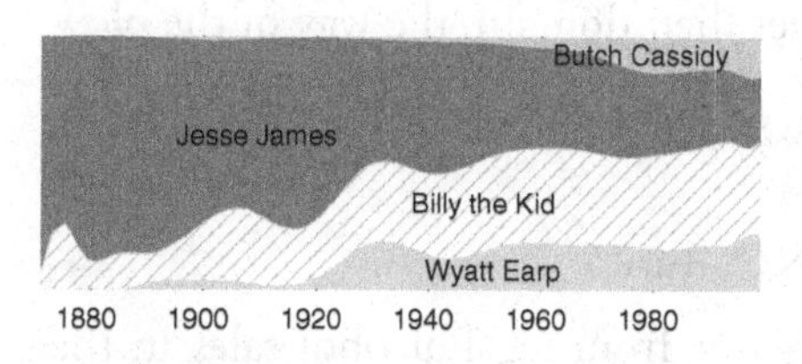

The Ngrams dataset lets us look back at the fame that Western figures had over time. It reveals that only *Jesse James (1847–1882)* [413] received much attention during his lifetime. *Billy the Kid (1859–1881)* [761] springs suddenly into popular consciousness during the silent film era. Lawman *Wyatt Earp (1848–1929)* [858], the sole survivor of the shootout at the O.K. Corral, reemerges shortly before his death, when he was recognized as one of the last survivors of this colorful era. His legend bloomed in Hollywood. Similarly, outlaw *Butch Cassidy (1866–1908)* [4230] was rediscovered in the 1950s, and hit the big time after being played by *Paul Newman (1925–2008)* [2970] in the film *Butch Cassidy and the Sundance Kid.*

Top Outlaws, Lawmen, and Showmen

Sig	Person	Dates	C/G	Description
413	Jesse James	(1847–1882)	C G	Western bank robber (James gang)
761	Billy the Kid	(1859–1881)	C G	Old west gunman and outlaw
788	Ned Kelly	(1854–1880)	C G	Irish Australian bushranger
858	Wyatt Earp	(1848–1929)	C G	Western gambler / lawman (O.K. Corral)
1344	Wild Bill Hickok	(1837–1876)	C G	Old West lawman and folk hero
2147	Doc Holliday	(1851–1887)	C G	American gambler, gunfighter (O.K. Corral)
3876	Calamity Jane	(1852–1903)	C G	American frontierswoman
4230	Butch Cassidy	(1866–1908)	C G	American train and bank robber
4608	Frank James	(1843–1915)	C G	Famous American outlaw (Jesse's brother)
4618	John Wesley Hardin	(1853–1895)	C G	American outlaw

Jesse and his brother *Frank James (1843–1915)* [4608] shared the leadership of the James gang, and show similar prominence in the early book Ngrams data. Their fates sharply diverged after Jesse got shot by the bounty hunter *Robert Ford (1862–1892)* [12757]. Frank surrendered to the authorities five months later, but was never convicted and lived the rest of his life as a free man. Jesse's afterlife proved considerably larger, because his colorful demise played much better on the silver screen.

Ned Kelly (1854–1880) [788] is an Australian outlaw/folk hero, who lived and died at roughly the same time as *Billy the Kid (1859–1881)* [761]. The closing of the frontier around 1890 marked the end of the outlaw. Once a critical mass of settlers took over their domain, the way of the plow triumphed over the way of the gun.

15.1.2 GANGSTERS

Prohibition (1919–1933) marked a hiatus from legal alcohol sales in the United States, but illegal bootleggers worked hard to satisfy the persistent demand. Gangsters like *Al Capone (1899–1947)* [646] consolidated the market, accruing great wealth, power, and notoriety in the process. The Great Depression led others to desperate measures, with bank robbers like *John Dillinger (1903–1934)* [2679] and *Baby Face Nelson (1908–1934)* [9276] becoming folk heros before being hunted down. Lawmen like FBI director *J. Edgar Hoover (1895–1972)* [1201] made their reputation by capturing the gangsters of the early 1930s.

Most Significant Gangsters

Sig	Person	Dates	C/G	Description
646	Al Capone	(1899–1947)	C G	American Prohibition-era gangster
2679	John Dillinger	(1903–1934)	C G	American Depression-era bank robber
2944	Lucky Luciano	(1897–1962)	C G	Pioneering Italian mobster
5318	Meyer Lansky	(1902–1983)	C G	American organized crime figure
6830	Pablo Escobar	(1949–1993)	C G	Colombian drug lord
7061	Bugsy Siegel	(1906–1947)	C G	American gangster (Las Vegas)
7374	John Gotti	(1940–2002)	C G	American mobster (Dapper Don)
8360	Frank Costello	(1891–1973)	C G	Boss of the Genovese crime family
8579	Dutch Schultz	(1902–1935)	C G	German-Jewish mobster and bootlegger
8814	Carlo Gambino	(1902–1976)	C G	Boss of the Gambino crime family
9243	Ma Barker	(1873–1935)	C G	American criminal (mother of Barker gang)
9276	Baby Face Nelson	(1908–1934)	C G	Bank robber
10599	Arnold Rothstein	(1882–1928)	C G	Jewish racketeer / gambler
11155	Pretty Boy Floyd	(1904–1934)	C G	American bank robber
11730	Machine Gun Kelly	(1895–1954)	C G	American gangster

A new generation of Italian (*Lucky Luciano (1897–1962)* [2944]) and Jewish (*Meyer Lansky (1902–1983)* [5318]) gangsters emerged after Prohibition, as organized crime syndicates took advantage of new business opportunities. American fascination with the Mafia continues in film and the popular media. However, Mafia influence has deteriorated since the conviction of *John Gotti (1940–2002)* [7374], the last American organized crime figure to capture the popular imagination. The rest of the world is not so fortunate, and drug lords like *Pablo Escobar (1949–1993)* [6830] have stepped up to replace them.

15.1.3 MASS KILLERS

The specter of large-scale terrorism is a relatively recent development in human history. The leader of the September 11 attacks on the World Trade Center, *Osama bin Laden (1957–2011)* [765], registers as a substantial historical figure, for the magnitude of the event and its consequences.

Most Significant Serial / Mass Killers

Sig	Person	Dates	C/G	Description
166	Jack the Ripper	(?–?)	C G	English serial killer (London)
765	Osama bin Laden	(1957–2011)	C G	Founder of al-Qaeda (Twin Towers attack)
3040	Gilles de Rais	(1404–1440)	C G	Breton knight (serial killer of children)
3404	Ted Bundy	(1946–1989)	C G	American serial killer and rapist
6049	Jeffrey Dahmer	(1960–1994)	C G	American serial killer and cannibal
7083	John Wayne Gacy	(1942–1994)	C G	American serial killer of boys
7379	Timothy McVeigh	(1968–2001)	C G	American terrorist (Oklahoma Federal Building)
7896	Charles Whitman	(1941–1966)	C G	American killer (Texas Tower Sniper)

But there is evidence that the perpetrators of more typical attacks quickly recede from memory. Oklahoma City bomber *Timothy McVeigh (1968–2001)* [7379] killed 166 people, yet his image has faded to the point where he now sits behind others guilty of horrific but ultimately smaller crimes.

The popular notion of a "serial killer" is quite recent, emerging in the Ngram records only around 1980. Indeed, the most significant serial killers in this table emerge from this time period, with two significant exceptions.

Jack the Ripper [166] was responsible for at least five murders in London in 1888, but his case remains unsolved today. He shocked the world with the brutal nature of the slayings, as well as his brazenness: he wrote letters to

Person	Sig	Dates	Victim	Sig	Dates
John Wilkes Booth	451	(1838–1865)	Abraham Lincoln	5	(1809–1865)
Marcus Brutus	763	(?–42 B.C.)	Julius Caesar	15	(100–44 B.C.)
Lee Harvey Oswald	1435	(1939–1963)	John F. Kennedy	61	(1917–1963)
Gavrilo Princip	2814	(1894–1918)	Franz Ferdinand	573	(1863–1914)
Charles J. Guiteau	3587	(1841–1882)	James A. Garfield	286	(1831–1881)
Leon Czolgosz	5566	(1873–1901)	William McKinley	167	(1843–1901)
Giuseppe Zangara	7238	(?–1933)	Anton Cermak	7679	(1873–1933)
Nathuram Godse	8910	(1910–1949)	Mohandas Gandhi	44	(1869–1948)
Udham Singh	11659	(1899–1940)	Michael O'Dwyer	19985	(1864–1940)
Sirhan Sirhan	12621	(1944–)	Robert F. Kennedy	1260	(1925–1968)
Dan White	16453	(1946–1985)	Harvey Milk	3386	(1930–1978)
James Earl Ray	19760	(1928–1998)	Martin Luther King Jr.	205	(1929–1968)
Yigal Amir	36317	(1970–)	Yitzhak Rabin	2523	(1922–1995)

FIGURE 15.1. Most significant assassins and their victims.

the police enclosing body parts of his victims. The mystery of his identity has kept his name alive: more than 100 published theories propose different suspects. However, the Ngrams data show that a different killer, *Gilles de Rais (1404–1440)* [3040], cut a higher profile than the Ripper over much of the past century. De Rais was a serial killer of children, and served as the inspiration for Bluebeard: a French folktale of a nobleman who repeatedly married and then murdered his wives.

15.1.4 ASSASSINS AND THEIR PREY

Political assassination offers perhaps the best opportunity for ordinary people to exert great influence over history. *Gavrilo Princip (1894–1918)* [2814] provoked World War I by assassinating *Archduke Franz Ferdinand of Austria (1863–1914)* [541]. That particular killing seems almost to have been fated. Princip was part of a failed conspiracy to bomb the Archduke's car at a procession in the morning, but then encountered the car again by chance an hour later and shot him.

How does the fame of the assassin grow with the size of his prey? We rank history's most significant assassins with their victims in Figure 15.1.

There is a strong correlation between the significance of assassin and victim. The killers of the four American presidents rank in nearly the same order as their victims. The out of order pair is explained by motive. *Charles J. Guiteau (1841–1882)* [3587] is famous as a “disappointed office seeker,” while *Leon Czolgosz's (1873–1901)* [5566] motives as an “anarchist” seem far more remote today.

The exceptions prove the rule. *Giuseppe Zangara's (?–1933)* [7238] fame comes not from shooting Chicago mayor *Anton Cermak (1873–1933)* [7958], but from missing his intended target *Franklin D. Roosevelt (1882–1945)* [43]. Only one assassin here is deemed more important than his victim. *Udham Singh (1899–1940)* [11659] was a revolutionary, whose killing of a British official responsible for a notorious massacre in the Punjab helped crystallize India's independence movement.

15.2 Law and Order

Residing somewhere between devils and angels are the security and legal personnel responsible for maintaining order in the world. Glory does not accrue to real police officers the way it does on television. Prohibition agent *Eliot Ness (1903–1957)* [7727] may be the most famous law enforcement officer to ever actually hit the street, but that is because he was the inspiration for television's *The Untouchables*. New York Police Department detective *Frank Serpico (1936–)* [37112] became famous as an honest cop, heroically testifying to corruption in the ranks. But it helped Serpico's reputation to be played by *Al Pacino (1940–)* [2673] in the movie version of the story.

The top brass rank somewhat higher than the flatfoots. Several major political figures got their start as chief of police, including New York's *Theodore Roosevelt (1858–1919)* [23]. Recent police chiefs widely recognized for their work include Los Angeles's *Daryl Gates (1926–2010)* [38518] and New York's *Raymond Kelly (1941–)* [62784].

15.2.1 FOR THE PROSECUTION

Far more political careers have been launched by prosecuting attorneys than those for the defense, because a "tough on crime" image always plays well among the electorate. President *William Howard Taft (1857–1930)* [153] got his start as a prosecutor. More recently, *Rudy Giuliani (1944–)* [2657] parlayed his convictions of important Mafia figures to become mayor of New York.

The attorney general is the highest position in the U.S. Department of Justice. The most prominent recent figure here is *Alberto Gonzales (1955–)* [8601], but he sits behind a dozen of his predecessors. Many held other important offices, serving as senators, governors, and justices of the Supreme Court.

Most Significant U.S. Attorney Generals

Sig	Person	Dates	C/G	Description
1253	Roger B. Taney	(1777–1864)	C G	5th chief justice of the U.S. Supreme Court
1355	Robert F. Kennedy	(1925–1968)	C G	Slain senator, presidential candidate
1910	Edwin M. Stanton	(1814–1869)	C G	Lincoln's cabinet official (Sec. of War)
2055	Edmund Randolph	(1753–1813)	C G	U.S. governor of Virginia
4407	John J. Crittenden	(1787–1863)	C G	U.S. governor of Kentucky
4551	Frank Murphy	(1890–1949)	C G	Supreme Court justice / governor of Michigan
5656	Harlan F. Stone	(1872–1946)	C G	12th chief justice of Supreme Court
6143	Robert H. Jackson	(1892–1954)	C G	Supreme Court justice, Nuremberg prosecutor
6334	Tom C. Clark	(1899–1977)	C G	U.S. attorney general / Supreme Court justice
6641	William M. Evarts	(1818–1901)	C G	American lawyer

The revelation here is how often attorney generals used to be promoted to the Supreme Court. Taney, Stanton,[1] Stone, Jackson, and Clark all became justices, a career path that seems closed off today. The highly charged political nature of the court has, paradoxically, served to disqualify candidates who have held elective or appointed office. This robs the court of practical experience in the functioning of other branches of government.

15.2.2 DEFENSE ATTORNEYS

Leading defense attorneys are more omnipresent in the news than prosecutors, who are bound to the jurisdiction they serve. Thus, they must catch a break to have a notorious case occur on their watch. By contrast, defense attorneys are hired guns, and the most prominent ones actively seek cases in the limelight. Two of the top defense attorneys in our rankings were associated with the *O. J. Simpson (1947–)* [3464] trial: *Alan Dershowitz (1938–)* [10078] and *Johnnie Cochran (1937–2005)* [22272].

Successful defense attorneys generally have limited political futures, but *John Adams' (1735–1826)* [61] early political career was helped by winning acquittals for most of the British troops responsible for the Boston Massacre.

Clarence Darrow (1857–1938) [1807] proves to be the most significant defense attorney by a large margin, particularly since his nearest rivals

1 *Edwin M. Stanton (1814–1869)* [1910] died four days after being confirmed a justice by the Senate, but never took the oath of office.

Most Significant Defense Attorneys

Sig	Person	Dates	C/G	Description
1807	Clarence Darrow	(1857–1938)	C G	Prominent American lawyer (Scopes Monkey trial)
2539	Alger Hiss	(1904–1996)	C G	American lawyer, Soviet spy
5019	Robert G. Ingersoll	(1833–1899)	C G	American lecturer and agnostic
7995	John Grisham	(1955–)	C G	Author of legal thrillers
10078	Alan Dershowitz	(1938–)	C G	American lawyer and political commentator
22272	Johnnie Cochran	(1937–2005)	C G	American lawyer (O. J. Simpson)
41840	F. Lee Bailey	(1933–)	C G	Criminal defense lawyer (Boston Strangler)

earned their real fame in other domains: *Alger Hiss (1904–1996)* [2539] as a spy, and *John Grisham (1955–)* [7995] as a mystery writer. Darrow's skills got several murderers off with life instead of the electric chair, including the notorious thrill killers Leopold and Loeb. Yet his reputation rises above that of today's hired guns, because he served as the defense attorney in the famous Scopes Monkey trial, concerning the teaching of evolution.

15.3 Peacemakers

The Book of *Saint Matthew (?–A.D. 34)* [848] states, "blessed are the peacemakers, for they will be called sons of God." Here, we consider two broadly recognized groups of peacemakers: the recipients of the Nobel Peace Prize and the secretary-generals of the United Nations.

15.3.1 NOBEL PEACE PRIZE WINNERS

Recall our graphs of the significance of Nobel Prize winners in Figure 13.3. The prominence of the recipients in the science categories has plunged since the early days of the awards, but the Peace Prize category remains quite robust. Scientific discoveries have become increasingly esoteric, but there remains no shortage of human conflicts in need of resolution.

The most significant winners of the Peace Prize prove to be a less overtly noble bunch than might be supposed. We see four U.S. presidents, plus other figures whose accomplishments in war seem to exceed that of peace, including *Henry Kissinger (1923–)* [864] and *Yasser Arafat (1929–2004)* [1486].

There are three types of recipients near the top of our rankings. *Martin Luther King (1929–1968)* [221] and *Nelson Mandela (1918–)* [356] broke

Top Nobel Peace Prize Laureates

Sig	Person	Dates	C/G	Description
23	Theodore Roosevelt	(1858–1919)	C G	26th U.S. president (Progressive Movement)
47	Woodrow Wilson	(1856–1924)	C G	28th U.S. president (World War I)
111	Barack Obama	(1961–)	C G	44th and current U.S. president
221	Martin Luther King	(1929–1968)	C G	American civil rights leader
356	Nelson Mandela	(1918–)	C G	Post-apartheid president of South Africa
462	Jimmy Carter	(1924–)	C G	39th president (Iranian hostage crisis)
623	Al Gore	(1948–)	C G	U.S. vice president / almost president
637	Mikhail Gorbachev	(1931–)	C G	Last Soviet premier (fall of USSR)
820	Mother Teresa	(1910–1997)	C G	Catholic nun / missionary in Calcutta
864	Henry Kissinger	(1923–)	C G	American diplomat (Realpolitik)
1204	Lester B. Pearson	(1897–1972)	C G	Prime minister of Canada (Suez treaty)
1256	Jane Addams	(1860–1935)	C G	Nobel Peace Prize laureate (Hull House)
1279	George Marshall	(1880–1959)	C G	American military leader (Marshall Plan)
1286	Linus Pauling	(1901–1994)	C G	Nobel Peace Prize–winning biochemist
1486	Yasser Arafat	(1929–2004)	C G	Palestinian leader (PLO) / Nobelist

the back of institutional racism in their nations, the United States and South Africa, respectively. Social workers like *Mother Teresa (1910–1997)* [820] and *Jane Addams (1860–1935)* [1256] were recognized for feeding and clothing the poor, from the slums of Calcutta to the wards of Chicago. Finally, we see inspired government officials like *George Marshall (1880–1959)* [1279], whose plan rebuilt Europe after the destruction of World War II, thus restoring peaceful, prosperous democracies.

15.3.2 LEADERS OF THE UNITED NATIONS

The secretary-general of the United Nations is presumably the man who logs the most miles in quest of peace each year. Part of being a diplomat is not drawing attention to one's self. Generally they have succeeded, achieving significance rankings akin to those of the less distinguished vice presidents of the United States. Secretary-generals of the UN are elected for five-year terms, with most serving two terms.

Dag Hammarskjöld (1905–1961) [5638] was martyred in a plane crash while serving on a peace mission in the Congo. This, along with his posthumous Nobel Peace Prize, helps explain his high level of significance. *Kofi Annan (1938–)* [3150] is the highest-ranking secretary-general, apparently because he served ten years in office and also received the Nobel Prize.

Top United Nations Leaders

Sig	Person	Dates	C/G	Description
3150	Kofi Annan	(1938–)	C G	Ghanaian UN Secretary-General / Nobelist
5133	Ban Ki-moon	(1944–)	C G	Current UN Secretary-General
5638	Dag Hammarskjöld	(1905–1961)	C G	Swedish UN Secretary-General / Nobelist
6881	U Thant	(1909–1974)	C G	Burmese UN Secretary-General
7287	Kurt Waldheim	(1918–2007)	C G	Austrian UN Secretary-General / suspected Nazi
9656	Boutros Boutros-Ghali	(1922–)	C G	Egyptian UN Secretary-General
13386	Trygve Lie	(1896–1968)	C G	First UN Secretary-General
19340	Javier Pérez de Cuéllar	(1920–)	C G	Peruvian UN Secretary-General

15.4 Social Activists

Social movements revolve around charismatic leaders, who tap into some previously unexpressed discontent in the Zeitgeist. The labor movement required brave leaders like *Samuel Gompers (1850–1924)* [1575] to organize the masses, at great personal risk. The Civil Rights movement was led by *Martin Luther King (1929–1968)* [221] and a host of other courageous figures. The history of the women's rights movement reads as a tale of starchy suffragettes followed by liberated intellectuals.

This book is being written at the moment when the issue of gay marriage is tipping from the unthinkable to the inevitable. The amazing thing about this process is how it has happened without a broadly visible gay rights leader. When the history texts record this story, who will be highlighted? The most prominent gay leader to date has been *Harvey Milk (1930–1978)* [3680], but he was not an active part of this process. The movement's leadership is conspicuous by its absence. Its success is perhaps related to this invisibility: the quiet role of influential members of the gay community silently greasing the skids of a cause whose time has finally come.

In this section we will explore the history of social activism, trying to better understand how leaders emerge, are identified, and are finally canonized.

15.4.1 WOMEN'S RIGHTS

A monument to three women's suffrage leaders sits in the Rotunda of the U.S. Capitol building in Washington: *Susan B. Anthony (1820–1906)* [432], *Elizabeth Cady Stanton (1815–1902)* [734], and *Lucretia Mott*

(1793–1880) [2208]. All were long dead by the time the right to vote was granted in 1920. So who deserves the credit for making it happen?

Most Significant Suffragists

Sig	Person	Dates	C/G	Description
432	Susan B. Anthony	(1820–1906)	C G	American women's rights / civil rights leader
734	Elizabeth Cady Stanton	(1815–1902)	C G	Woman's suffrage leader
1093	Harriet Tubman	(1820–1913)	C G	African American abolitionist
1135	Helen Keller	(1880–1968)	C G	Inspirational deaf-blind author
1231	Sojourner Truth	(1797–1883)	C G	African American abolitionist
2208	Lucretia Mott	(1793–1880)	C G	American Quaker and abolitionist
2738	Lucy Stone	(1818–1893)	C G	American abolitionist and suffragist
3783	Ida B. Wells	(1862–1931)	C G	African American journalist
4832	Victoria Woodhull	(1838–1927)	C G	American suffragist and free love advocate
5216	Frances Willard	(1839–1898)	C G	American educator and women's suffragist
6037	Carrie Chapman Catt	(1859–1947)	C G	American women's suffrage leader
6784	Alice Paul	(1885–1977)	C G	American suffragist (National Women's Party)

The end game of the suffrage movement proved somewhat messy. Two opposing suffrage organizations were founded after the Civil War, by *Lucy Stone (1818–1893)* [2738] and Susan B. Anthony. They differed in tactics and goals before merging in 1890. A mix of insider activity by the resulting National American Woman Suffrage Association's (NAWSA) president *Carrie Chapman Catt (1859–1947)* [6037] and more radical protest by *Alice Paul's (1885–1977)* [6784] National Woman's Party ultimately granted women the vote. The NAWSA eventually became the League of Women Voters following the passage of the Nineteenth Amendment in 1920.

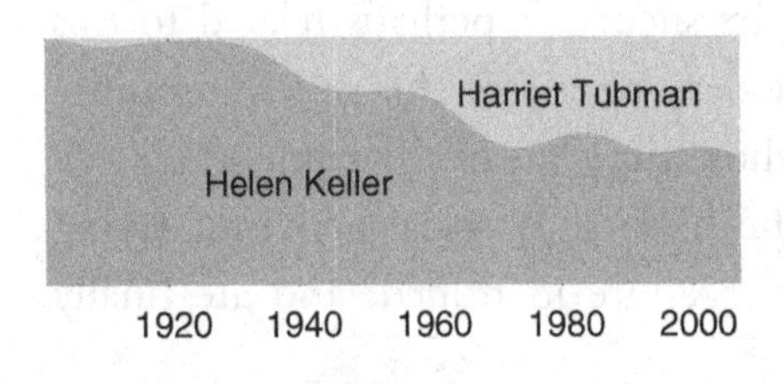

Women's leaders who were more strongly aligned with other causes also supported the suffrage movement. *Helen Keller's (1880–1968)* [1135] struggle to overcome deafness and blindness made her the most prominent advocate for the disabled, yet she also embraced women's fight for the vote. *Harriet Tubman (1820–1913)* [1093] was an abolitionist, yet saw no reason why the vote should be granted to black men but not women after the Civil War. The image of *Helen Keller (1880–1968)* [1135] appears to be receding in the Ngrams data relative to Tubman, perhaps because vaccination, cochlear implants, and other medical technology have sharply reduced the prevalence of deafness among today's youth.

The women's rights movement has fought many battles since gaining the vote. Early advocates sought increased opportunities in education, and demonstrated that women could succeed in fields such as medicine if granted the same rights as men.

Other Top Women's Rights Advocates

Sig	Person	Dates	C/G	Description
1315	Emma Goldman	(1869–1940)	C G	Exiled American anarchist
2453	Elizabeth Blackwell	(1821–1910)	C G	First U.S. woman to receive a medical degree
2672	Margaret Sanger	(1879–1966)	C G	American sex educator (birth control)
3815	Gloria Steinem	(1934–)	C G	American feminist activist
4289	Julia Ward Howe	(1819–1910)	C G	Poet / abolitionist ("Battle Hymn of the Republic")
5553	Betty Friedan	(1921–2006)	C G	American writer, feminist

Others pursued causes far more radical for their day. *Emma Goldman's (1869–1940)* [1315] outspokenness led to her deportation to Russia. *Margaret Sanger's (1879–1966)* [2672] efforts to promote family planning and access to contraception provided women with the freedom to pursue life paths previously denied to them.

The leading women's activists of the 1960s (*Gloria Steinem (1934–)* [3815] and *Betty Friedan (1921–2006)* [5553]) are substantial historical figures, yet they resonate less strongly than the leading black activists of the period. For example, both are outranked by *Coretta Scott King (1927–2006)* [3481], the widow of *Martin Luther King (1929–1968)* [221].

15.4.2 THE LABOR MOVEMENT

The notion that workers had the right to fair pay, a safe workplace, and an eight-hour day did not come easily. It took a war, fought with and against the law. Early union conflicts often turned violent, with management's hired Pinkerton agents combating union activities through infiltration and thuggery. Turn-of-the-century union leaders like Big *Bill Hayward (1868–1947)* [192701] and Mother *Mary Harris Jones (1837–1930)* [2398] fought back, and went to prison on a mix of real and trumped-up charges. Socialist union leader *Eugene V. Debs (1855–1926)* [1340] gathered 913,664 write-in votes for president from his jail cell in 1920.

Cigar maker *Samuel Gompers (1850–1924)* [1575] served as the president of the American Federation of Labor (AFL) from its founding in 1886 almost continuously until his death in 1924. The AFL was the first truly successful labor union in the United States, organizing the craft/trade unions while avoiding the radical socialism of earlier efforts. *John L. Lewis (1880–1969)* [8358] founded the Congress of Industrial Organizations (CIO) to organize workers in the growing mass production industries. Both were federations of smaller unions, that stood together to increase their clout with management. The two major federations merged in 1955 to form the AFL-CIO.

Most Significant Labor Leaders

Sig	Person	Dates	C/G	Description
1340	Eugene V. Debs	(1855–1926)	C G	Socialist presidential candidate
1575	Samuel Gompers	(1850–1924)	C G	American labor union leader (AFL)
2398	Mary Harris Jones	(1837–1930)	C G	American labor organizer
3552	Jimmy Hoffa	(1913–1975)	C G	Disappeared Teamster union leader
3948	César Chávez	(1927–1993)	C G	Co-founder of United Farm Workers
5936	Bill Haywood	(1869–1928)	C G	Radical workers union leader
6144	A. Philip Randolph	(1889–1979)	C G	African American union / civil rights leader
8358	John L. Lewis	(1880–1969)	C G	Union leader (United Mine Workers)
15150	Walter Reuther	(1907–1970)	C G	American labor union leader (UAW)

Pride of place in our rankings of labor leaders generally lies with the founding figures of the movement. They rank ahead of the establishment figures who led labor at its high watermark after World War II, such as longtime AFL-CIO president *George Meany (1894–1980)* [33502].

Teamsters leader *Jimmy Hoffa (1913–1975)* [3552] played with a rough crowd, and his organized crime connections led to his expulsion from the AFL-CIO in 1957. These friends presumably were responsible for his mysterious disappearance as well.

Some organizers founded unions whose successes were critical to the Civil Rights movement. *César Chávez's (1927–1993)* [3948] campaign on behalf of migrant farm workers turned him into the most prominent Hispanic leader in America, while *A. Philip Randolph's (1889–1979)* [6144] leadership of the Brotherhood of Sleeping Car Porters made him a godfather to the black Civil Rights movement.

15.5 Civil Rights Leaders

The fight for black civil rights is one of the most stirring episodes in American history. *Martin Luther King (1929–1968)* [221] and *Malcolm X (1925–1965)* [1106] formed the yin and yang of the movement in the 1960s: one preaching peaceful protest, with the other advocating more forceful means of black self-determination.

Most Significant Civil Rights Leaders

Sig	Person	Dates	C/G	Description
221	Martin Luther King	(1929–1968)	C G	American civil rights leader
964	Rosa Parks	(1913–2005)	C G	African American civil rights activist
1106	Malcolm X	(1925–1965)	C G	Black Muslim activist
1589	Jesse Jackson	(1941–)	C G	Black civil rights leader
1707	Thurgood Marshall	(1908–1993)	C G	First African American Supreme Court Justice
3481	Coretta Scott King	(1927–2006)	C G	MLK widow, civil rights leader
4700	James Baldwin	(1924–1987)	C G	Black novelist (*Go Tell It on the Mountain*)
6144	A. Philip Randolph	(1889–1979)	C G	African American union / civil rights leader
6589	Al Sharpton	(1954–)	C G	Black activist and talk show host
6690	Medgar Evers	(1925–1963)	C G	Slain black civil rights activist

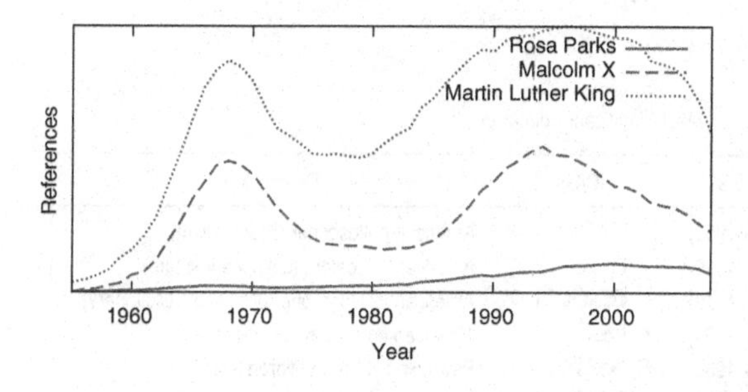

The book Ngrams data show that their reputations have moved roughly in parallel since the mid-1950s: rising rapidly to a peak a few years following their respective assassinations, sharply declining in the 1970s to set the stage for a resurgence in the 1990s. The first peak reflects the power of their movements as contemporary events, the later when the history of the movement began to be written. By contrast, the reference frequency for *Rosa Parks (1913–2005)* [964] has increased steadily over this period. We project that Malcolm's mindshare will decline with respect to King and Parks, as the story of the movement becomes simplified with time and distance.

Jesse Jackson (1941–) [1589] became the most prominent black leader after Dr. King's death, but it is not clear that he will ever have a meaningful successor. The emergence of black president *Barack Obama (1961–)* [111] marked a transition into the political mainstream that seems difficult to reverse. We cannot identify any major American figure today who purports to represent a particular cultural or ethnic group. This seems to mark a maturation of our political system, where constituencies are no longer based on single-dimensional categories of color, ethnicity, or gender.

15.6 The Education Movement

The history of education in America is one of growth and increasing access, from the little, one-room schoolhouse to the comprehensive public and private schools of today.

Several of the most significant educators were theorists with novel ideas of how children learn. *John Dewey (1859–1952)* [417] was the leading proponent of Progressive education, the notion that schools served to teach how to live, in addition to general knowledge and skills. *Maria Montessori (1870–1952)* [3332] emphasized sensory exploration and manipulatives, inspiring the Montessori schools that remain popular today. *Jean Piaget (1896–1980)* [1590] was a developmental psychologist who studied how children learn.

Most Significant Educators

Sig	Person	Dates	C/G	Description
417	John Dewey	(1859–1952)	C G	American philosopher (Pragmatism)
525	Booker T. Washington	(1856–1915)	C G	American educator (Tuskegee Institute)
1132	Noah Webster	(1758–1843)	C G	American lexicographer (Webster's Dictionary)
1193	Horace Mann	(1796–1859)	C G	American education reformer
1590	Jean Piaget	(1896–1980)	C G	Psychologist (how children learn)
3332	Maria Montessori	(1870–1952)	C G	Italian educator (Montessori schools)
5756	Anne Sullivan Macy	(1866–1936)	C G	Instructor of Helen Keller
6451	Catharine Beecher	(1800–1878)	C G	American educator (sister of Harriet and Henry)
6779	Daniel Coit Gilman	(1831–1908)	C G	American educator (president of Johns Hopkins)
7113	Thomas Hopkins Gallaudet	(1787–1851)	C G	Inventor (American Sign Language)

Others worked to widen access to education. *Horace Mann (1796–1859)* [1193] created a state-wide system of teaching in Massachusetts, with the important principle of progressing students through grades based on age. This marked the end of the one-room schoolhouse. *Booker T. Washington's (1856–1915)* [525] Tuskegee Institute provided new educational opportunities for Southern blacks after the Civil War. *Daniel Coit Gilman (1831–1908)* [6779] served 25 years as the founding president of Johns Hopkins University, and deserves credit for establishing the modern model of postgraduate education in the United States.

Several college presidents built substantial reputations in other fields. For example, President *Dwight D. Eisenhower (1890–1969)* [110] served as president of Columbia University before serving as president of the United States. We have removed these figures from our rankings of educators here.

APPENDIX A

Ranking Methodology

Here we provide greater technical detail concerning our historical significance ranking methods. We will discuss (1) exactly what data we used, (2) how the data was normalized, and (3) the statistical methodology we used to develop our rankings. Interested readers should begin with the high-level overview of our methods that we presented in Chapter 2.

A.1 Feature Set

We used the October 11, 2010 Wikipedia release (`http://dumps.wikimedia.org/enwiki/20101011/`) as the basis for our analysis. We extracted six input variables from this distribution and other available datasets:

- *Page Rank (NPR and PPR)* – Wikipedia is a hyperlinked document, defining a network with vertices corresponding to articles and directed edges (x, y), meaning that article x references article y. The well-established PageRank measure of network centrality [Brin and Page, 1998] computes a significance score based on the number and strength of the in-links to each node. We compute two forms of PageRank, based on two different graphs derived from Wikipedia. The first contains all Wikipedia pages, while the second consists only of the pages corresponding to people. Page annotations were determined using Freebase (`http://www.freebase.com/`), a collaborative knowledge database. We employed the Cloud9 map-reduce library (`https://github.com/lintool/Cloud9`) to compute Page-Rank.

- *Page Hits (PH)* – Web logging data reveal how often each Wikipedia page is viewed. More famous/significant entities should have their pages read more frequently. We analyzed six months of log data, collected immediately prior to the date of our Wikipedia release. In order to reduce the large monthly variance in readership owing to news events, we used the median monthly frequency as our measure of page hits.
- *Number of Page Revisions (NR)* – The collaborative model that underlies Wikipedia permits thousands of authors to contribute their knowledge to every single article. More famous/important people would be expected to have more refined articles than lesser personages. We considered a variety of measures of page edit frequency, including the number of distinct editors, the number of changes, and the number of edit reversions. All correlated quite strongly with our primary measure, which is simply the number of distinct revisions. The residual components of these variables, after accounting for the number of distinct revisions, also proved uninformative.
- *Article Length (AL)* – The length of a given entity's Wikipedia page (as measured in words) provides a natural measure of fame: more significant entities typically have longer articles written about them. For our purposes, we use the length of the body of the article, removing references and tables that appear in the document. This eliminates certain biases, such as overvaluing sports figures and performing artists with excessively long tables of events or recordings.
- *News Frequency (NF)* – Famous/important people are expected to appear regularly in the news. We used data derived from an independent analysis of more than 500 U.S. daily newspapers from between November 2004 and October 2010, totaling in excess of one terabyte of text. This analysis was performed using our *Lydia* news/blog analysis system [Bautin et al., 2010], deriving frequencies of mention that measure the media coverage of these entities.

  This proved to be our least informative variable, which is why it was not discussed in Chapter 2. Part of the difficulty arose from resolving which proper reference names are associated with a given person, a problem we

	FPR	PPR	PH	NR	AL	NF
FPR	1.00	0.68	0.67	0.63	0.27	0.30
PPR		1.00	0.50	0.49	0.24	0.29
PH			1.00	0.71	0.27	0.37
NR				1.00	0.34	0.36
AL					1.00	0.20
NF						1.00

FIGURE A.1. Correlation matrix of the major input variables to our significance analysis methods. All pairs show substantial correlation, yet none are so high as to be redundant.

discussed in Section 4.3 with respect to Google Ngrams. Also, contemporary people are disproportionally represented in the news relative to significant figures of earlier times. This variable ultimately contributes very little to our final significance scores.

We also considered several other measures as potential input variables for our analysis, but ultimately rejected them. One was the estimated number of web pages returned by Google and Bing search engine queries. We found these counts largely meaningless, because their algorithms estimate name frequency from the individual word components making up the name. These counts correlated poorly with the other variables. Other properties of the Wikipedia graph, like the out-degree of each entity's page, were subsumed by PageRank. Sentiment analysis from our news dataset also proved to be largely uninformative as to significance.

Figure A.1 presents the pairwise correlation among the six variables we employed in our analysis, after employing the normalization procedures we will describe in the next section.

A.1.1 NORMALIZATION OF VARIABLES

None of our six variables were normally distributed. Most resembled either half-normal (that is, one tail of a normal distribution), or log half-normal (where taking the logarithm of the variable yields a half-normal distribution) distributions. Some had long, power law tails.

We sought to convert these variables to normally distributed standard scores for analysis. However, standard z-scores did not result in normally distributed data, and produced oddities with extreme values when applied to the power law distributions.

We instead performed a normal re-expression of each input variable, a commonly used technique for the analysis of ordinal variables [Mosteller and Tukey, 1977]. Each variable is rank ordered, and then a value is converted to a normal score based on its rank. The resulting standardized variables are, by definition, as normally distributed as possible given the discretization of the input, and prove more amenable to subsequent analysis.

Processing artifacts occasionally resulted in missing data values for a given person. When one variable was missing, its value was imputed using a linear regression model, based on the other five variables. When two or more variables were missing, the person was discarded from subsequent analysis.

A.2 Ranking Methodology

Exploratory factor analysis [Harman, 1976; Pearson, 1901; Spearman, 1904] is a statistical methodology related to principal component analysis [Eckart and Young, 1936; Hotelling, 1933], intended to explain observed variables by a linear combination of unobserved (latent) variables. Exploratory factor analysis can be contrasted with confirmatory factor analysis [Joreskog, 1969], in which an underlying structure of latent variables is assumed a priori and the accuracy (or fit) of the model then computed.

Specifically, we attempted to derive the latent factors that explain the variance in our six variables of interest: both types of PageRank, page hits, article length, number of article revisions, and the number of references in our news dataset. We conducted exploratory factor analysis using the statistical package *R*, and found that the majority (59%) of the variance in our six variables could be explained by two factors. Both factors explained roughly equal proportions of the variance (31% and 28%), meaning that these latent variables are of approximately equal importance.

Factor loadings reflect the extent to which each variable is associated with a factor. A factor loading can be thought of as analogous to a correlation coefficient. The square of the factor loading reflects the proportion of the variance of that variable that can be explained by the factor. The first factor loads highly on page hits and article revisions, while the second factor

Variable	F1 Loading	F2 Loading
Full PageRank	0.403	0.912
Person PageRank	0.401	0.630
Page Hits	0.697	0.485
# Revisions	0.829	0.395
Article Length	0.360	0.184
News Hits	0.376	0.167

FIGURE A.2. Loadings for the largest two factors extracted from the significance data.

loads highly on the two forms of PageRank. Neither substantially loads on article length or news volume, as shown in Figure A.2.

A factor score is the score of an observation (the input associated with a specific person) on a factor. The factor score of each individual can be estimated from the observed variables, in a variety of ways. We computed both factor scores for each person in our dataset using Bartlett maximum likelihood estimation [Bartlett, 1937]. The resulting factor scores are normally distributed.

Upon examining the individuals with high scores for each factor, it quickly becomes apparent that the first factor captures a notion of lurid fame, or "celebrity," while the second factor better captures a notion of substantive accomplishment, or "gravitas." Here we refer back to Section 2.3, where we show several well-known people with very high Celebrity and Gravitas factor scores.

Neither of these measures individually captured the sense of historical "significance" we were interested in measuring. However, the linear combination of these two factors yields a measure that proved relatively robust across domains, creating a unified representation of contemporary renown we call Fame. Because the Gravitas and Celebrity factors are both normally distributed, the linear combination of these two factors into our Fame score is also normally distributed.

A.3 The Decay of Fame

The need to compensate for the destructive nature of time on reputation becomes apparent on examination of our Fame measure, which wildly overstates the significance of contemporary figures such as *George W. Bush (1946–)* [1], *Barack Obama (1961–)* [3], *Bill Clinton (1946–)* [7], and

Rank	Raw Score	Person of Rank
1	7.48	Jesus
10	6.54	Thomas Jefferson
50	5.99	Oliver Cromwell
100	5.65	John Locke
200	5.33	Wilhelm II
500	4.85	Hatshepsut
1000	4.39	Georges Seurat
5000	3.30	Jerry Seinfeld
10000	2.80	Pau Gasol
50000	1.50	Catterino Cavos
100000	0.85	Sean O'Hagan
300000	–0.38	Sophie Blake
500000	–1.23	Guy U. Hardy
700000	–2.15	Masashi Kawakami
843790	–5.16	Sagusa Ryusei

FIGURE A.3. Representative mappings of raw significance score to rank, with the person occupying that rank

Michael Jackson (1958–2009) [9]: all presented here with their Fame rankings. Although all are important, none belong among the top ten figures in historical (as opposed to contemporary) significance.

Modeling the speed and extent to which reputation fades is essential to appropriately compare the historical significance of current individuals with those long dead. To correct for this, we built a model to systematically age contemporary figures, so as to predict what their reputation will be a given number of years from now.

There are two distinct processes at work here: first, the lapse from living memory inherent in the passage of generations, and second, a strictly contemporary bias from the recent authorship of Wikipedia. We correct for both factors in turn, to complete our model for quantifying historical significance.

In order to quantify the first effect, we turned to the Google Books Ngrams dataset [Michel et al., 2011], discussed in Chapter 4. This rich dataset covers a significant fraction of all books ever published in the English language, and is nearly a terabyte in size. This allows us to count the number of times each person, place, and thing appeared in English books during any given year, and provides an excellent means to track the appearance and decay of individual reputations across a two-century period.

We separated individuals into order-of-magnitude tiers, based on the relative frequency of references to them during their peak year, and then

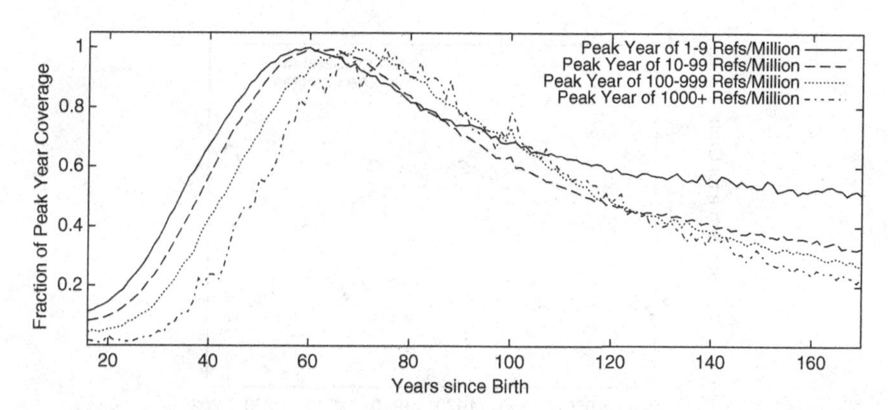

FIGURE A.4. Reputations peak roughly 70 years after birth. The peak is typically later the more famous a person is.

plotted the average trajectory of each group. These results are presented in Figure A.4. From this data, we fit a polynomial decay model for each tier of individuals.

This correspondence yields a decay schedule for each individual, which can be used to project their significance forward to the age of 170. We do not continue further beyond this, because insufficient reference data exist to evaluate a model on such a timescale. However, if the mechanisms of historical preservation involve living memory, we do not believe that the fame can continue to decay indefinitely into the future. Moreover, experiments with long-term decay did not further improve our rankings, as measured by the validation criteria discussed in Section 2.5.

Even after incorporating this reputational decay into our model, contemporary figures still appeared to be overvalued. As described in Section 2.4.2, the rate at which Wikipedia people are recorded as dying over the past 20 years far exceeds our baseline relative to human population, meaning more have been finding their way into Wikipedia than was the case for previous generations.

This recency effect exaggerates several variables we employ in our analysis for contemporary people, including page hits, revisions, and newspaper frequency. Entities of current interest must be further discounted in significance to compensate for this effect.

To do so, we extracted the dates mentioned in each individual's Wikipedia page. This provides us with an excellent measure of the prime

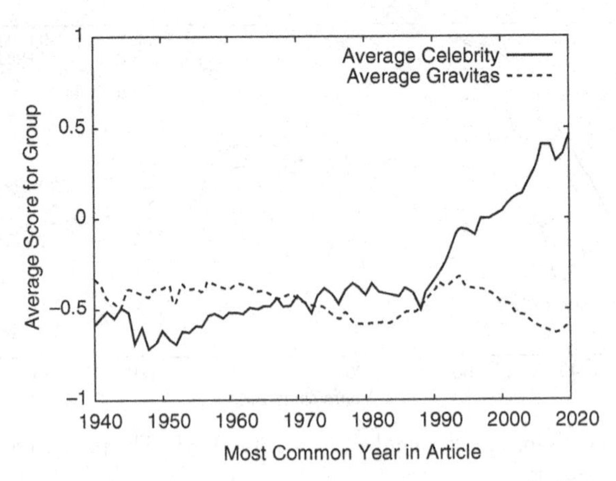

FIGURE A.5. The Wikipedia Effect on reputation. Contemporary figures have come to exhibit greater celebrity since the founding of Wikipedia.

of each individual's career. And indeed, when we examine entities active in recent years, we see a significant skew upward in overall celebrity. Figure A.5 plots the average celebrity and gravitas for all entities peaking each year.

The magnitude of this skew provides an estimate of the popularity over-inflation of recent figures. We subtracted out this skew from the decayed significance score for each person, based on their year of peak activity. This yields our final significance score. Figure A.3 shows how the overall significance rank, used throughout the book, corresponds to raw significance score values.

These corrections not only anecdotally improve the quality of rankings, but provide a substantial boost when compared to human-created rankings. Our evaluation procedure and results are discussed in Section 2.5.

APPENDIX B

Resources

B.1 Websites and Databases

We hope that some readers will be inspired by this book to perform further investigations into historical significance. We encourage you to visit our website at `http://www.whoisbigger.com`, where you can access our latest analysis for more than 800,000 different people. These rankings will be updated to reflect more recent editions of Wikipedia, including those of languages beyond English.

This website is still under active development, so you will have to visit to see exactly what information we present on each individual. But at the time of this writing, we include rosters of people with comparable significance, celebrity, and gravitas, to help you put their achievements into perspective. We also provide an interface to rank individuals within hundreds of thousands of Wikipedia categories, so you can see how they stack up in comparison with their peers.

We also encourage the reader to check out the Google books Ngram viewer, which Google makes available at `http://books.google.com/ngrams`. Although we have developed our own Ngram visualizations, which we make available at `http://www.whoisbigger.com`, the Google site has far more comprehensive data. This includes more general vocabularies, annotation by parts of speech, and even Ngrams in multiple languages. It is a fabulous resource to explore.

B.2 Birth/Death Calendars

In Section 7.2 of Chapter 7, we showed that life outcomes can be affected (at least to a minor extent) by the specifics of one's calendar birth date. It is natural to be curious about the historical figures which share your birth date: perhaps this kinship may reflect something of our own prospects for the future.

For fun, we have computed the most significant people born on each date of the year, and placed them here, in Figures B.1 and B.2. Take a peek at the calendars to find which major historical figure was born under your star. Steve is proud to share his birth date with *Franklin D. Roosevelt (1882–1945)* [43], while Charles has something nice in common with *Bill Gates (1955–)* [904].

Here are a few amusing curiosities to note as you peruse the calendars:

- *Albert Einstein (1879–1955)* [19] is the most significant person ever born on March 14, a festive date for math and science known as *Pi Day* because it is written as 3/14.
- The rarest birth date on the calendar, by far, is the leap day February 29. Yet its representative is adequately distinguished: the classical composer *Gioachino Rossini (1792–1868)* [390]. He is most famous for the Lone Ranger's *William Tell Overture.*
- The least significant person to command a birth date appears to be the Italian poet *Giacomo Leopardi (1798–1837)* [2909], born on June 29. Other possible candidates for this date might have been singer *Nelson Eddy (1901–1967)* [13795] and author/aviator *Antoine de Saint-Exupéry (1900–1944)* [6148]. Leopardi presents a tempting target, so if you were born on this date, you now have a new life goal to shoot for.
- The most famous birthday of all is Christmas, yet historical and calendrical uncertainties make it impossible to assign December 25 as the true birth date of *Jesus (7 B.C.–A.D. 30)* [1]. Instead, the most significant person born on that date was *Muhammad Ali Jinnah (1876–1948)* [542], the founder of Pakistan.

We have also constructed tables of the most significant historical figures by death date. In a nice bit of symmetry, we note that *Charles I (1600–1649)* [39] was beheaded on Steve's birthday. The most significant person

January

Date	Person	Sig
1-01	Huldrych Zwingli	484
1-02	James Wolfe	687
1-03	J. R. R. Tolkien	192
1-04	Isaac Newton	21
1-05	Shah Jahan	677
1-06	Richard II	527
1-07	Millard Fillmore	446
1-08	Elvis Presley	69
1-09	Richard Nixon	82
1-10	Rod Stewart	1646
1-11	John A. Macdonald	159
1-12	Edmund Burke	381
1-13	Salmon P. Chase	1164
1-14	Benedict Arnold	222
1-15	Martin Luther King	221
1-16	J. C. Breckinridge	2071
1-17	Benjamin Franklin	35
1-18	Montesquieu	268
1-19	Edgar Allan Poe	54
1-20	Charles III	707
1-21	Stonewall Jackson	283
1-22	Francis Bacon	81
1-23	John Hancock	616
1-24	Frederick the Great	163
1-25	Robert Burns	317
1-26	Douglas MacArthur	315
1-27	Wolfgang A. Mozart	24
1-28	Thomas Aquinas	90
1-29	Thomas Paine	113
1-30	F. D. Roosevelt	43
1-31	Franz Schubert	198

February

Date	Person	Sig
2-01	Isaac Asimov	916
2-02	Charles Darwin	12
2-03	Felix Mendelssohn	199
2-04	Charles Lindbergh	736
2-05	Robert Peel	584
2-06	Ronald Reagan	32
2-07	Charles Dickens	33
2-08	William T. Sherman	233
2-09	William H. Harrison	288
2-10	Bertolt Brecht	933
2-11	Thomas Edison	40
2-12	Abraham Lincoln	5
2-13	Thomas R. Malthus	265
2-14	Babur	587
2-15	Susan B. Anthony	432
2-16	Ernst Haeckel	630
2-17	Frederick Douglass	402
2-18	Mary I	126
2-19	Nicolaus Copernicus	74
2-20	Rihanna	1185
2-21	John Henry Newman	534
2-22	George Washington	6
2-23	George F. Handel	114
2-24	Charles V	84
2-25	P. Renoir	549
2-26	Victor Hugo	208
2-27	Constantine the G.	67
2-28	Michel de Montaigne	791
2-29	Gioachino Rossini	390

March

Date	Person	Sig
3-01	Frédéric Chopin	195
3-02	Pope Pius XII	294
3-03	Alexander G. Bell	106
3-04	Antonio Vivaldi	262
3-05	Henry II	170
3-06	Michelangelo	86
3-07	Maurice Ravel	812
3-08	Oliver W. Holmes	885
3-09	Amerigo Vespucci	407
3-10	Ferdinand II	454
3-11	Harold Wilson	1337
3-12	George Berkeley	776
3-13	Joseph Priestley	330
3-14	Albert Einstein	19
3-15	Andrew Jackson	66
3-16	James Madison	51
3-17	Toyotomi Hideyoshi	515
3-18	Grover Cleveland	98
3-19	David Livingstone	635
3-20	Henrik Ibsen	403
3-21	Johann S. Bach	48
3-22	Maximilian I	609
3-23	P. Laplace	659
3-24	William Morris	302
3-25	Elton John	1042
3-26	Robert Frost	644
3-27	Mariah Carey	636
3-28	Teresa of Ávila	1216
3-29	John Tyler	341
3-30	Vincent van Gogh	73
3-31	René Descartes	92

April

Date	Person	Sig
4-01	Otto von Bismarck	97
4-02	Charlemagne	22
4-03	Washington Irving	433
4-04	Muddy Waters	2374
4-05	Thomas Hobbes	167
4-06	Raphael	140
4-07	William Wordsworth	182
4-08	Timur	290
4-09	Isambard K. Brunel	371
4-10	Matthew C. Perry	645
4-11	Septimius Severus	662
4-12	Henry Clay	252
4-13	Thomas Jefferson	10
4-14	Christiaan Huygens	715
4-15	Leonardo da Vinci	29
4-16	Charlie Chaplin	295
4-17	Cap Anson	2342
4-18	David Ricardo	561
4-19	Roger Sherman	1029
4-20	Adolf Hitler	7
4-21	Elizabeth II	132
4-22	Immanuel Kant	59
4-23	William Shakespeare	4
4-24	William the Silent	690
4-25	Oliver Cromwell	50
4-26	David Hume	128
4-27	Ulysses S. Grant	28
4-28	James Monroe	220
4-29	Alexander II	385
4-30	Carl F. Gauss	207

May

Date	Person	Sig
5-01	Arthur Wellesley	174
5-02	Catherine the Great	108
5-03	Niccolò Machiavelli	168
5-04	Thomas Henry Huxley	365
5-05	Karl Marx	14
5-06	Sigmund Freud	44
5-07	P. I. Tchaikovsky	63
5-08	Harry S. Truman	94
5-09	John Brown	304
5-10	John Wilkes Booth	451
5-11	Salvador Dalí	1021
5-12	F. Nightingale	270
5-13	Pope Pius IX	334
5-14	Charles IV	826
5-15	Claudio Monteverdi	704
5-16	Janet Jackson	559
5-17	Edward Jenner	746
5-18	Pope John Paul II	91
5-19	Ho Chi Minh	779
5-20	John Stuart Mill	201
5-21	Philip II	87
5-22	Richard Wagner	62
5-23	Carl Linnaeus	31
5-24	Queen Victoria	16
5-25	Ralph Waldo Emerson	122
5-26	John Wayne	459
5-27	Ibn Khaldun	437
5-28	George I	289
5-29	John F. Kennedy	71
5-30	Benny Goodman	1800
5-31	Walt Whitman	160

June

Date	Person	Sig
6-01	Dante Alighieri	96
6-02	Thomas Hardy	378
6-03	Jefferson Davis	188
6-04	George III	58
6-05	John Maynard Keynes	469
6-06	Alexander Pushkin	420
6-07	Paul Gauguin	540
6-08	Frank Lloyd Wright	245
6-09	Peter the Great	107
6-10	Prince Philip	909
6-11	Ben Jonson	421
6-12	George H. W. Bush	363
6-13	Winfield Scott	787
6-14	Che Guevara	457
6-15	Edvard Grieg	711
6-16	Adam Smith	56
6-17	Edward I	155
6-18	Paul McCartney	447
6-19	Blaise Pascal	157
6-20	Jacques Offenbach	968
6-21	Jean-Paul Sartre	1003
6-22	George Vancouver	725
6-23	Oda Nobunaga	458
6-24	Robert Dudley	474
6-25	George Orwell	342
6-26	William Thomson	450
6-27	Charles S. Parnell	836
6-28	Henry VIII	11
6-29	Giacomo Leopardi	2909
6-30	Joseph D. Hooker	1151

FIGURE B.1. The most significant figure born on each day of the year, January through June

July

Date	Person	Sig
7-01	Diana	837
7-02	Thomas Cranmer	455
7-03	Franz Kafka	724
7-04	Nathaniel Hawthorne	227
7-05	Cecil Rhodes	706
7-06	George W. Bush	36
7-07	Gustav Mahler	364
7-08	John D. Rockefeller	172
7-09	Franz Boas	1053
7-10	Nikola Tesla	93
7-11	John Quincy Adams	135
7-12	Henry David Thoreau	131
7-13	John Dee	906
7-14	Gerald Ford	230
7-15	Rembrandt	189
7-16	Joshua Reynolds	1143
7-17	Ismail I	1358
7-18	Robert Hooke	216
7-19	Edgar Degas	422
7-20	Gregor Mendel	250
7-21	Ernest Hemingway	248
7-22	Philip I	1472
7-23	Bal Gangadhar Tilak	1791
7-24	Simón Bolívar	448
7-25	Arthur Balfour	1036
7-26	George Bernard Shaw	213
7-27	Triple H	1596
7-28	Marcel Duchamp	1245
7-29	Benito Mussolini	101
7-30	Henry Ford	148
7-31	Milton Friedman	700

August

Date	Person	Sig
8-01	Herman Melville	251
8-02	Francis Scott Key	1050
8-03	Stanley Baldwin	1416
8-04	Percy B. Shelley	329
8-05	Guy de Maupassant	1160
8-06	Alfred	301
8-07	Nathanael Greene	1025
8-08	Roger Federer	743
8-09	Whitney Houston	1191
8-10	Herbert Hoover	183
8-11	Hulk Hogan	1353
8-12	George IV	318
8-13	Alfred Hitchcock	439
8-14	Pope Pius VII	1441
8-15	Napoleon	2
8-16	Madonna	121
8-17	Davy Crockett	606
8-18	Franz Joseph I	398
8-19	Bill Clinton	115
8-20	Benjamin Harrison	339
8-21	William IV	377
8-22	Claude Debussy	380
8-23	Louis XVI	83
8-24	William Wilberforce	387
8-25	Ivan the Terrible	297
8-26	Antoine Lavoisier	276
8-27	Georg W. F. Hegel	142
8-28	J. W. von Goethe	88
8-29	John Locke	100
8-30	Mary Shelley	223
8-31	Caligula	210

September

Date	Person	Sig
9-01	Johann Pachelbel	1369
9-02	Liliuokalani	1088
9-03	Louis Sullivan	1719
9-04	Anton Bruckner	854
9-05	Louis XIV	26
9-06	Gilbert du Motier	217
9-07	Elizabeth I	13
9-08	Richard I	120
9-09	Leo Tolstoy	196
9-10	Charles S. Peirce	225
9-11	D. H. Lawrence	1166
9-12	Francis I	351
9-13	John J. Pershing	764
9-14	A. von Humboldt	520
9-15	Marco Polo	147
9-16	Henry V	497
9-17	Bernhard Riemann	1091
9-18	Samuel Johnson	141
9-19	Henry III	920
9-20	Upton Sinclair	1150
9-21	H. G. Wells	249
9-22	Michael Faraday	175
9-23	Kublai Khan	346
9-24	John Marshall	401
9-25	Qianlong Emperor	856
9-26	T. S. Eliot	436
9-27	Louis XIII	426
9-28	Georges Clemenceau	1085
9-29	Horatio Nelson	219
9-30	Jacques Necker	2355

October

Date	Person	Sig
10-01	Jimmy Carter	462
10-02	Mohandas K. Gandhi	46
10-03	Stevie Ray Vaughan	2312
10-04	Rutherford B. Hayes	325
10-05	Chester A. Arthur	499
10-06	Louis Philippe I	482
10-07	Heinrich Himmler	671
10-08	Jesse Jackson	1589
10-09	John Lennon	162
10-10	Giuseppe Verdi	228
10-11	Eleanor Roosevelt	517
10-12	Edward VI	206
10-13	Margaret Thatcher	271
10-14	D. D. Eisenhower	110
10-15	Friedrich Nietzsche	42
10-16	Oscar Wilde	77
10-17	Eminem	823
10-18	Pierre Trudeau	870
10-19	Thomas Browne	1568
10-20	John Dewey	417
10-21	Samuel T. Coleridge	280
10-22	Franz Liszt	116
10-23	Pelé	1418
10-24	Domitian	3035
10-25	Pablo Picasso	171
10-26	Hillary R. Clinton	575
10-27	Theodore Roosevelt	23
10-28	Bill Gates	904
10-29	Joseph Goebbels	810
10-30	John Adams	61
10-31	John Keats	305

November

Date	Person	Sig
11-01	Alfred Wegener	1146
11-02	Marie Antoinette	125
11-03	Aurangzeb	332
11-04	Sean Combs	1708
11-05	Eugene V. Debs	1340
11-06	Suleiman the M.	354
11-07	Leon Trotsky	278
11-08	Bram Stoker	651
11-09	Benjamin Banneker	1797
11-10	George II	340
11-11	Fyodor Dostoyevsky	244
11-12	Auguste Rodin	574
11-13	Augustine of Hippo	72
11-14	William III	143
11-15	William Herschel	719
11-16	John Bright	2678
11-17	Louis XVIII	491
11-18	W. S. Gilbert	744
11-19	Charles I	39
11-20	Alexander Hamilton	45
11-21	Voltaire	64
11-22	Charles de Gaulle	396
11-23	Franklin Pierce	427
11-24	Zachary Taylor	299
11-25	Andrew Carnegie	104
11-26	Tina Turner	1189
11-27	Jimi Hendrix	338
11-28	William Blake	102
11-29	C. S. Lewis	327
11-30	Winston Churchill	37

December

Date	Person	Sig
12-01	A. of Denmark	878
12-02	Britney Spears	689
12-03	George B. McClellan	554
12-04	Francisco Franco	291
12-05	Martin Van Buren	277
12-06	Henry VI	599
12-07	Gian L. Bernini	680
12-08	Mary	127
12-09	John Milton	165
12-10	Emily Dickinson	194
12-11	Hector Berlioz	419
12-12	Frank Sinatra	357
12-13	Henry IV	238
12-14	Tycho Brahe	362
12-15	Nero	656
12-16	Jane Austen	139
12-17	William L. M. King	668
12-18	Joseph Stalin	18
12-19	Philip V	749
12-20	Robert Menzies	1514
12-21	Thomas Becket	264
12-22	Diocletian	181
12-23	Alexander I	404
12-24	John	136
12-25	Muhammad Ali Jinnah	542
12-26	Mao Zedong	151
12-27	Louis Pasteur	112
12-28	Woodrow Wilson	47
12-29	Andrew Johnson	105
12-30	Rudyard Kipling	259
12-31	Jacques Cartier	306

FIGURE B.2. The most significant figure born on each day of the year, July through December

January

Date	Person	Sig
01-01	Basil of Caesarea	1044
01-02	James Longstreet	1105
01-03	Josiah Wedgwood	1720
01-04	T. S. Eliot	436
01-05	Calvin Coolidge	370
01-06	Theodore Roosevelt	23
01-07	Nikola Tesla	93
01-08	Marco Polo	147
01-09	Victor Emmanuel II	897
01-10	Carl Linnaeus	31
01-11	Thomas Hardy	378
01-12	Maximilian I	609
01-13	James Joyce	406
01-14	George Berkeley	776
01-15	Vasco N. d. Balboa	1136
01-16	Edward Gibbon	573
01-17	Rutherford B. Hayes	325
01-18	Rudyard Kipling	259
01-19	P. Proudhon	1076
01-20	John Ruskin	550
01-21	Vladimir Lenin	75
01-22	Queen Victoria	16
01-23	William Pitt the Y.	409
01-24	Winston Churchill	37
01-25	Al Capone	646
01-26	Edward Jenner	746
01-27	Ali	89
01-28	Henry VIII	11
01-29	George III	58
01-30	Charles I	39
01-31	Guy Fawkes	392

February

Date	Person	Sig
02-01	Mary Shelley	223
02-02	Bertrand Russell	253
02-03	Woodrow Wilson	47
02-04	Septimius Severus	662
02-05	Thomas Carlyle	733
02-06	Charles II	78
02-07	Pope Pius IX	334
02-08	Peter the Great	107
02-09	Fyodor Dostoyevsky	244
02-10	Montesquieu	268
02-11	René Descartes	92
02-12	Immanuel Kant	59
02-13	Richard Wagner	62
02-14	James Cook	60
02-15	H. H. Asquith	782
02-16	Charles Theodore	4959
02-17	Wilfrid Laurier	543
02-18	Michelangelo	86
02-19	Timur	290
02-20	Frederick Douglass	402
02-21	Baruch Spinoza	309
02-22	Amerigo Vespucci	407
02-23	John Quincy Adams	135
02-24	Robert Fulton	954
02-25	Christopher Wren	600
02-26	Roger II	2591
02-27	Ivan Pavlov	946
02-28	Henry James	408
02-29	Ludwig I	2174

March

Date	Person	Sig
03-01	Leopold II	1365
03-02	John Wesley	359
03-03	Robert Hooke	216
03-04	Saladin	138
03-05	Joseph Stalin	18
03-06	Ayn Rand	486
03-07	Thomas Aquinas	90
03-08	William Howard Taft	153
03-09	William I	769
03-10	Harriet Tubman	1093
03-11	Charles Sumner	1094
03-12	Pope Gregory I	394
03-13	Benjamin Harrison	339
03-14	Karl Marx	14
03-15	H. P. Lovecraft	713
03-16	Tiberius	272
03-17	William II	2711
03-18	Ivan the Terrible	297
03-19	Ibn Khaldun	437
03-20	Henry IV	585
03-21	Pocahontas	428
03-22	J. W. von Goethe	88
03-23	Paul I	828
03-24	Elizabeth I	13
03-25	Claude Debussy	380
03-26	Walt Whitman	160
03-27	M. C. Escher	1749
03-28	D. D. Eisenhower	110
03-29	Robert Falcon Scott	808
03-30	Rudolf Steiner	775
03-31	Isaac Newton	21

April

Date	Person	Sig
04-01	E. of Aquitaine	476
04-02	Pope John Paul II	91
04-03	Johannes Brahms	204
04-04	Martin Luther King	221
04-05	Douglas MacArthur	315
04-06	Richard I	120
04-07	Henry Ford	148
04-08	Pablo Picasso	171
04-09	Francis Bacon	81
04-10	Joseph L. Lagrange	683
04-11	Llywelyn the Great	1714
04-12	F. D. Roosevelt	43
04-13	Jean de La Fontaine	1447
04-14	George F. Handel	114
04-15	Abraham Lincoln	5
04-16	Francisco Goya	366
04-17	Benjamin Franklin	35
04-18	Albert Einstein	19
04-19	Charles Darwin	12
04-20	Bram Stoker	651
04-21	Mark Twain	53
04-22	Richard Nixon	82
04-23	William Shakespeare	4
04-24	Daniel Defoe	551
04-25	Anders Celsius	1510
04-26	John Wilkes Booth	451
04-27	Ralph Waldo Emerson	122
04-28	Benito Mussolini	101
04-29	Alfred Hitchcock	439
04-30	Adolf Hitler	7

May

Date	Person	Sig
05-01	David Livingstone	635
05-02	Leonardo da Vinci	29
05-03	Mehmed II	496
05-04	Josip Broz Tito	562
05-05	Napoleon	2
05-06	Henry David Thoreau	131
05-07	Otto I	762
05-08	John Stuart Mill	201
05-09	Friedrich Schiller	564
05-10	Stonewall Jackson	283
05-11	William Pitt	881
05-12	John Dryden	806
05-13	Georges Cuvier	772
05-14	Henry IV	238
05-15	Emily Dickinson	194
05-16	Joseph Fourier	1542
05-17	John Jay	411
05-18	Gustav Mahler	364
05-19	Anne Boleyn	154
05-20	C. Columbus	20
05-21	Hernando de Soto	473
05-22	Constantine the G.	67
05-23	John D. Rockefeller	172
05-24	Nicolaus Copernicus	74
05-25	Pope Gregory VII	774
05-26	Bede	308
05-27	John Calvin	99
05-28	Noah Webster	1132
05-29	W. S. Gilbert	744
05-30	Voltaire	64
05-31	Joseph Haydn	129

June

Date	Person	Sig
06-01	James Buchanan	237
06-02	Giuseppe Garibaldi	352
06-03	William Harvey	468
06-04	Wilhelm II	200
06-05	Ronald Reagan	32
06-06	John A. Macdonald	159
06-07	Robert the Bruce	470
06-08	Muhammad	3
06-09	Charles Dickens	33
06-10	Frederick I	456
06-11	George I	289
06-12	William C. Bryant	1779
06-13	Miyamoto Musashi	940
06-14	Benedict Arnold	222
06-15	James K. Polk	240
06-16	John Churchill	879
06-17	John III Sobieski	1157
06-18	Roald Amundsen	1109
06-19	Maximilian I	798
06-20	William IV	377
06-21	N. Machiavelli	168
06-22	Judy Garland	1031
06-23	Vespasian	588
06-24	Grover Cleveland	98
06-25	Michael Jackson	180
06-26	George IV	318
06-27	Giorgio Vasari	902
06-28	James Madison	51
06-29	Henry Clay	252
06-30	Moctezuma II	928

FIGURE B.3. The most significant figure who died on each day of the year, January through June

July

Date	Person	Sig
07-01	Harriet B. Stowe	449
07-02	J. Rousseau	80
07-03	Jim Morrison	1311
07-04	Thomas Jefferson	10
07-05	Stamford Raffles	1634
07-06	Henry II	170
07-07	Edward I	155
07-08	Percy B. Shelley	329
07-09	Zachary Taylor	299
07-10	Hadrian	214
07-11	George Gershwin	792
07-12	Alexander Hamilton	45
07-13	Rashi	658
07-14	Philip II	655
07-15	Anton Chekhov	523
07-16	Pope Innocent III	598
07-17	Adam Smith	56
07-18	Jane Austen	139
07-19	Petrarch	326
07-20	Pope Leo XIII	424
07-21	Robert Burns	317
07-22	William L. M. King	668
07-23	Ulysses S. Grant	28
07-24	Martin Van Buren	277
07-25	Samuel T. Coleridge	280
07-26	Sam Houston	498
07-27	Bob Hope	1098
07-28	Johann S. Bach	48
07-29	Vincent van Gogh	73
07-30	Otto von Bismarck	97
07-31	Andrew Johnson	105

August

Date	Person	Sig
08-01	Anne	256
08-02	Alexander G. Bell	106
08-03	Joseph Conrad	544
08-04	Hans C. Andersen	435
08-05	Marilyn Monroe	347
08-06	Ben Jonson	421
08-07	Rabindranath Tagore	397
08-08	Trajan	298
08-09	Ernst Haeckel	630
08-10	Leoš Janáček	3244
08-11	Andrew Carnegie	104
08-12	William Blake	102
08-13	H. G. Wells	249
08-14	William R. Hearst	565
08-15	Stephen I	1280
08-16	Elvis Presley	69
08-17	Frederick the Great	163
08-18	Pope Alexander VI	625
08-19	Augustus	30
08-20	Pope Pius X	571
08-21	Leon Trotsky	278
08-22	Richard III	344
08-23	William Wallace	296
08-24	Henry VII	2097
08-25	Friedrich Nietzsche	42
08-26	William James	442
08-27	Titian	319
08-28	Augustine of Hippo	72
08-29	Brigham Young	471
08-30	J. J. Thomson	514
08-31	Henry V	497

September

Date	Person	Sig
09-01	Louis XIV	26
09-02	J. R. R. Tolkien	192
09-03	Oliver Cromwell	50
09-04	Robert Dudley	474
09-05	Mother Teresa	820
09-06	Suleiman the M.	354
09-07	John G. Whittier	1698
09-08	Richard Strauss	586
09-09	William the C.	70
09-10	Mary Wollstonecraft	654
09-11	Nikita Khrushchev	416
09-12	Johnny Cash	900
09-13	Philip II	87
09-14	Dante Alighieri	96
09-15	Isambard K. Brunel	371
09-16	James II	137
09-17	Dred Scott	883
09-18	Leonhard Euler	185
09-19	James A. Garfield	285
09-20	Jean Sibelius	1309
09-21	Charles V	84
09-22	Guru Nanak Dev	414
09-23	Sigmund Freud	44
09-24	Dr. Seuss	893
09-25	Pope Clement VII	988
09-26	Daniel Boone	501
09-27	Edgar Degas	422
09-28	Louis Pasteur	112
09-29	Gustav I	1063
09-30	Jerome	320

October

Date	Person	Sig
10-01	Pierre Corneille	2024
10-02	Samuel Adams	467
10-03	Francis of Assisi	275
10-04	Rembrandt	189
10-05	Charles Cornwallis	481
10-06	Alfred	301
10-07	Edgar Allan Poe	54
10-08	Franklin Pierce	427
10-09	Pope Pius XII	294
10-10	Orson Welles	729
10-11	Huldrych Zwingli	484
10-12	Robert E. Lee	76
10-13	Claudius	211
10-14	Harold Godwinson	570
10-15	Andreas Vesalius	754
10-16	Marie Antoinette	125
10-17	Frédéric Chopin	195
10-18	Thomas Edison	40
10-19	John	136
10-20	Herbert Hoover	183
10-21	Horatio Nelson	219
10-22	Paul Cézanne	389
10-23	W. G. Grace	849
10-24	Tycho Brahe	362
10-25	Geoffrey Chaucer	173
10-26	Alfred the Great	267
10-27	Akbar	186
10-28	John Locke	100
10-29	Walter Raleigh	313
10-30	Charles Tupper	1270
10-31	Harry Houdini	621

November

Date	Person	Sig
11-01	Ezra Pound	507
11-02	George Bernard Shaw	213
11-03	Umar	234
11-04	Felix Mendelssohn	199
11-05	Casimir III the G.	1563
11-06	P. I. Tchaikovsky	63
11-07	Eleanor Roosevelt	517
11-08	John Milton	165
11-09	Charles de Gaulle	396
11-10	Mustafa K. Atatürk	360
11-11	Søren Kierkegaard	400
11-12	Cnut the Great	355
11-13	Gioachino Rossini	390
11-14	Georg W. F. Hegel	142
11-15	Johannes Kepler	156
11-16	Louis Riel	463
11-17	Mary I	126
11-18	Chester A. Arthur	499
11-19	Franz Schubert	198
11-20	Leo Tolstoy	196
11-21	Franz Joseph I	398
11-22	John F. Kennedy	71
11-23	Roald Dahl	1177
11-24	John Knox	425
11-25	Upton Sinclair	1150
11-26	Isabella I	236
11-27	Clovis I	666
11-28	Washington Irving	433
11-29	Hans Holbein the Y.	555
11-30	Oscar Wilde	77

December

Date	Person	Sig
12-01	Henry I	438
12-02	John Brown	304
12-03	Diocletian	181
12-04	Thomas Hobbes	167
12-05	Claude Monet	178
12-06	Saint Nicholas	331
12-07	William Bligh	1255
12-08	John Lennon	162
12-09	Sigismund	712
12-10	Avicenna	190
12-11	Llywelyn the Last	2090
12-12	Robert Browning	608
12-13	Samuel Johnson	141
12-14	George Washington	6
12-15	Walt Disney	337
12-16	Camille Saint-Saëns	842
12-17	Simón Bolívar	448
12-18	J. Lamarck	479
12-19	Emily Brontë	2008
12-20	John Steinbeck	576
12-21	George S. Patton	526
12-22	George Eliot	590
12-23	Thomas R. Malthus	265
12-24	Vasco da Gama	281
12-25	Samuel de Champlain	205
12-26	Harry S. Truman	94
12-27	Gustave Eiffel	807
12-28	Maurice Ravel	812
12-29	Thomas Becket	264
12-30	Robert Boyle	310
12-31	John Wycliffe	579

FIGURE B.4. The most significant figure who died on each day of the year, July through December to die on our Charles's birthday is *John Locke (1632–1704)* [100]. You can check out yours in Figures B.3 and B.4.

B.3 The Who's Bigger Game

The inspiration for this book goes back to long family car rides that I (Steve) took while growing up. To pass the time, my brothers Len and Rob and I took turns posing pairs of actors, athletes, and other celebrities to our parents, asking them to decide who was bigger. We felt we won whenever Mom and Dad came up with conflicting answers, or gave up when challenged with a ridiculous or obscure-enough pairing, say, singer *Tony Martin (1913–2012)* [20800] and actor *Gene Raymond (1908–1998)* [47144], with whom my father served in World War II.

Inspired by this, we have created a free *Who's Bigger* app for the Apple iPhone and iPad. Cast your vote as to which is the more significant or famous person in a head-to-head battle between two historical figures. We use the significance analysis methods detailed in this book to determine the "right" answer. You gain points when you agree with our selection, and lose them when you disagree. Higher levels become more challenging, involving increasingly obscure figures and battles between more closely matched people.

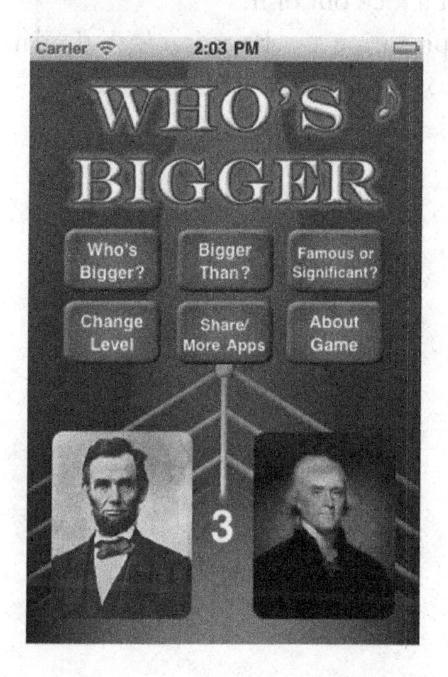

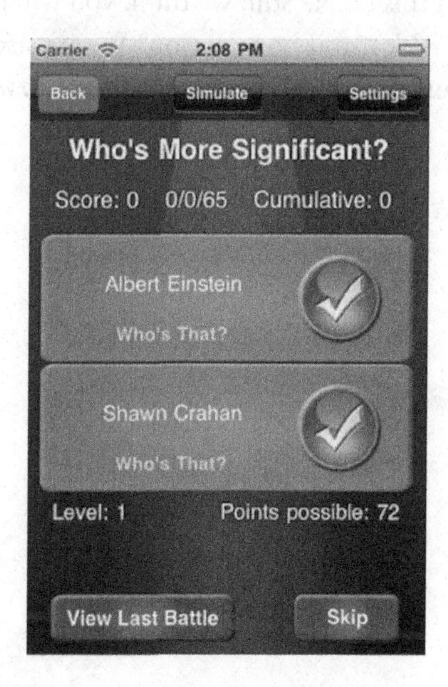

This App includes a database with 50,000 of the most significant historical figures, enough to challenge even the greatest history buff. If you cannot remember why you have heard of someone, click *Who's That* for a brief description and *Read More* for a link to their full Wikipedia article. We give each person's historical fame rank at the conclusion of each battle, so you can see exactly where they fit into the world's pecking order. *Simulate* mode gives you the opportunity to watch battles and get the hang of how the game is played.

As the reader of this book should know by now, significance and fame measure two different things. *Albert Einstein (1879–1955)* [19] clearly is more historically significant, while a celebrity like *Justin Bieber (1994–)* [8633] likely is more famous than he is important. You choose between your favorite measure of merit, although we believe significance makes for a more interesting game. The *Settings* menu permits you to move to harder questions if the initial settings prove too easy. Upgrading to the full version grants access to two more game variants: one asking whether a particular figure is more famous or significant, the other letting you specify that all questions involve your favorite historical figure.

We warn that this App was produced using an earlier version of our rankings, so we make no promise that they jibe exactly with what you see in this book. Still, we think you will get a kick out of it.

Not an iPhone person? We are also preparing a web version with similar functionality. Check it out at `www.whoisbigger.com/game`.

APPENDIX C

Biographical Dictionary

Here we give a brief description of the 100 most significant people in history, so you can refresh your memory on any of them who seem fuzzy. But first a confession, to give credit where properly due. These definitions have been edited down from the first paragraph of each person's Wikipedia article. We have tried to capture the essence of each article in a text short enough to be sent over Twitter, i.e., at most 140 characters.

Adams, John [61] (1735–1826) was the second president of the United States; he also served as an American lawyer, statesman, diplomat, and political theorist.

Alexander the Great [9] (356–323 B.C.), the King of Macedonia and conqueror of the Persian Empire, is one of the greatest military geniuses of all times.

Ali [89] (598–661) was the cousin and son-in-law of the Islamic prophet Muhammad, and ruled over the Islamic Caliphate from 656 to 661.

Alighieri, Dante [96] (1265–1321) Durante Alighieri, commonly known as Dante, was a major Italian poet of the Middle Ages.

Aquinas, Thomas [90] (1225–1274) was an Italian theologian who is remembered for his attempt to reconcile faith and reason in a comprehensive theology.

Arc, Joan of [95] (1412–1431) was a French heroine and military leader inspired by religious visions to organize resistance against the England.

Aristotle [8] (384–322 B.C.) was a Greek philosopher, a student of Plato, and teacher of Alexander the Great.

Augustus [30] (63 B.C.–A.D. 14) is considered the first emperor of the Roman Empire, which he ruled alone from 27 B.C. until his death in A.D. 14.

Bach, Johann Sebastian [48] (1685–1750) was a Baroque composer; he is considered to be one of the greatest composers of all time.

Bacon, Francis [81] (1561–1626) was an English philosopher, statesman, scientist, lawyer, jurist, author, and father of the scientific method.

Beethoven, Ludwig van [27] (1770–1827) was a German composer and pianist. He remains one of the most famous and influential composers of all time.

Bismarck, Otto von [97] (1815–1898) was a German-Prussian statesman of the late nineteenth century, and he is a dominant figure in world affairs.

Buddha, Gautama [52] (563–483 B.C.) was a spiritual teacher from the Indian subcontinent; his teachings were the foundation of Buddhism.

Bush, George W. [36] (1946–) was an American politician who served as the 43rd president of the United States, from 2001 to 2009.

Caesar, Julius [15] (100–44 B.C.) was a Roman general and politician. He played a critical role in the transition of the Roman Republic into the Roman Empire.

Calvin, John [99] (1509–1564) was an influential French theologian and pastor during the Protestant Reformation.

Charlemagne [22] (742–814) was the king of the Franks, first sovereign of the Christian Empire of the West.

Charles I [39] (1600–1649) was the King of England from 1625 to 1649; he was the only English king to be executed due to power struggle.

Charles II [78] (1630–1685) was the king of England who was invited to return to Britain after his 9-year exile in mainland Europe due to civil war.

Charles V [84] (1500–1558) was the ruler of the Holy Roman Empire best known for his role in opposing the Protestant Reformation.

Churchill, Winston [37] (1874–1965) was a British politician and statesman known for his leadership of the United Kingdom during World War II.

Cicero [79] (106–43 B.C.) was a Roman statesman. He established a model for Latin prose and he was a supporter of Pompey against Julius Caesar.

Cleveland, Grover [98] (1837–1908) was the 22nd and 24th U.S. president. Cleveland is the only president to serve two nonconsecutive terms.

Columbus, Christopher [20] (1451–1506) was an Italian explorer, whose voyages across the Atlantic Ocean led European awareness of the New World.

Constantine the Great [67] (272–337) was the first Christian Roman Emperor who ruled from 306 to 337.

Cook, James [60] (1728–1779) was a British explorer, navigator, and cartographer who eventually rose to the rank of captain in the Royal Navy.

Copernicus, Nicolaus [74] (1473–1543) was a Renaissance astronomer and the first person to formulate a comprehensive heliocentric cosmology.

Cromwell, Oliver [50] (1599–1658) was an English military and political leader best known in England for his overthrow of the English monarchy.

Darwin, Charles [12] (1809–1882) was a scientist who laid the foundations of the theory of evolution and changed our perspective about the natural world.

David [57] (1040–970 B.C.) was the second king of the united Kingdom of Israel according to the Hebrew Bible.

Descartes, René [92] (1596–1650) was a French philosopher, mathematician, and writer who is considered to be the father of modern philosophy.

Dickens, Charles [33] (1812–1870) was an English novelist, also regarded as the greatest author of the Victorian period.

Edison, Thomas [40] (1847–1931) was an American inventor who developed many devices (phonograph, lightbulb, etc.) that profoundly influenced the world.

Einstein, Albert [19] (1879–1955) was a theoretical physicist who discovered the theory of relativity, which laid the foundation of modern physics.

Elizabeth I [13] (1533–1603) was the fifth and last monarch of the Tudor dynasty whose reign is famously known as the Elizabethan era.

Franklin, Benjamin [35] (1706–1790) was a noted polymath from Pennsylvania in the colonial era, and one of the Founding Fathers of the United States.

Freud, Sigmund [44] (1856–1939) was a prominent Austrian neurologist who founded the discipline of psychoanalysis.

Galilei, Galileo [49] (1564–1642) was an Italian physicist, astronomer, mathematician, and philosopher who played a major role in the Scientific Revolution.

Gandhi, Mohandas Karamchand [46] (1869–1948) was the political and ideological leader of India during the Indian independence movement.

George III [58] (1738–1820) was the third British monarch of the House of Hanover who is remembered as "The King Who Lost America."

Goethe, Johann Wolfgang von [88] (1749–1832) was a German writer and polymath who is considered the supreme genius of modern German literature.

Grant, Ulysses S. [28] (1822–1885) was the 18th U.S. president as well as military commander during the Civil War and Reconstruction periods.

Hamilton, Alexander [45] (1755–1804) was the first U.S. Secretary of the Treasury, a Founding Father, economist, and political philosopher.

Henry VIII [11] (1491–1547) of England was the second monarch of the House of Tudor who separated the Church of England from papal authority in Rome.

Hippo, Augustine of [72] (354–430) was a Christian bishop whose writings were very influential in the development of Western Christianity.

Hitler, Adolf [7] (1889–1945) was an Austrian-born German politician and the leader of the Nazi Party during World War II.

Jackson, Andrew [66] (1767–1845) was the 7th president of the United States who destroyed the national bank and relocated most Indian tribes to the west.

James I [41] (1566–1625) of England was an English king known mainly for his advocacy of the divine right of kings and his authorization of the King James Bible.

Jefferson, Thomas [10] (1743–1826) was the third president of the United States and the principal author of the Declaration of Independence.

Jesus [1] (7 B.C.–A.D. 30) of Nazareth, also referred to as Jesus Christ or Jesus, is the central figure of Christianity.

Kant, Immanuel [59] (1724–1804) is one of the most influential philosophers in the history of Western philosophy.

Kennedy, John F. [71] (1917–1963) was the 35th president of the United States, serving from 1961 until his assassination in 1963.

Khan, Genghis [38] (1162–1227) was the founder of the Mongol Empire, which became the largest contiguous empire in history after his death.

King Arthur [85] (?–?) was a legendary British leader of Late Antiquity who led the defense of Britain against Saxon invaders in the early sixth century.

Lee, Robert E. [76] (1807–1870) was a career military officer who is best known for having commanded the Confederate Army in the American Civil War.

Lenin, Vladimir [75] (1870–1924) was a Russian Marxist revolutionary, creator of the Soviet Communist Party, and leader of the October Revolution.

Leonardo da Vinci [29] (1452–1519) was an Italian Renaissance polymath who epitomized the Renaissance humanist ideal.

Lincoln, Abraham [5] (1809–1865) was the 16th president of the United States who preserved the Union during the Civil War and abolished slavery.

Linnaeus, Carl [31] (1707–1778) was the "Father of Taxonomy" whose system for naming, ranking, and classifying organisms is still in use today.

Locke, John [100] (1632–1704) was an English philosopher and physician regarded as one of the most important Enlightenment thinkers.

Louis XIV [26] (1638–1715) of France was the longest-reigning and most powerful king in European history who built the lavish Palace of Versailles in France.

Louis XVI [83] (1754–1793) was the king of France and Navarre from 1774 who was guillotined after the outbreak of the French Revolution in 1793.

Luther, Martin [17] (1483–1546) was a German priest and professor of theology who initiated the Protestant Reformation.

Madison, James [51] (1751–1836) was an American politician and political philosopher who served as the 4th president of the United States.

Marx, Karl [14] (1818–1883) was a German philosopher, political economist, and revolutionary socialist, who developed the theory of Marxism.

Michelangelo [86] (1475–1564) was an Italian Renaissance sculptor and architect who exerted a tremendous influence on the development of Western art.

Mozart, Wolfgang Amadeus [24] (1756–1791) was a prolific and influential Austrian composer of the Classical era.

Muhammad [3] (570–632) is regarded as the founder of Islam, and is considered by Muslims to be a messenger and prophet of God.

Napoleon [2] (1769–1821) Bonaparte was a military and political leader during the latter stages of the French Revolution.

Newton, Isaac [21] (1643–1727) was an English physicist, mathematician, astronomer, natural philosopher, alchemist, and theologian.

Nietzsche, Friedrich [42] (1844–1900) was a German philosopher whose writings challenged the foundations of Christianity and traditional morality.

Nixon, Richard [82] (1913–1994) was the 37th U.S. president; he became the first president to resign owing to his involvement in the Watergate scandal.

Paul the Apostle [34] (A.D. 5–67) was one of the most influential Christian missionaries, whose writings form a considerable portion of the New Testament.

Paul, Pope John II [91] (1920–2005) was the second-longest serving pope in history who reigned as pope of the Catholic Church from 1978 until his death.

Peter, Saint [65] (?–?) was an early Christian leader and one of the twelve apostles of Jesus who is featured prominently in the New Testament Gospels and the Acts of the Apostles.

Philip II [87] (1527–1598) of Spain was a monarch in sixteenth-century Europe under whose rule Spain reached the height of its influence and power.

Plato [25] (427–347 B.C.) was a Classical Greek philosopher, who also founded the first institution of higher learning in the Western world.

Poe, Edgar Allan [54] (1809–1849) was an American author, poet, editor, and literary critic, considered part of the American Romantic Movement.

Presley, Elvis Aaron [69] (1935–1977) was an American singer, musician, and actor. He was one of the most popular American singers of the twentieth century.

Reagan, Ronald [32] (1911–2004) was the 40th president of the United States, the 33rd governor of California, and an actor before his political career.

Roosevelt, Franklin D. [43] (1882–1945) was the 32nd U.S. president and a central world figure during the time of worldwide economic crisis and world war.

Roosevelt, Theodore [23] (1858–1919) was the 26th president who played an important role in the Spanish-American War.

Rousseau, Jean-Jacques [80] (1712–1778) was a philosopher who influenced the

American and French Revolutions with his writings on education and politics.

Shakespeare, William [4] (1564–1616) was an English poet and playwright, regarded as the greatest writer in the English language.

Smith, Adam [56] (1723–1790) was a Scottish economist whose work *The Wealth of Nations* is considered to be the first modern work of economics.

Smith, Joseph [55] (1805–1844) was an American religious leader and the founder of the Latter Day Saint movement, which gave rise to Mormonism.

Socrates [68] (469–399 B.C.) was a Classical Greek philosopher. His teaching was known chiefly through the writings of his student Plato.

Stalin, Joseph [18] (1878–1953) was the unrivaled dictator of the Soviet Union, who ruled from 1928 until his death in 1953.

Tchaikovsky, Pyotr Ilyich [63] (1840–1893) was a composer of the Romantic era who is considered to be one of the most important composers of all time.

Tesla, Nikola [93] (1856–1943) was an inventor and electrical engineer. He was an important contributor to the birth of commercial electricity.

Truman, Harry S. [94] (1884–1972) was the 33rd U.S. president; he played a significant role in WWII, the establishment of the UN, and the Marshall Plan.

Twain, Mark [53] (1835–1910) was an American author whose writings such as *The Adventures of Tom Sawyer* have had worldwide influence.

Van Gogh, Vincent [73] (1853–1890) was a post-Impressionist painter whose work had a big influence on modern art for its vivid colors and emotional impact.

Victoria, Queen [16] (1819–1901) was the longest reigning monarch of the United Kingdom from 1837 until her death.

Voltaire [64] (1694–1778) was a French Enlightenment philosopher famous for his advocacy of civil liberties, including freedom of religion and free trade.

Wagner, Richard [62] (1813–1883) was a German composer, conductor, theater director, and essayist, primarily known for his operas.

Washington, George [6] (1732–1799) was the dominant military and political leader of the United States from 1775 to 1799.

Wilde, Oscar [77] (1854–1900) was an Irish writer and poet famous for his sophisticated, brilliantly witty plays.

William the Conqueror [70] (1027–1087) was the first Norman king of England from Christmas 1066 until his death.

Wilson, Woodrow [47] (1856–1924) was the 28th U.S. president who played an important role in WW I and the establishment of the League of Nations.

Bibliography

Applegate, D. 2006. *The Most Famous Man in America: The Biography of Henry Ward Beecher*. Doubleday.

Astor, P. 2010. The Poetry of Rock: Song Lyrics Are Not Poems but the Words Still Matter; Another Look at Richard Goldstein's Collection of Rock Lyrics. *Popular Music* 29:143–48.

Axelrod, S., C. Roman, and T. Travisano, editors. 2012. *The New Anthology of American Poetry: Vol. III: Postmodernisms 1950–Present*. Rutgers University Press.

Bartlett, M. S. 1937. The Statistical Conception of Mental Factors. *British Journal of Psychology* 28:97–104.

Bautin, M. C. Ward, A. Patil, and S. Skiena. 2010. Access: News and blog analysis for the social sciences. In *19th Int. World Wide Web Conference (WWW 2010)*, Raleigh, NC.

Brin, S. and L. Page. 1998. The Anatomy of a Large-Scale Hypertextual Web Search Engine. In *Proc. 7th Int. Conf. on World Wide Web (WWW)*, pp. 107–17.

Broder, A., A. Kirsch, R. Kumar, M. Mitzenmacher, E. Upfal, and S. Vassilvitskii. 2009. The Hiring Problem and Lake Wobegon Strategies. *SIAM Journal of Computing* 39:1233–55.

Brooks, K. 2012. Lost Leonardo Da Vinci Painting Worth £100 Million May Have Been Found in a Scottish Farmhouse. *Huffington Post*, August 7.

Burt, D. 2000. *The Literary 100: A Ranking of the Most Influential Novelists, Playwrights, and Poets of All Time*. Checkmark Books.

CDC. 2012. Vital statistics data available online. http://www.cdc.gov/nchs/data_access/Vitalstatsonline.htm.

Chafets, Z. 2009. *Cooperstown Confidential: Heros, Rogues, and the Inside Story of the Baseball Hall of Fame*. Bloomsbury.

Cohen, N. 2011. Define Gender Gap? Look Up Wikipedia's Contributor List. *New York Times*, January 30.

Collins, A., R. O'Doherty, and M. C. Snell. 2006. Rising stars, superstars and dying stars: Hedonic explorations of autograph prices. Discussion Papers 0603, University of the West of England, Department of Economics, March. URL http://ideas.repec.org/p/uwe/wpaper/0603.html.

Croce, L. E. 1978. The Hall of Fame for Great Americans. In D. Wallechinsky and I. Wallace, editors, *The People's Almanac 2*, pp. 1050–56. William Morrow and Co.

Danilov, V. 1997. *Hall of Fame Museums: A Reference Guide*. Greenwood Publishing Group.

Dawkins, Richard. 1990. *The Selfish Gene*. Oxford University Press.

Desser, A., J. Monks, and M. Robinson. 1999. Baseball Hall of Fame Voting: A Test of the Customer Discrimination Hypothesis. *Social Science Quarterly* 80:591–603.

Dickert-Conlin, S. and A. Chandra. 1999. Taxes and the Timing of Births. *Journal of Political Economy* 107:161–177.

Duignan, B. 2009. *The 100 Most Influential Philosophers of All Time*. Britannica Educational Publishing.

Eckart, C. and G. Young. 1936. Approximation of one matrix by another of lower rank. *Psychometrika* 1:211–218.

Findlay, D. and C. Reid. 1997. Voting Behavior, Discrimination, and the National Baseball Hall of Fame. *Economic Inquiry* 35(3):562–578. ISSN 1465-7295. doi: 10.1111/j.1465-7295.1997.tb02033.x.

Findlay, D. and C. Reid. 2002. A Comparison of Two Voting Models to Forecast Election into the National Baseball Hall of Fame. *Managerial and Decision Economics* 23:99–113.

Findlay, D. and J. Santos. 2012. Race, Ethnicity, and Baseball Card Prices: A Replication, Correction, and Extension of Hewitt, Muñoz, Oliver, and Regoli. *Economic Journal Watch* 9(2):122–40.

Franklin, B. 1818. *Memoirs of the life and writings of Benjamin Franklin. Edited by William Franklin*. T.S. Manning, Philadelphia.

Giles, J. Internet Encyclopaedias Go Head to Head. *Nature* 438:900–01.

Gladwell, M. 2008. *Outliers*. Little, Brown.

Goldstein, R. 1969. *The Poetry of Rock*. Bantam.

Gottlieb, Agnes Hooper, Henry Gottlieb, Barbara Bowers, Brent Bowers, and Agnes Gottlieb. 1998. *1,000 Years, 1,000 People: Ranking the Men and Women Who Shaped the Millennium*. Kodansha America.

Grahame-Smith, S. and J. Austen. 2009. *Pride and Prejudice and Zombies*. Quirk Classics.

Hafner, K. 2007. Seeing Corporate Fingerprints in Wikipedia Edits. *New York Times*, August 19.

Harman, H. H. 1976. *Modern Factor Analysis*. University of Chicago Press.

Hart, M. H. 1992. *The 100: A Ranking of the Most Influential Persons in History*. Citadel Press.

Hochschild, A. 1999. *King Leopold's Ghost: A Story of Greed, Terror, and Heroism in Colonial Africa.* Mariner Books.

Hotelling, H. 1933. Analysis of a Complex of Statistical Variables into Principal Components. *Journal of Educational Psychology* 24:417–41, 498–520.

Infoplease. 2007. Life Expectancy by Age, 1850–2004. http://www.infoplease.com/ipa/A0005140.html.

James, B. 1985. *The Bill James Historical Baseball Abstract.* Villard.

James, B. 1995. *Whatever Happened to the Hall of Fame: Baseball, Cooperstown, and the Politics of Glory.* Fireside Press.

Jewell, R., R. Brown, and S. Miles. 2002. Measuring Discrimination in Major League Baseball: Evidence from the Baseball Hall of Fame. *Applied Economics* 34:167–77.

Johnson, R. 1935. *Your Hall of Fame.* New York University Press.

Jones, S. 2005. *Afterlife as Afterimage: Understanding Posthumous Fame.* Peter Lang.

Joreskog, K. G. 1969. A General Approach to Confirmatory Maximum Likelihood Factor Analysis. *Psychometrika,* 34:183–202.

Klepper, M. and R. Gunther. 1996. *The Wealthy 100: From Benjamin Franklin to Bill Gates – A Ranking of the Richest Americans, Past and Present.* Citadel Press.

Lehman, D., editor. 2006. *The Oxford Book of American Poetry.* Oxford University Press.

Levoy, M., K. Pulli, B. Curless, S. Rusinkiewicz, D. Koller, L. Pereira, M. Ginzton, S. Anderson, J. Davis, J. Ginsberg, and J. Shade. 2000. The Digital Michelangelo Project: 3D Scanning of Large Statues. In *Proceedings of the 27th Annual Conference on Computer Graphics and Interactive Techniques,* SIGGRAPH '00, pages 131–144. ISBN 1-58113-208-5. doi: 10.1145/344779.344849.

Lin, Y., J. B. Michel, E. Aiden, J. Orwant, W. Brockman, and S. Petrov. 2012. Syntactic Annotations for the Google Books Ngram Corpus. In *Proc. 50th Meeting for the Association of Computational Linguistics,* pp. 169–174.

MacCracken, H. M. 1901. *The Hall of Fame.* G. P. Putnam's Sons.

Life Magazine. 2000. Life Magazine Top 100 People of the Millennium. www.lifemag.com/Life/millennium/people/01.html.

Mamet, D. 2007. *Bambi vs. Godzilla: On the Nature, Purpose, and Practice of the Movie Business.* Pantheon.

Matheson W. and R. Baade. 2003. The Death-Effect on Collectible Prices. Department of Economics, Williams College Working Papers. URL `http://ideas.repec.org/p/wil/wileco/2003-12.html`.

McConnell, J. 2010. *Cooperstown by the Numbers: An Analysis of Baseball Hall of Fame Elections.* McFarland and Company.

Merry, R. 2012. *Where They Stand: The American Presidents in the Eyes of Voters and Historians.* Simon & Schuster.

Michel, J., Y. Shen, A. Aiden, A. Veres, M. Gray, Google Books Team, J. Pickett, D. Hoiberg, D. Clancy, P. Norvig, J. Orwant, S. Pinker, M. Nowak, and E. Aiden.

2011. Quantitative Analysis of Culture Using Millions of Digitized Books. *Science* 331:176–82.

Morello, T. 1977. *Great Americans: A Guide to the Hall of Fame for Great Americans.* New York University Press.

Mosteller, Frederick, and John W. Tukey. 1977. *Data Analysis and Regression: A Second Course in Statistics.* Addison-Wesley.

Mullin, C. and L. Dunn. 2002. Using Baseball Card Prices to Measure Star Quality and Monopsony. *Economic Inquiry* 40:620–32.

Musch, J. and S. Grondin. 2001. Unequal Competition as an Impediment to Personal Development: A Review of the Relative Age Effect in Sport. *Developmental Review* 21:147–67.

New York Times. 2006. How Common Is Your Birthday? December 19.

Pearson, K. 1901. On Lines and Planes of Closest Fit to Systems of Points in Space. *Philosophical Magazine* 2(6):559–72.

Pemmaraju, S. and S. Skiena. 2003. *Computational Discrete Mathematics: Combinatorics and Graph Theory with Mathematica.* Cambridge University Press.

Peterson, R. 1992. *Only the Ball Was White: A History of Legendary Black Players and All-Black Professional Teams.* Oxford University Press.

Phillips, D., C. Voorhees, and T. Ruth. 1992. The Birthday: Lifeline or Deadline? *Psychosomatic Medicine* 54:532–42.

Pinker, S. 2011. *The Better Angels of Our Nature: Why Violence Has Declined.* Viking.

Plato. *Laws.*

Pollock, R., P. Stepan, and M. Valimaki. 2010. The Value of the EU Public Domain. www.rufuspollack.org.

Polsby, N., M. Gallaher, and B. Rundquist. 1969. The Growth of the Seniority System in the U.S. House of Representatives. *American Political Science Review* 63:787–807.

Qi, Lin. 2012. Li Keran Bucks Market Trend. *China Daily*, August 8.

Renneboog, L. and C. Spaenjers. 2012. Buying Beauty: On Prices and Returns in the Art Market. *Management Science.*

Saffro, R. 2009. *The Sanders Autograph Price Guide.* Autograph Media/Odyssey, 7th edition.

Seelye, K. 2005. A Little Sleuthing Unmasks Writer of Wikipedia Prank. *New York Times*, December 11.

Simmons, J. 2000. *The Scientific 100: A Ranking of the Most Influential Scientists.* Citadel Press.

Skiena, S. 2001. *Calculated Bets: Computers, Gambling, and Mathematical Modeling to Win.* Cambridge University Press.

Skiena, S. 2008. *The Algorithm Design Manual*, Springer-Verlag, 2nd edition.

Skiena, S. and M. Revilla. 2003. *Programming Challenges: The Programming Contest Training Manual.* Springer-Verlag.

Spearman, C. 1904. General Intelligence Objectively Determined and Measured. *American Journal of Psychology* 15:201–93.

Surowiecki, J. 2004. *The Wisdom of Crowds: Why the Many Are Smarter Than the Few and How Collective Wisdom Shapes Business, Economies, Societies and Nations.* Doubleday.

U.S. News. 2011. U.S. News Best Colleges. http://colleges.usnews.rankingsandreviews.com/best-colleges.

Wales, J. 2009. Wikipedia. http://www.wikipedia.org.

Wikipedia. 2011. World Population. http://en.wikipedia.org/wiki/World_population.

Wikipedia. 2012a. Historical Rankings of Presidents of the United States. http://en.wikipedia.org/wiki/Historical_rankings_of_Presidents_of_the_United_States.

Wikipedia. 2012b. Wikipedia: List of Freedom Indices. http://en.wikipedia.org/wiki/List_of_freedom_indices.

Index

Printed in the United States
by Baker & Taylor Publisher Services